Science, technology and society for a post-truth age

Comparative dialogues on reflexivity

Edited by
Emine Öncüler Yayalar
Melike Şahinol

Series in Philosophy of Science
VERNON PRESS

www.vernonpress.com

In the Americas:
Vernon Press
1000 N West Street, Suite 1200
Wilmington, Delaware, 19801
United States

In the rest of the world:
Vernon Press
C/Sancti Espiritu 17,
Malaga, 29006
Spain

Series in Philosophy of Science

Library of Congress Control Number: 2023946580

ISBN: 978-1-64889-788-7

Cover design by Vernon Press. Background image by Daniele Levis Pelusi on Unsplash.

Next, the backdrop and method of the dissident method of "living in truth" to oppose ideological manipulation; the multifaceted structure of Havel's truth claims, and the devolution of their currency in the post-communist period are examined by Kilburn in "Living in Post-Truth: Anti-Politics and the Power of the Powerless in the Twenty-First Century". Kilburn discusses the challenges posed by a post-truth society and how the case of socialist Czechoslovakia, particularly Václav Havel's "living in truth" philosophy, may offer insights for navigating the current crisis. The chapter makes an important contribution to the understanding of the post-truth condition by providing a detailed analysis of how Havel's moral positioning may have inadvertently contributed to the distrust of authority and expertise and the rise of right-wing populism, which led to the corrosion of truth.

The main goal of the final chapter of part two, entitled "Re-conceptualizing Cross-disciplinary Expertise in International Criminal Investigations: An STS Perspective," is to create, articulate, and disseminate a normatively coherent conceptual framework of transdisciplinary expertise as it is used in international criminal justice. McGregor delves into the realm of Science and Technology Studies (STS) to address challenges related to scientific and technological expertise, particularly in the context of international criminal justice and applies theories of expertise to a real-world case study, aiming to provide a theoretical foundation for addressing ontological and epistemological tensions arising from multidisciplinary scientific and technological advancements. McGregor's analysis hints at the potential for STS to contribute to resolving these difficulties.

In the third part, the contributions focus on local knowledge structures and questioning of singular knowledge claims as well as on participatory research/science. It starts with Diamary and Kumar's case study of grassroots innovations in India and proposes a potential model for improving the understanding of communication processes in informal knowledge systems. Drawing on the case study of Shodytara, the authors argue that an engagement with local communities and an awareness of place-based evidence and experiential knowledge may be crucial in moving beyond the public understanding of the science model. An investigation of alternative knowledge production and new forms of science communication are proposed as possible ways to overcome the erosion of trust in science.

In the next chapter, "Reflexive philosophical dialogue about science, technology and society: tacking the post-truth condition," Dunlop and Stubbs claim that in a post-truth era, public philosophy, particularly philosophical discussion, can create conditions for critical thinking through reflexive philosophical inquiry into science, technology, and society. According to the authors, it is important to recognize the limitations of scientific decision-making and the need for interdisciplinary collaboration and public engagement in

In “New Way of Critique in the Post-Truth Era”, Lee asks if post-truth simply means the end and failure of the age of truth and explains a common misconception about STS, thereby underlining the significance of the STS as a framework for raising important concerns based on reflexivity. Through the concept of ‘critique,’ Lee (re)historicizes the post-truth regime in order to better comprehend the contemporary period. According to Lee, the demise of expertise and the rejection of objective truth are important drivers of the post-truth era. After problematizing the attacks on postmodernism for paving the way for the destabilization of truth, Lee deploys the theories of Boltanski and Foucault to offer a new perspective based on the adoption of reflexivity and the genealogical method. The author concludes that a productive relationship between praxis and critique with a reflexive attitude is necessary to address the challenges of the Post-truth era.

In “Social Reflexivity as a Means for Critical Life Science Research on Human Diversity”, Plümecke focuses on life science research dealing with human diversity and race/ethnicity and gender categories by reconsidering the concept of reflexivity. He provides an empirical analysis of how reflexivity is employed in the knowledge production process. According to Plümecke, the concept of reflexivity is not just used as a means to develop a critical awareness of how one’s embodied presence impacts their intellectual output but also as a set of research practices and a normative requirement. In this chapter, the concept of reflexivity is presented as a panacea to the debates on post-truth. Plümecke emphasizes the collective and transformative nature of reflexivity as a possible tool for understanding the social dimensions of science and making a difference in the world. The chapter aligns with calls to reclaim the original mission of STS by highlighting the importance of reflexivity as a tool for improving science and ensuring its ethical and social responsibility.

The second part examines the rise of post-factual politics from two different perspectives. The articles in this section help us understand the complexity of the post-truth moment by analyzing the conditions that have enabled the corrosion of trust in expert knowledge. In the first chapter of the second part, “The spread of the ‘gender ideology’ and the peace referendum in Colombia: post-truth age's digital and infrastructural materialities”, Guerrero Castro sheds light on Colombia's post-conflict and post-truth politics by examining the relationship between social media and political rhetoric. The chapter views “gender ideology” as part of the post-truth era, where facts and expertise are publicly contested, and fake news threatens democratic institutions. The author highlights that post-truth is not a new development but has a long political and intellectual history that goes hand in hand with modern liberal democracy.

problem from the perspective of the deficit model of science communication, the contributors to the volume emphasize the importance of the co-constructed nature of knowledge and expertise. Instead of trying to re-claim truth and inform the public through one-way communication, the authors recognize the underlying social, political, and economic factors that shape the existing knowledge claims and highlight the possibility of participatory epistemologies. This approach recognizes that a more reflexive approach to knowledge production is necessary, one that acknowledges the role of social actors in shaping scientific knowledge and that recognizes the inherent value judgments that underpin all scientific work.

The debates about the post-truth condition have focused on various concepts of STS, including symmetry, social constructivism, and "epistemic democracy" which have all been accused of being co-opted and utilized in pushing forward the post-truth agenda (Lynch 2017). According to some scholars, STS should be held responsible for instigating anti-science currents through its appeals to question the objectivity of scientific knowledge production (Latour 2004; Sismondo 2017; Collins, Evans, and Weinel 2017; Fuller 2018). This ongoing conversation has resulted in an impasse with STS being vilified for providing the tools of anti-scientific thinking. The volume provides a contemporary engagement with these debates to emerge from this impasse. In the first part of the edited volume, we propose "Rethinking STS in the post-truth era," taking reflexivity into account. In doing so, we gain an important understanding of STS concepts. This provides the foundation for a conceptually informed discussion about the post-truth era.

In the first article of this part, "(Good) post-truth, (bad) alternative facts, (fake) fake news and its ideological bubbles: a dialogue", Restrepo Forero and Ashmore critically reflect on the concept of alternative facts and fake news and argue that STS needs to embrace post-truth in order to move forward. Their use of the dialogue format is reminiscent of Freire's arguments on moving beyond an understanding of the dialogue as a mere technique. It allows the reader to engage with new ways of thinking about epistemological relationships fostered by the use of the dialogue format. The authors provide a more nuanced STS-inspired version of post-truth that is supported by their analysis of current events and their reception in the United States and Colombia. The article cautions against uncritically accepting the low credibility of post-truth. Drawing from recent discussions within the discipline, the authors argue that the democratizing project of STS persists, resulting in a post-truth world where the construction of facts is recognized as a contingent and complex process. Following Latour's call, they emphasize the importance of moving our attention from "matters of fact to matters of concern" and paying attention to the messy and entangled realities that shape our world.

Introduction

Emine Öncüler Yayalar and Melike Şahinol

When the Oxford Dictionary designated "post-truth" as the word of the year in 2016, it was perceived as an indication of the prevailing epistemic crisis characterized by the dissemination of misinformation, the rise of anti-science movements, and the proliferation of conspiracy theories. Subsequently, there has been a profusion of scholarly articles, books, and reports endeavoring to examine the emergence of the post-truth era. This concern intensified during COVID-19 as researchers sought to explain the accompanying infodemic.

Amidst these debates, the interdisciplinary field of Science, Technology and Society (STS) played a central role. STS is an interdisciplinary domain that investigates the mutual shaping of science, technology, and society (Jasanoff 2010). As the concept of post-truth gained more prominence, practitioners of STS found themselves embroiled in a fervent controversy regarding the discipline's responsibility in engendering such a phenomenon and the appropriate stance it should adopt in response.

Confronted with this backdrop, the purpose of this volume is twofold. On the one hand, the book provides an intervention to ongoing debates around post-truth by reclaiming the main tenets of STS and suggesting new visions for the development of the field. This collection fosters a constructive dialogue aimed at reimagining the potential of Science, Technology, and Society in effectively addressing the challenges arising in the widely acknowledged era of post-truth. Secondly, the volume contributes to the expanding body of literature on post-truth. The prevailing emphasis on objectivity and the pursuit of truth has garnered significant attention, yet the field of Science, Technology, and Society (STS) underscores the necessity of acknowledging the contextual factors that shape the production of knowledge. In line with this observation, the volume accentuates the significance of both theoretical and empirical approaches to the increasing tendency to disregard scientific knowledge and factual evidence from an STS perspective.

The volume moves away from a clear distinction between facts and values. Instead, following Latour, it sets the stage for embracing the hybrid entanglements of the "objective" and the "subjective". The rise of post-truth is placed in its socio-technical context and is understood to be an effect of a widespread understanding of truth as unassailable. Rather than framing the

Acknowledgements

We would like to express our heartfelt gratitude to all those who contributed to this project.

When we released the Call for Papers for this book, during a time when the world was in the midst of the unprecedented challenges presented by the COVID-19 pandemic, we were surprised by the extraordinary global response. The sheer number of submissions from across the world not only filled us with a deep sense of pride but also highlighted the critical need for discourse on the nexus of science, technology, and society in this increasingly complex "post-truth era."

We would like to express our foremost thanks to the authors who entrusted us with their valuable contributions. Their scholarly rigor and intellectual generosity have undoubtedly enriched this collection. They conducted the necessary adjustments and revisions with great dedication. Their willingness to heed the reviewers' suggestions and recommendations has been instrumental in the excellence of this volume. They engaged with the editorial process with exceptional commitment, ensuring their work not only met but exceeded our expectations.

Our heartfelt thanks go out to the diligent reviewers, whose meticulous insights and extensive expertise have significantly enhanced the articles in this collection. Their contributions during the double-blind peer review process have been crucial in elevating the quality of this scholarly work.

We are also deeply grateful to the editors at Vernon Press, especially Blanca Caro and Argiris Legatos, who weathered numerous challenges and versions alongside us with remarkable patience and professionalism. Their unwavering support has been a cornerstone of this project's success.

Our families also deserve special recognition for their patience and understanding; their unwavering support has been our harbor amidst the tempest of deadlines and revisions, and for this, we are immeasurably thankful.

We are overwhelmed by the support and collaboration that has accompanied this project from the outset. Without the contributions, dedication, and professionalism of all involved, this volume would not exist in its current form.

Thank you to all who have contributed to this book. It is our sincere hope that the ideas and research presented therein will help advance the discussion of the challenges and opportunities in the post-truth era.

Emine Öncüler Yayalar and Melike Şahinol
Ankara/ Istanbul (Türkiye) 2023

List of tables

List of figures

Table of contents

shaping research trajectories and policy decisions. By promoting public reflexive philosophical inquiry, we may be able to treat the post-truth condition, particularly in relation to climate change and geoengineering (large-scale intervention in the Earth's climate). They argue that philosophical dialogue develops scientific and political criticality and engagement, which is required to construct and deconstruct arguments and communicate sensitively through challenge, disagreement, and disjuncture based on empirical, interdisciplinary work in chemistry/philosophy/education.

Vidmar's contribution "Towards the PERIpatetic Approach: Epistemology of Close Encounters Within Contemporary STS" outlines major characteristics of an evolving shift in research philosophy in STS, which involves closer proximity to participants through longer-term contact and embeddedness in diverse contexts of Participatory Action Research. The text also emphasizes the significance of the researcher's positionality and makes the case that a closer relationship based on meaningful engagement and embeddedness in the studied environment while maintaining critical analytical distance is crucial for the post-truth age. This approach of the "uninformed insider" fosters greater confidence among stakeholders and increases confidence in the research process. This multi-layered process incorporates principles of reflexivity as well as participatory and longitudinal analysis, which may be useful in overcoming the paralysis presented by the post-factual period.

Finally, the fourth part brings together key empirical examples that address changing practices in the post-truth era. In doing so, the authors enrich the volume with examples of practices in art, education, and work. Da Veiga's chapter on "Post-fake artivism: how activism and art can break reflexivity" focuses on the Science Gallery in Dublin's "Fake" show as a case of how the dismantlement of the unprecedented curtain of clichés becomes urgent in a reality guided by the cognitive overlay of a barrage of buzzwords – from the simpler like friend, or tag to the more complex Internet of behaviors, extended reality, or enhanced connectivity – produced and controlled by dominant cultures to inculcate habits and norms and to consolidate power. Da Veiga concludes that post-fake artivism may be a step in the right direction.

Aldous Arantes' chapter on "Comparative dialogues on reflexivity: What do pre-service teachers need to understand about 'objectivity' and 'evidence' in their emergent workplace, during a post-truth age?" offers both a theoretical and an empirical approach, drawing on the findings of the author's PhD, the Apps in Australian Classrooms Project. In her analysis, Arantes explores the various challenges that teachers face in the post-truth era of technology and argues that STS perspectives could be beneficial in addressing these issues. She emphasizes that we need to shift our focus from approaching edtech as a neutral learning tool to addressing its role in the commercialization of education and the potential biases amplified by the extensive use of these tools. The article underscores the importance of understanding the political implications of commercial data in education and advocates for a critical stance towards using edtech. By highlighting the importance of the sociotechnical context, Arantes emphasizes the need for teachers to develop a reflexive understanding of 'objectivity' and 'evidence' in their emergent workplace.

In "Exploring the Masculine Roots of Scientific Tradition: A Comparative Analysis of Two Cases from Sweden and Turkey," Pehlivanlı-Kadayıfçı gives us insights into current discussions of gender in STEM departments in Lund University and Middle East Technical University (Ankara, Turkey) in the wider context of STS and post-truth. The article provides a critique of the pipeline model, which focuses on increasing the number of underrepresented groups in the STEM fields without addressing privilege and questioning the objectivity of science itself. This results in the persistence of gender disparities in the physical sciences, with differences in representation, inclusion, and pay equity. Pehlivanlı bases her analysis at the backdrop of the post-truth rhetoric and anti-gender movements and highlights the significance of contextual factors in the creation of scientific knowledge.

In the final chapter, "Techno-social equities and inequities in the post-truth age. The case of remote work in Greece during the COVID-19 pandemic" Papazafeiropoulou stresses that the rise of remote working necessitates a reconsideration of the STS paradigm so that its tools can aid in the understanding of the interplay of techno-social elements not just in the sector of consumption, as has been stressed to date, but also in the sphere of production. In light of the foregoing, the presented case study focuses on distant working conditions in Greece in 2020 to explore the discourses of working groups that are primarily influenced by such changes.

The contributors to the volume provide a comprehensive exploration of the post-truth era with insights from the interdisciplinary field of Science, Technology, and Society (STS). By emphasizing the co-constructed nature of knowledge and expertise, the volume not only challenges the prevailing emphasis on objectivity and the pursuit of truth but it also presents the significance of the unique contributions of STS in thinking beyond the post-truth period. Doing so advocates for a more reflexive approach to knowledge production that acknowledges the social, political, and economic factors shaping existing knowledge claims.

While some scholars have accused STS of contributing to the erosion of trust in expert knowledge, the authors argue for the continued relevance of STS in addressing the challenges of the post-truth era. They advocate for embracing post-truth as an opportunity for reimagining the potential of Science,

Technology, and Society rather than dismissing it as a threat. By exploring concepts such as reflexivity, critique, and social reflexivity, the volume offers innovative perspectives for understanding and navigating the complexities of the post-truth condition. The empirical case studies presented in the book shed light on various aspects of the post-truth era, including its impact on political rhetoric, grassroots innovations, philosophical inquiry, and changing practices in art, education, and work. Collectively, these contributions contribute to the expanding body of literature on post-truth while highlighting the importance of interdisciplinary approaches and participatory research in addressing the challenges of our increasingly complex information landscape.

References

Collins, Harry, Robert Evans, and Martin Weinel. 2017. "STS as Science or Politics?" *Social Studies of Science* 47(4): 580–86. (https://doi.org/10.1177/0306312717710131).

Fuller, Steve. 2018. *Post-truth: Knowledge as a power game.* New York: Anthem Press.

Jasanoff, Sheila. 2010. *States of knowledge: The co-production of Science and Social Order.* London: Routledge.

Latour, Bruno. 2004. "Why Has Critique Run out of Steam? From Matters of Fact to Matters of Concern." *Critical Inquiry* 30(2): 225–48. https://doi.org/10.1086/421123.

Lynch, Michael. 2017."STS, Symmetry and Post-Truth." *Social Studies of Science* 47(4): 593-99. https://doi.org/10.1177/0306312717720308.

Sismondo, Sergio. 2017."Post-Truth?" *Social Studies of Science* 47(1) 3–6. https://doi.org/10.1177/0306312717692076.

PART 1.
Rethinking STS in the post-truth era

Chapter 1

(Good) post-truth, (bad) alternative facts, (fake) fake news and its ideological bubbles: a dialogue[1]

Olga Restrepo Forero and Malcolm Ashmore

Abstract: Despite the 'linguistic panics' over the prominence of the hyper-recent tropes of 'post-truth' (P-T), 'alternative facts' (AFs), and 'fake news', along with the tendency to treat the second and third as instances of the first, we argue that these terms denote rather different, and even opposed, phenomena. We show that P-T is a mature, well-known, and cuddly beast that those doing STS should honor. Even when we construe P-T as relevant only within the discourse of politics, we argue that its effects are benign, permitting a much larger range of voices and framings - a wider and healthier democracy - than is the case in a world of the Singular Truth, with its strictly accredited spokespersons.

AFs, on the other hand, are phenomena with equally pernicious effects on their proponents and their critics. Instead of promoting democracy, they encourage authoritarianism. As we argue in relation to the events surrounding the coinage of the term, AFs are performative admissions of argumentative defeat, succeeding only in increasing their users' 'incredibility' by opening a new avenue for scornful dismissals.

To complete the Trumpian trio, we look briefly at 'fake news' and find that this is a term that, when used as an accusation, can be countered simply by turning it against itself: 'fake news' is fake. Less easy to shrug off, though, is the way it

[1] This text is a hybrid, being both a scholarly paper and an authorial dialogue. As the former, it uses the classic fact-writing devices of footnotes (and here we would like to footnote Anthony Grafton's *The Footnote* (1997); but we cannot, as we are already inside one, and meta-footnotes are not allowed), headings, citations, and Bibliography. As the latter (within the constraints of the former) it is performatively designed, as befits its precursor's existence as a live presentation at the Second Colombian National Colloquium of Social Studies of Science and Technology, (Restrepo Forero and Ashmore 2017).

accentuates the perilous phenomenon of political polarization in the form of ideological bubbles – 'ours' as well as 'theirs'.

Our article takes the form of a lively dialogue, as befits an examination of such contentious and ideologically divisive issues.

Keywords: Post-truth; alternative-facts; fake-news; ideological-bubbles; dialogue

***

Getting started

Malcolm – We have three (and a half) distinct topics…

Olga – …clearly linked though they are.

M – Indeed. And of course, we hope that these topical connections will add some coherence to this piece…

O – …of writing…

M – …and reading…

O – …and performing…

M – …and attending…

O – Let's just say 'wrighting' (Ashmore 1989).

M – Let's just get going. So, everybody, the order of events is as our title suggests: first, post-truth, then alternative facts, thirdly, fake news incorporating some thoughts on ideological bubbles.

O – We'll see if it works that neatly.

M – Unlikely, I agree, so…

O – We need to avoid Ian Hacking's "elevator words" (Hacking 1999:22). You know, those Capitalized Terms which tend to operate at meta levels; to be circularly defined; and to have remarkable histories of change, despite the sacredly-intoned rhetorics of inevitability and timelessness in which they get used.

M – You mean like 'Truth' and 'Fact'?

O – Quite. Which is another benefit of talking instead of 'post-truth' and 'alternative facts.'

M – Because they are not so elevated?

O – They seem radically incapable of elevation.

M – Except that Oxford Dictionaries (2016), as is well-known, chose 'post-truth' as its word of the year in 2016, defining it thus: "Relating to or denoting circumstances in which objective facts are less influential in shaping public opinion than appeals to emotion and personal belief."

O – Well, that is less than useful…

M – And strangely limited: "public opinion," which is treated as a solid entity and not as the product of public opinion surveys (Osborne and Rose 1999) and of political discourses attempting to tame it, shape it, and define it.

O – "[E]motion and … belief", which always has been the stuff of politics and political action.

M – And, of course, "objective facts" as the contrasted Good Persuader to the Bad of the personal...

O – …which somewhat gives the game away. For those guardians of the elevated at the Oxford Dictionaries, 'post-truth' is a Bad Thing which we'd be much better off without.

M – Back to Objective Facts: naked facts, singular facts, incontestable facts, sacred facts.

O – Back to simplicity; no artifacts, factions, factishes (Latour 2010) nor even factoids.

M – Back to clarity…

O – …to certainty…

M – …when we all knew what was what.

O – What was which?

M – And that too!

O – So, back to the world we have lost, the stable world of tradition, of the pre-modern…

M – …well, certainly not the post-modern...

O – …and not the modern we have never been, neither; although we have always believed that we were, at least until Bruno Latour's (1993) bombshell… But wait! It all depends on what is this *we* and where 'it' is *situated* (and you and I always run into this trouble), but let's leave this for now.

M – OK, but on the proviso that we will come back to it.

O – Oh, we will.

M – Good. So now we know what we are doing…

O – …we do?

M – Certainly! Post-truth the alternative version. Proudly unelevated and properly grounded. And released from Oxford Dictionaries sub-political ghetto…

O – …though we will have to head into the arena of politics even so.

M – Of course! But later. Instead, why don't we do…

A brief disciplinary history: STS and/as PostTruth

O – STS[2] 101?

M – Not quite. There are many such overviews (e.g., Ashmore 1989:1-25; Shapin 1982; Shapin 1995; Lynch 1993:39-158; Lynch 2016). No, let's just deal with some of the most startling early stuff, the work that was so exciting and radical and that has provided the whole field of STS – however greatly expanded and differentiated it has since become – with nothing less than its conscience and its soul.

O – You sound almost religious! But I'll play along if we can start with Fleck; Ludwik Fleck, a Polish medical doctor and, in the 1930s, the modest purveyor of a revolutionary mode of epistemics. As we know through Thomas Kuhn's account, the title of Fleck's (1935) "almost unknown monograph" (Kuhn 1962)[3] was all he needed to recognize "that a book with that title was likely to speak to my own concerns" (Kuhn 1979:viii). When published in English for the first time in 1979, the title became *Genesis and development of a scientific fact.*

M – Didn't he articulate a similar distinction to that made in Latour's brilliant *Science in Action* (1987) between Ready Made Science (what I later called in reviewing Annemarie Mol's (2002) equally brilliant book, the Left Hand Side), and Science in the Making (the Right Hand Side)? In my review· (Ashmore 2005), I made this division, and its many cognates, the founding division of SSK.

O – Absolutely. Ready Made Science provides us with facts, which are supposed to be, according to Fleck´s understanding, "definite, permanent, and independent of any subjective interpretation by the scientist" (Fleck 1979: xxvii). Latour's Ready Made Science is what Fleck called "vademecum science"[4], while the Right Hand Side, Science in the Making, "journal science," is full of claims, and counterclaims, and controversies, that carry the marks of the "provisional, uncertain and personally colored" (Fleck 1979:118-119). And Fleck also invented genetic, social, and comparative epistemology well before these expressions became widely used in the 1970s and 80s, mainly through

[2] Science and Technology Studies; or – now retro – Science, Technology and Society.

[3] Fleck's 1935 title is *Entstehung und Entwicklung einer wissenschaftlichen.*

[4] 'Handbook science' or 'textbook science'.

Kuhn's ([1962]2012) massively influential *Structure*.[5] But, please, don't let me carry on talking about the Fleck-Kuhn connection.

M – Hasn't that been dealt with by Steve Fuller?

O – Hmmmm... not really, as all Fuller says in his book on Kuhn, in a footnote, is that Fleck is "Kuhn's most celebrated mythical progenitor" (Fuller 2000:295,n.76).

M – OK, though we must deal with Fuller's pronouncements on 'post-truth' later. For now, we'll leap forward from the 1930s to the 1970s and David Bloor's Strong Programme, with its strategy of undermining the solid distinction between (proper) Knowledge and (mere) Belief, through contrasting the robustness of the social with the weakness of the idiosyncratic individual. The Strong Programme's four tenets became the grounding for SSK.[6] Of these rules, the most long-lasting and consequential has been the symmetry principle[7], which, as Steve Woolgar (2017) reminds us, has been extended and expanded from Bloor and the Edinburgh School's[8] original use of it to undermine classical epistemology's most treasured distinction: that between everything TRASP and everything trash – as Harry Collins almost put it in 1981 with his own radical programme for SSK.[9] Extended symmetry is evident in later developments,

[5] *The Structure of Scientific Revolutions:* 2nd edition in 1970, 3rd edition in 1996, 4th edition in 2012 (50th anniversary). It is a truism in STS that this text was influential on the early sociology of scientific knowledge. (In)famously, its author has regretted that he had anything to do with this misbegotten and mistaken mode of analysis (Kuhn 1992). The historian of science, Paolo Rossi, originally in his introduction to the Italian translation of Fleck's book in 1983, rejects the idea of an "epistemological revolution" brought about by almost the sole influence of Kuhn (Rossi 1990). A recent book by the Taiwanese scholar, Daiwei Fu (2019) also questions the 'influence' truism, especially given Kuhn's own repudiation of SSK's understanding of his work. On this, see Song and Gao 2021.

[6] 1. Causality; 2. Impartiality; 3. Symmetry; 4. Reflexivity (Bloor 1976:4-5; 1991:7). SSK is the sociology of scientific knowledge.

[7] The strong programme in the sociology of knowledge would be "symmetrical in its style of explanation. The same types of cause would explain, say, true and false beliefs" (Bloor 1976:5; 1991:7)

[8] A group based in the Science Studies Unit at Edinburgh University, originally headed by David Edge (1923-2003) and including Bloor, Barry Barnes, Donald MacKenzie and the early careers of Brian Wynne, Andrew Pickering and Steven Shapin.

[9] TRASP = **T**rue, **RA**tional, **S**uccessful, **P**rogressive (Collins 1981). The "radical programme" (Collins 1981) transmuted into "an empirical relativist programme" (Collins 1983) and was later finalised by being acronymised as EPOR – **E**mpirical **P**rogramme **O**f **R**elativism.

particularly ANT's[10] double redistribution of agency between Humans and Non-Humans, the reflexivists' (Mulkay 1985; Woolgar 1988; Ashmore 1989;) refusal of any special authorial authority, and Sheila Jasanoff (2004) and the co-productionists' two-way street between knowledge and social order.

O – Yes, that last being at the center of Latour's *We Have Never Been Modern* (1993), a book title that many in Colombia would understand as a description of the country's failed attempts to achieve Modernity despite all its modernizing projects. It is more usually understood, however, as a description of Western Modernity in general and its failed attempts to purify, classify, order, and eliminate hybrids so as to produce clear-cut distinctions between Nature and Culture (or Society, or Politics) and of course, between Humans and Non-Humans.

M – So, Fleck on the production-history of one particular fact; Kuhn on how science periodically trashes its truths and adopts new ones; Bloor on the necessity of agnosticism (and relativism) in the study of 'whatever counts as' true or false; and Latour, in *Science in Action* (1987) on the Janus-faced character of science: stuffed with incontrovertible facts on the one face and teeming with disputes and uncertainties on the other.

O – And let's not forget *Laboratory Life* (Latour and Woolgar 1979; 1986) and its deconstruction of the textual economy of fact-making…

M – …and *Changing Order* (Collins [1985]1992) which taught us about the importance of studying the resolution of contemporary controversies[11] for a grandstand view of truth-making, and also about the deep circularity at the heart of novel experimental practice, which Collins memorably calls "the experimenters' regress."

O – Yes, capital-T Truth and capital-F Fact had certainly taken a battering! And in the 1990s, the bruised defenders of the cultural authority of these elevated terms hit back hard in that bizarre episode known as the science wars[12]…

M – …when our quiet corner of the academy was suddenly accused of undermining civilization and poisoning the minds of the young.

---

[10] **A**ctor-**N**etwork **T**heory; as developed by Bruno Latour, Michel Callon, John Law, … Classics: Callon 1986; Latour 1992. For a late introduction: Latour 2005b; and an introduction to the late: Law and Hassard 1999.

[11] In Collins' case no matter how long these last, sometimes as if waiting for the long Aha! (Collins 2004; 2010; 2017).

[12] Major warriors included Gross and Levitt 1994; Sokal 1996; The Editors of *Lingua Franca* 2000; and Koertge 1998.

O – Yes, a weird case of flattery by insult. In an utterly negative fashion, we were granted a degree of influence and visibility we had no idea we possessed!

M – But of course, the science warriors refused to see that we were not anti-science irrationalists.

O – But they got it right in one way: we *were* engaged in a project involving a certain flattening, a kind of democratizing. The Fact (naked, singular, incontestable, sacred) was deconstructed, firstly by turning 'It' into 'them,' and then by close attention to *their* making, *their* building, *their* production, *their* construction; in short, to the manufacture of knowledge, to quote Karin Knorr-Cetina's perfectly-titled book (1981). Facts are made (or supported, or undermined, or destroyed) through the contingent practices involved in the production of knowledge-claims. Claims that achieve the stability required for factual status are relatively rare. 'Fact' is an honorific and fragile socio-epistemic status. Facts exist, of course, they do, but only by virtue of being constructed, enacted, and performed.

M – Thus, any and every fact Could Have Been Otherwise. And this democratizing project continues through the temporal conceptual choreography of STS: practices, multiplicities, complexities, heterogeneities, assemblages, enactments, performances... And in the process, a post-Truth world has been built; of course, with allies from elsewhere (e.g., The 2nd January Group 1986; Edwards et al. 1995; Vattimo 2011).

O – But what good is this nice and cuddly 'post-Truth world' of yours in our harsh, all-too-stern and thoroughly depressing world: the world of far too few posts (never-ending wars, never-ending inequality, never-ending austerity) along with far too many: post nine eleven, post-financial crash, post-2016: the rolling madness of Brexit in the UK, the no-then-yes peace plebiscite in Colombia, and of course the 2016 election in the USA. Although, admittedly, we are all of us, at least for now, post-Trump. Any temptation to breathe a sigh of relief, however, is somewhat tempered by the global disaster of the COVID-19 pandemic.

M – "Somewhat tempered"? You sound more British than me! But you asked me what good are the democratized epistemics and the 'it could be otherwise' ethics, that we have described. So, this is Steve's question again, as posed in his contribution to our colloquium entitled "It could be otherwise, in times of trouble" (Woolgar 2017): in times like these, do we not need a bit more elevation? Do we not need to draw on the rhetoric of the Capitalized? To claim a bit of that old-time Truthiness for ourselves? I hope not; I really do. Our kind of post-truth, our kind of epistemic democratizing, is not only nothing to be frightened of, and not only what sustains our corner of academia, but is positively...

O – …useful? Really? In the Real World? Was that what you were going to say? But think of the recent efflorescence of the 'Post-Truth' phrase which Oxford Dictionaries was duly noting: its use by the media, in public intellectual commentary, and by disappointed politicians and upset citizens. Think of the way it is used by those on the losing sides of referenda, plebiscites, and elections, who feel betrayed by the lies they believe they were told during those campaigns. This, surely, is more difficult to handle. Because, on the one hand, we would like to say to them: "Come on, grow up, when were politicians truthful, when were political campaign promises kept? Please, don't try to convince us now that the mainstream media, the 24-hour news channels, the embedded journalists have ever been fountainheads of The Truth! Remember the 'dodgy dossier' and Saddam's WMDs?[13] Remember the non-closing of Guantanamo?[14] But, on the other hand, we, as disappointed citizens ourselves, don't we want to join in and denounce the blatant lies, the false campaign promises, and the deceptive slogans ('take back control'[15]; 'subsidized impunity for guerrillas, punishment for the army'[16]; the imposition of 'gender ideology' in Colombian schools[17]; 'drain the swamp'[18]; 'strong and stable'[19]).

M – Well, yes, we do. But that is a luxury we really should refuse, and not just for the sake of reflexive consistency. I think this argument can be best pursued

---

[13] *Iraq – Its Infrastructure of Concealment, Deception and Intimidation* often known as the Dodgy Dossier, was a February 2003 briefing document for the UK government used as a pretext for going to war in Iraq. It concerned Saddam Hussein and his 'weapons of mass destruction' (WMDs). It was later found to be plagiarised, 'sexed up', and false…or was it? It took a long time to establish this as a fact in media reports, from some of the first allegations/proofs of the "10 ways to sex up a dossier" (MacAskill and Norton-Taylor 2003) to the 'final' confirmation more than a decade later, after horrible loss of life, and through the very long and painstaking process of a Public Inquiry: 'The Iraq Inquiry', chaired by Sir John Chilcot, was set up in 2009 to look at decision-making in relation to the invasion of Iraq. The findings were not published until 2016. See "Chilcot report: The flawed chain of intelligence that led to war" (Taylor 2016).

[14] Before his first presidential term, Obama promised to close Guantanamo, the controversial illegal US military base/prison in Cuba. This was never done.

(M: But this promise was impossible to fulfill! O: You mean "it turned out" to be impossible.)

[15] Used, successfully, in the UK Brexit referendum, June 2016. 'Leave' - 52%.

[16] Used, successfully, in the Colombian peace plebiscite, October 2016. 'No' - 50.2%

[17] Used, successfully, in the Colombian peace plebiscite, October 2016. See Chapter 4.

[18] Used, successfully, in the US presidential election, November 2016. Trump – 304 electoral votes; Clinton – 227.

[19] Used, unsuccessfully, in the UK general election, June 2017. Conservatives - 318 seats, and loss of majority. 326 needed. The slogan was amusingly, and almost successfully, turned on its head: "weak and wobbly".

by attending directly to the theme of the colloquium that gave this text its birth: 'making peace in a post-truth world' (Segundo Coloquio Nacional de Estudios Sociales de las Ciencas y las Tecnologías September 2017). The implication of this phrase, I think, is that the practice of peace-making is made more difficult in conditions of post-truth.

O – Yes, that could be seen as the implication, but that is exactly what we want to contest, isn't it? That's why we, and the other organizers, thought of this precise theme for the colloquium. For we as citizens need to be more vigilant, more aware, more engaged in the "matters of concern" that affect us all rather than leaving them in the less-than-secure hands of experts, and Fact Producers of all kinds, both in politics and in science. We should be aware, though, with Latour, that the results of representation are neither automatic nor given in advance: "Transparency and immediacy are bad for science as well as for politics; they would make both suffocate. ... No unmediated access to agreement; no unmediated access to the facts of the matter. ... If politics is earthly, so is science" (Latour 2005a: 21-22).

M – Which implies that what we must deal with now are not the few as-of-now incontestable facts carefully crafted in laboratories through 'secluded research,' but with knowledge produced in "hybrid forums" by a range of unpredictable and heterogeneous actors; in the process enhancing "technical democracy" (Callon et al. 2009: passim.).

O - All this has been brilliantly encapsulated by Latour (2004) in his call to move our attention from "matters of fact to matters of concern," that is, from the controlled and clear-cut (and increasingly rare) Fact to the entangled and messy (and proliferating) Thing. In uncertain times, when science is produced not only In Action but also On Display, it is better for citizens to pay attention, to demand more and to give more of our own doubts, our own uncertainties, and to carry on building our imperfect agora... our imperfect assemblies...

M – Yes, indeed. But the crucial opening of Latour's exploration of critique that we have been drawing on is: "Wars. So many wars" (2004: 225). There was one here...

O – ...actually, several at the same time.

M – It, OK, they lasted for 52 years. Now officially over.

O – Though some Colombians insisted that we were never actually 'at war'. More importantly, not all our many-sided wars have yet ended.

M – Nevertheless, we accepted a chance of peace in Colombia at the second time of asking. Except that peace is more than what automatically 'takes place' after the combatants have signed their accords.

O – That's important to bear in mind when there is so much talk of post-conflict...

M – Post-war, post-conflict, is not the same as Peace. As the ending of recent conflicts in the world has taught us, Peace – yes, very much with a capital-P – is hard-won by hard work. The work that has Peace as its goal has often included formal proceedings of various kinds, most notably 'Truth and Reconciliation' commissions[20], as pioneered in post-apartheid South Africa. In these proceedings, success, the establishment of Peace, relies on Reconciliation taking precedence over Truth and the future taking precedence over the past, with numerous local acts of forgiveness and remorse being performed. If Reconciliation is sacrificed for Truth, on the other hand, then vengeance is the more likely result. Here, then, post-truth is exactly what is needed. It seems more than possible that the peace negotiations with the guerrillas here in Colombia, with their hopeful future-oriented dynamics, have incorporated this understanding; and that it will survive any single political cycle's...

O – ...you mean President's[21]...

M – ...attempts to destroy it.

O – Amen to that. Hopefully, we will learn to Remember and also learn to Forget, as both are part of the same process of reassembling social worlds. Understanding Truth and Reconciliation and Remembering and Forgetting, not as absolutes but as hard-worked results of a painful disassemblage of so many horrors, is going to take much time and patient effort. Will we have them, I wonder?

M – The question of memory, of all our memories, is also messy, as is the accountability for all the losses and all the victims of these Colombian wars (Mora Gámez 2016).

O – Memor*ies*... this multiplicity must be kept for as long as possible, especially if we are not going to fall into the tra(s)p of old certainties. We need to contest attempts to produce a single narrative, an official history. In the peace accords, the victims were rightly placed at the center, but as the negotiations were being achieved only with the FARC guerrillas, the overwhelming number of victims, of the disappeared and displaced, at the hands of paramilitaries, the Colombian armed forces, other guerrilla groups, and neo-paramilitaries, have not been as

[20] Wikipedia lists 42, including most of South America: Argentina, Bolivia, Brazil, Chile, Colombia, Ecuador, Panama, Paraguay, Peru, Uruguay. https://en.wikipedia.org/wiki/List_of_truth_and_reconciliation_commissions. Accessed 25 November 2021.

[21] President Iván Duque (2018-2022). As of this writing (January 2023), Colombia has elected Gustavo Petro as his successor. Petro's now five-month-old presidency is seemingly successful, seemingly popular...

visible in mainstream discourse as those of the FARC, despite the intense and valuable work done by *El Centro Nacional de Memoria Histórica.*[22] Unfortunately, since the appointment of its new director, this organization seems to be in the process of disassembling itself; on the model, perhaps, of similar US governmental organizations under Trump. This new director, like one unnameable Colombian ex-president, sometimes denies the existence of the Colombian Armed Conflict while others complain of it becoming some sort of "official truth" (Aguirre 2020)[23]. So, yes, currently, perhaps permanently, contested memories. This imbalance was painfully obvious, especially during the plebiscite campaigns. Also, in the recent debates about the guerrilla's declared or undeclared riches and whether these should be used to compensate their victims, the massive land grab perpetrated by paramilitaries[24] and the very slow process of compensating their victims has taken second place. (The compensation process, begun in 2005, has had very scant results in the subsequent seventeen years.) So, yes, post-truth means understanding and undertaking the hard work that goes into producing agreements and in crafting provisional and revisable versions of events. Post-truth has nothing to do with producing a white-and-black world of Good Guys and Bad Hombres. Now, what might we call a production of that kind?

M – Trump World, what else! We had to end up there! But after dealing with all these very complex and painful realities, entering the tawdry world of Trump by turning, as we must, to alternative facts seems rather tasteless. Inevitably. But don't worry, we will not be staying there for long.

O – I hope not! Anyway, the main thing we want to do is examine briefly the origins of this strange term. For followers of recent US politics, the story, at least in its outlines, is well-known.

'Alternative facts' – its inglorious origin

On 21 January 2017, the second day of Trump's presidency, Sean Spicer, the newly appointed press secretary, at what would ordinarily have been his first press conference, chose instead to lecture the assembled journalists and to take no questions. The lecture, which told off the assembled press for minimizing the size of the crowds for yesterday's inauguration, included the claim, or rather the declaration, that "this was the largest audience to ever witness an inauguration, *period!*" (CNN 2017:00:47).

---

[22] https://centrodememoriahistorica.gov.co/ Accessed 23 November 2021.

[23] Alvaro Uribe (2002-2010).

[24] See for a thorough, though contested, discussion, "Right-wing paramilitarism in Colombia". https://en.wikipedia.org/wiki/Right-wing_paramilitarism_in_Colombia. Accessed 6 December 2021.

M – Immediately, the media went to work demolishing this assertion and its associated evidence, including Spicer's use of comparisons of the Washington Metro 'ridership' figures on Obama's (2013) and Trump's inauguration days: "we know that 420,000 people used the DC metro public transit yesterday which actually compares to 317,000 that used it for President Obama's last inaugural" (CNN 2017). According to the Metro system's own figures, tweeted on January 20th, Spicer was comparing Trump's whole-day numbers with Obama's up-to-11 a.m.

O – However, the *Washington Post* used different numbers for their debunking efforts. Unfortunately, the origin of these latter numbers is not immediately clear, which reminds us that the work of debunking (Ashmore 1993) is seldom straightforward...

M – ...unless you can use an image (Ashmore and Guillot 2009) like Figure 1.1's side-by-side photographs of the crowd on the National Mall for Trump's inaugural (right) and for Obama's first in 2009 (left) (Business Insider 2017).

Figure 1.1. Photographic debunking

Source: Business Insider: 2017
Authors' photograph of a processed screenshot from the author's computer screen

O – This swift debunking had worked so well that by the time the NBC News "Meet the Press" program was broadcast that evening, the presenter, Chuck Todd, was able to ask his guest, Kellyanne Conway, special counsel to the

President, at least three times "why did the President send out his press secretary…for the very first time in front of that podium to utter a provable falsehood…?" (NBC News 2017: 00:20) Conway responds to this formulation of Todd's question with an attack of her own on the falsehood-mongering character of the press. This echoed not only Spicer's lecture but Trump's regular characterization of the media as purveying 'fake news'[25], at one point saying: "that's flat-out false" (NBC News 2017: 01:08) initiating a bad-tempered argument with Todd, or as we might want to say, constructing a usually effective diversion in the form of the 'reverse attack' rhetorical device. But Todd repeats his "you have not answered the question"; Conway claims she has; Todd repeats the question for the third time. And Conway answers thus: "Don't be so over-dramatic about it, Chuck, you're saying it's a falsehood [but Spicer] ((nod of head, flick of hair)) gave ((pause)) alternative facts ((pause)) to that…." Overlapping, Todd laughs, not believing his good fortune, and exclaims, "alternative facts?" (NBC News 2017: 01:56-02:06)

M – So that's the story. And it's not a story about another regrettable manifestation of a post-truth world at all. Rather, it shows the lengths defenders of Trump's 'fake news' agenda will go to in the effort to deny that the lie-telling, the fakery, the spin, the bullshit, is coming largely from their own side. When asked to explain Spicer's uttering "a provable falsehood" (and one indeed, to the satisfaction of the press[26] at least, already thoroughly proved), Conway went on the attack accusing the accusers of the self-same crime: uttering falsehoods. Losing the battle for accusation rights, she is faced for the third time with Todd's 'why' question, and rather than give in, she proceeds to dig an enormous rhetorical hole for herself: the one called 'alternative facts.'

O – Oh, I like the 'rhetorical hole device'…

M – …the RHD!

O – But note how Todd's glee and subsequent definitive statement, "alternative facts are not facts - they're falsehoods," shows how their expression has allowed media spokespersons suddenly to become truth teachers reveling in their occupation of an improbable moral high ground. These new Defenders of the Fact, however, are interested only in unmediated and spectacularized Facts, not well-built lower-case facts. They are nothing but born-again Fundamentalist Factistas!

M – Finally, and going back to our STS field, all this chatter around post-truth in the media, mainly in the USA and in the UK, has preoccupied some of our colleagues. Was the field's embrace of a kind of post-truth on the one hand, and

---

[25] See Section Four below.

[26] The mainstream media, the liberal press, the *New York Times*, the *Guardian*…

on the other, post-truth's association with Trump and the Brexit campaign, about to unleash a new version of the 'Science Wars' of the 1990s? Sergio Sismondo, the editor of *Social Studies of Science*, produced a worried editorial on post-truth (2017).

O – Yes. He seems most concerned with the idea that "embracing epistemic democratization" (Sismondo 2017: 3) – which chimes nicely with our take on post-truth – amounts to accepting the cheap relativism of 'anything goes,' or 'everything's equal' or 'alternative facts'…

M – Hang on! 'Alternative facts' are *not* relativist. Remember "Death and Furniture" (Edwards et al. 1995)?[27]

O – But the whole point is that they are *taken* to be! Anyway, Sismondo also mentions the wrong idea that claiming that "it could be otherwise," citing Woolgar as the foremost proponent of ICBO, is equivalent to saying that "it could *easily* be otherwise." But we already know this, don't we? We are entirely on Sismondo's side as he points out what is required for epistemic change: "other possible infrastructures, efforts, ingenuity and validation structures." And then he ruins it by adding: "That doesn't look at all like post-truth" (Sismondo 2017:3).

M – But of course it does!

O – But of course, the question is, whose post-truth? The media's, the moral entrepreneurs', the science warriors'? Or the version we have been arguing for here, our home-grown STS version? Here, we find ourselves agreeing with Steve Fuller, who has called for STSers (Fuller 2017; 2018: 53-68), along with scientists (Fuller 2016) and sociologists (Fuller 2018:53-68), actively to embrace post-truth as partly our own creation, to walk the P-T walk as well as, on more frequent occasions, merely talking its talk.

M – The problem, though, is that he sees Trump-world post-truth as a descendent of our original, arguing that we should take the Trumpistas' appropriation of our creation as a flattering endorsement.

O – But this all is so self-centered or field-centered, I should say. The argument is that the "basket of deplorables"[28], the what-have-you deniers, the

---

[27] A no-holds-barred attack on Realisms, in the form of a full-blooded defence of Relativisms.

[28] An unfortunate phrase used by Hilary Clinton in her unsuccessful 2016 election campaign to describe Trump fans and voters; but not all of them, apparently. "I'd say you can take Trump supporters and put them in two big baskets" …. "There are what I would call the deplorables -- you know, the racists and the haters[,] and [in the other basket are] the people who are drawn because they think somehow he's going to restore an America that no longer exists." For this quote and other candidate instances of the Original 'BoDs', see Kelsey 2016.

Q-Anonsensers, have taken their lead from sophisticated STS. That is as hard to believe as the idea that "politics" and the search for power, and for how to hold on to power, were based on Objective Facts before we entered the corrupt reign of appeals to Personal Emotion and all such despicable efforts to manipulate innocent "public opinion."

M – Which is where we started, of course. With quotations from Oxford Dictionaries definition of post-truth. Are we running around in circles? Let's move on...

Fake news: an ideological bubble machine

O – I thought we had already addressed this, though perhaps not adequately. Let's start by saying that the liberal media are not entirely blameless in this situation they (and we) are facing, with the "Fake you! No, Fake you!! NO, Fake you again, motherFaker!!!" crossfire. They have been very keen to present themselves as innocent bystanders when the Trump bombshell exploded on their screens and newspapers as if all by itself.[29]

M – Well, they had it coming, if that's what you mean, with their cozy relationships with governments and their increasing embeddedness with warmongering before, after, and during the invasions of Iraq and Afghanistan. In countries like the UK, where public media such as the BBC have been so important, it has become increasingly difficult to keep intact their traditional role of holding power to account. Instead, their role has become more about producing a "balanced" view that leaves everybody unhappy.[30]

O – Their traditional role of holding power to account is very similar to the traditional understanding (pre-STS) of scientists' speaking truth to power. In both cases, a self-flattering description of rather more complicated

---

[29] For the media's role in building Trump's candidacy through their ever more obsessive attention to his every move, see Confessore and Yourish 2016. In an analysis of the "free media" earned by Trump during his 2016 presidential campaign, they say "Mr. Trump earned $400 million worth of free media [in February to March 2016], ...[and] is strong in every type of earned media, from television to Twitter. Over the course of the campaign, he has earned close to $2 billion worth of media attention, about twice the all-in price of the most expensive presidential campaigns in history."

[30] And, of course, the jury, even within the BBC itself, is constantly out on the topic of 'Balance' or 'Bias'. See, for example, the description of the topic and contents of a BBC Two *Newsnight* program (Edwards and Cromwell 2007). Though of course, at this point the Corporation could be understood as trying to restore some (reflexive, meta) balance following the findings of an academic study, which had concluded, against the claims of Alastair Campbell, Tony Blair's director of communications, that the BBC was not biased against (the government line on) the war (Wells 2003).

relationships. So, the impression has been growing that the media, the so-called Fourth Estate, has really become more like the fourth branch of government.

M – And this was really clear in the build-up to the war in Iraq when Tony Blair continually complained of having hostile media coverage, even though the BBC kept organizing balanced debates for and against the invasion and the war in Iraq, even when majority public opinion[31] was clearly opposed.

O – But despite much of the UK media opposing the prospect of the war, they seem to have been the first to rally around in favor of it once the war started, ahead, indeed, of Parliament and Public.[32]

M – Oh yes, and in the US, as you mentioned, embedded journalists, or in-bedded as they were popularly called, initially spent more efforts in promoting the 'war effort' than in later denouncing the false WMD claims or the hype of the US military being welcomed with open arms.

O – They also had their cameras wide shut on all the behind-the-scenes preparations for the toppling of the statue of Saddam Hussein in Firdos Square, Baghdad. If you remember, we were shown over and over (on the BBC or on CNN) close-up images of the happy people who somehow had managed to pull down a gigantic bronze statue of Hussein[33], symbolizing the sense of relief, liberation, and triumph that the fall of the dictator had brought to the Iraqis, only a fortnight or so after the beginning of the ground war. But no media at the time[34] showed images of the US soldiers assisting or of the US M-88 armored recovery vehicle that helped with the task. It was much better to stage it as a spontaneous event captured by the cameras as if by a lucky chance. Like so many other events and their iconic images...

---

[31] We make this strong assertion alongside our endorsement of Osbourne and Rose (1999).

[32] This according to the detailed analysis conducted by James Strong (2017).

[33] Ordinary people can, though, engage in spontaneous statue-disassembly, in Bristol, if not Baghdad (Siddique and Skopeliti 2020). Unless of course this too was prearranged...

[34] Years later, some reports started appearing of what went on and how the whole thing had been largely staged for and with the patient attention of the media. See, for example, Maas 2011, and more recently von Tulzenman 2021. The former author makes the media directly accountable: "The media have been criticized for accepting the Bush Administration's claims, in the run-up to the invasion, that Iraq had weapons of mass destruction. The W.M.D. myth, and the media's embrace of it, encouraged public support for war. The media also failed at Firdos Square, but in this case it was the media, rather than the government, that created the victory myth".

M – …such as the (in)famous 'raising the flag at Iwo Jima' photograph, images of which still proliferate on the internet, or its strange reflection in the immediate aftermath of 9/11 (Schaible 2010).

O – But back to the present. The real shock and wake-up call to the liberal media in the US came when, as reported in *The New York Times*, Donald Trump, in a visit to the CIA, "used his first full day in office" to "unleash a remarkably bitter attack on the news media" in which he described journalists as "among the most dishonest human beings on earth" (Davis and Rosenberg 2017).[35]

M – That is when the truly unprecedented thing happened. Later, we would see repeated attempts at (re)producing the initial shock that came after this new president of the USA, the country so used to lecturing the world about the importance of democracy, about the need for press freedom, about the necessity of electoral integrity…

O – …not that Trump put a stop to the lectures, but at this point, he was far more interested in his inaugural crowd size and happy to open what would be a very long-lasting tug-of-words with the media, forever after calling them Fake News. This created a new political reality.

M – It did, and one which has not disappeared upon Trump's loss of the presidency. The 'fake news' trope is, unlike 'post-truth' or even 'alternative facts,' sharply recipient-designed for consumption by Trump's own followers. It says to them, "The liberal, mainstream, conventional media can and should be ignored and treated with contempt. They are not for you. All you need is Fox, Breitbart, and the rest,[36] together with your favorite social media sites." The accusation of fakery is clearly aimed at what we have called the 'liberal media.' It is thus one method among others for the inflating of distinct ideological bubbles (here, on the Right; we, of course, occupy our own – nobody is innocent), the members of which talk only to each other, never putting their heads above the parapet to see, and hear, their enemies' critiques.

O – Their enemies being us, I suppose?

M – Well, yes, in a way. We clearly occupy different sides in the increasingly canyon-like 'political divide' evident throughout EuroAmerica. Though some intrepid missionaries, or observers, or ethnographers have attempted the crossing; for example, Hochschild (2016).

O – Only from the one side?

---

[35] This was the same event at which Trump, as his main example of the media's deplorableness, first made the infamous complaint, reported and repeated by press secretary Spicer, about the reporting of his inauguration crowd-size, as we detail in Section 3, above.

[36] For a comprehensive listing of US right wing media, see *The Righting* (N.d.)

M – As far as I know, yes. But, of course, that ignorance is just another effect of the ideological bubbling. Where I am, where we…

O – We? Who're we?

M – Oh, you know – our side. Politically left/liberal, educated, *The Guardian, The New York Times,* perhaps writing, definitely reading, culturally on the ball, 'Good Taste' in all things, self-admiring…

O – Really?

M – Oh yes. Because we're the winners, we're on top…

O – Right. Which is why the other side is trying so hard to topple us, to undo all our 'progressivenesses,' to regain all the White rights…

M – …all the Manly privileges…

O – …and to contest all our spurious claims of Expertise.

M – Climate change is a hoax!

O – Abortion is murder! Women belong in the bedroom and the kitchen!

M – Blacks are ignorant brutes!

O – Voting is a privilege for the Chosen Few!

M – And that's US! White Men! Patriots!! Proud Boys!!! Republicans!!!![37]

O – Not women…

M – …blacks…

O – …jews…

M – …Mexican rapists…

O – …Colombian drug dealers…

M – …Polish plumbers…

O – …French philosophers…

M – …German engineers…

O – …Turkish academics.

Them – All of them! Taking our jobs and our self-respect. We hate them all!

Us – But what about the multi-billionaires with their trillion-dollar corporations?

Them – Post-Truth!

Us – The decades of 'austerity' and the hollowing out of the public sphere?

---

[37] These last two categories are becoming indistinguishable – at least in Miami (Mazzei and Feuer 2022).

Them – Alternative Facts!!

Us –The blatant corruption? The lies, the lying, the liars! They're laughing in your face!

Them – Fake News, Fake News, Fake News!!!

[[at this point of high drama and verbal fisticuffs, the dialogue breaks off]]

[and is replaced by]

The conclusion

The piece begins stutteringly (1. Getting Started), eventually laying out its three and a half topics and the connections between them. The authors note that post-truth was Oxford Dictionaries 2016 word of the year and proceed to critique the OD's naïve understanding of politics, public opinion, and, of course, post-truth. As a grounding for the authors' more sophisticated understanding of this last, we are then treated to section 2. A Brief Disciplinary History: STS and/as PostTruth. All their heroes appear here: Fleck, Bloor, Collins, Latour, the usual suspects. From this, the authors proceed to examine recent political events, mainly in the United States and in Colombia, to show the validity and utility of their STSish version of post-truth, taking in such large themes as truth and reconciliation, remembering and forgetting, and the perils of post-conflict thinking. Arriving, as if by accident, at the figure of Donald Trump, we proceed in section 3. 'Alternative Facts' – Its Inglorious Origin, to examine the very different status of AFs through another telling of their first appearance. The authors conclude that using this phrase is rhetorically suicidal and consequently that P-T and AFs cannot coherently be treated as 'two sides of the same coin,' as, they note, has become a regular feature of contemporary political discourse. Returning to STS and academia, we note that any too-easy acceptance of the junk status of 'post-truth' could find us defending our enterprise in a 'science wars mark 2'. Finally, in section 4. Fake News: An Ideological Bubble Machine, the authors complete their commentaries on Trump-speak through an examination of 'fake news.' We note that FN is far more narrowly applicable than either P-T or AFs, being precisely recipient-designed for Trumpistas as a descriptor of the (liberal) media - "nothing to see/read/hear here." Thus, we/they conclude that using the FN trope is "one method among others for the inflating of distinct ideological bubbles," leading to, as is immediately performed, the polarization of the most extreme and vituperative kind. Insults. Shouting. Scuffles. Whatever next?

This (Section 5. The Conclusion). Now concluded.

References

Aguirre, María Clara Calle. 2020. "Colombia: las ambigüedades del director del Centro de Memoria sobre el conflicto armado." *FRANCE 24*, 5 February. Retrieved July 29 2023. https://www.france24.com/es/20200205-colombia-existencia-conflicto-memoria-controversia

Ashmore, Malcolm. 1989. *The Reflexive Thesis: Wrighting Sociology of Scientific Knowledge.* Chicago: The University of Chicago Press.

Ashmore, Malcolm. 1993. "The Theatre of the Blind: Starring a Promethean Prankster, a Phoney Phenomenon, a Prism, a Pocket, and a Piece of Wood." *Social Studies of Science,* 23:67-106.

Ashmore, Malcolm. 2005. "Book Review: The Life inside/the Left-Hand Side." *Social Studies of Science* 35(5):827–30. doi: 10.1177/0306312705056053.

Ashmore, Malcolm, and Javier Guillot. 2009. "Seeing Double and Finding Fraud: Hwang Woo Suk and his Stem Cell Photographs." Paper presented at the annual meeting of 4S (Society for Social Studies of Science), October.

Edwards, David and David Cromwell. 2007. "BBC's Iraq Coverage - Biased or Balanced?" *Talk About Newsnight Blog,* March 19. https://www.bbc.co.uk/blogs/newsnight/2007/03/bbcs_iraq_coverage_bia sed_or_balanced.html.

Bloor, David. [1976]1991. *Knowledge and Social Imagery.* Chicago: The University of Chicago Press; London: Routledge.

Business Insider. 2017. "Trump Press Secretary Disputes Reports of Low Turnout at Inauguration." Retrieved July 29, 2023 https://www.youtube.com/watch?v=5IyGwxeK5fA

CNN. 2017. "Spicer: Inauguration Had Largest Audience Ever." Retrieved July 29, 2023 https://www.youtube.com/watch?v=PKzHXelQi_A

Callon, Michel. 1986. "Some Elements of a Sociology of Translation: Domestication of the Scallops and the Fishermen of St. Brieuc Bay." Pp. 196-223 in *Power, Action and Belief: A new sociology of knowledge?,* edited by J. Law. London: Routledge and Kegan Paul.

Callon, Michel, Pierre Lascoumes, and Yannick Barthe. 2009. *Acting in an Uncertain World: An Essay in Technical Democracy.* Cambridge: MIT Press.

Collins, Harry. 1981. "What is TRASP? The Radical Programme as a Methodological Imperative." *Philosophy of the Social Sciences* 11: 215-224.

Collins, Harry. 1983. "An Empirical Relativist Programme in the Sociology of Scientific Knowledge."Pp. 85-113 in *Science Observed: Perspectives on the Social Study of Science.* edited by Karin Knorr-Cetina and Michael Mulkay. London: Sage.

Collins, Harry. [1985]1992. *Changing Order: Replication and Induction in Scientific Practice.* Chicago: The University of Chicago Press.

Collins, Harry. 2004. *Gravity's Shadow: The Search for Gravitational Waves.* Chicago: The University of Chicago Press.

Collins, Harry. 2010. *Gravity's Ghost: Scientific Discovery in the Twenty-First Century.* Chicago: The University of Chicago Press.

Collins, Harry. 2017. *Gravity's Kiss: The Detection of Gravitational Waves.* Cambridge: The MIT Press.

Confessore, Nicholas, and Karen Yourish. 2016. "$2 Billion Worth of Free Media for Donald Trump." *The New York Times,* 15 March. Retrieved 25 April 2022. https://www.nytimes.com/2016/03/16/upshot/measuring-donald-trumps-mammoth-advantage-in-free-media.html.

Davis, Julie Hirschfeld, and Matthew Rosenberg. 2017. "With False Claims, Trump Attacks Media on Turnout and Intelligence Rift." *The New York Times,* 21 January. Retrieved 23 November 2021. https://www.nytimes.com/2017/01/21/us/politics/trump-white-house-briefing-inauguration-crowd-size.html.

Edwards, Derek, Malcolm Ashmore, and Jonathan Potter. 1995. "Death and Furniture: The Rhetoric, Politics and Theology of Bottom Line Arguments Against Relativism." *History of the Human Sciences* 8: 25-49.

Fleck, Ludwik. 1935. *Entstehung und Entwicklung einer wissenschaftlichen Tatsache.* Basel.

Fleck, Ludwik. 1979. *Genesis and Development of a Scientific Fact.* Chicago: The University of Chicago Press.

Fu, Daiwie. 2019. 傅大為, STS de yuan qi yu duo zhong jian gou - heng kan jin dai ke xue de yi zhong bian zhi yu da zao STS 的緣起與多重建構—橫看近代科學的一種編織與打造 [A Genealogical History of STS and Its Multiple Constructions: To Weave an Extensive Network for Gazing Upon the Modern Sciences] Taiwan: National Taiwan University Press.

Fuller, Steve. 2000. *Thomas Kuhn: A Philosophical History for Our Times.* Chicago: The University of Chicago Press.

Fuller, Steve. 2016. "Science Has Always Been a Bit 'Post-Truth'." *The Guardian,* 15 December. Retrieved 7 December 2021. https://www.theguardian.com/science/political-science/2016/dec/15/science-has-always-been-a-bit-post-truth.

Fuller, Steve. 2017. "Is STS All Talk and No Walk?" *EASST Review* 36(1). Retrieved 16 June 2017. https://easst.net/article/is-sts-all-talk-and-no-walk/.

Fuller, Steve. 2018. *Post Truth: Knowledge as a Power Game.* London: Anthem Press.

Grafton, Anthony. 1997. *The Footnote: A curious history.* Cambridge: Harvard University Press and London: Faber and Faber.

Gross, Paul R., and Norman Levitt. 1994. *Higher Superstition: The academic left and its quarrels with science.* Baltimore: Johns Hopkins University Press.

Hacking, Ian. 1999. *The Social Construction of What?* Cambridge: Harvard University Press.

Hochschild, Arlie Russell. 2016. *Strangers in Their Own Land: Anger and Mourning on the American Right.* New York: The New Press.

Jasanoff, Sheila, ed. 2004. *States of Knowledge: The Co-Production of Science and the Social Order.* London: Routledge.

Kelsey, Adam. 2016. "Behind Hillary Clinton's 'Basket of Deplorables' Line." *ABC News,* 13 September. Retrieved 25 April 2022. https://abcnews.go.com/Politics/hillary-clintons-basket-deplorables-line/story?id=42069200.

Knorr-Cetina, Karin. 1981. *The Manufacture of Knowledge.* London: Pergamon.

Koertge, Noretta, ed. 1998. *A House Built on Sand: Exposing Postmodernist Myths About Science.* Oxford: Oxford University Press.

Kuhn, Thomas S. [1962] 2012. *The Structure of Scientific Revolutions.* Chicago: The University of Chicago Press.

Kuhn, Thomas S. 1979. "Foreword." Pp. vii-xii in *Genesis and Development of a Scientific Fact,* by L. Fleck. Chicago: The University of Chicago Press.

Kuhn, Thomas S. 1992. "The Trouble With the Historical Philosophy of Science." Robert and Maurine Rothschild Distinguished Lecture, 19 November 1991, An occasional publication of the Department of the History of Science. Cambridge: Harvard University Press.

Latour, Bruno. 1987. *Science in Action.* Milton Keynes: Open University Press.

Latour, Bruno. 1992. "Where are the Missing Masses? The Sociology of a Few Mundane Artifacts." Pp. 225-258 in *Shaping Technology/Building Society: Studies in Sociotechnical Change.* edited by W. Bijker and J. Law. Cambridge: The MIT Press.

Latour, Bruno. 1993. *We Have Never Been Modern.* Hemel Hempstead: Harvester Wheatsheaf.

Latour, Bruno. 2004. "Why Has Critique Run Out of Steam? From Matters of Fact to Matters of Concern." *Critical Inquiry* 30: 225-248.

Latour, Bruno. 2005a. "From Realpolitik to Dingpolitik: Or How to Make Things Public." Pp. 21-22 in *Making Things Public: Atmospheres of Democracy* edited by B. Latour and P. Weibel. Cambridge: The MIT Press.

Latour, Bruno. 2005b. *Reassembling the Social: An Introduction to Actor-Network Theory.* Oxford: Oxford University Press.

Latour, Bruno. 2010. *On the Modern Cult of the Factish Gods.* Durham: Duke University Press.

Latour, Bruno and Steve Woolgar. 1979. *Laboratory Life: The Social Construction of Scientific Facts.* London: Sage.

Latour, Bruno and Steve Woolgar. 1986. *Laboratory Life: The Construction of Scientific Facts.* Princeton: Princeton University Press.

Law, John, and John Hassard, eds. 1999. *Actor Network Theory and After.* Oxford/Keele: Blackwell Publishers/The Sociological Review.

Lynch, Michael. 1993. *Scientific Practice and Ordinary Action: Ethnomethodology and Social Studies of Science.* Cambridge: Cambridge University Press.

Lynch, Michael. 2016. Social constructivism in science and technology studies. *Human Studies* 39: 101–112.

Maas, Peter. 2011. "The Toppling: How the Media Inflated a Minor Moment in a Long War." *The New Yorker,* 10 January. Retrieved 23 November 2021. http://www.newyorker.com/magazine/2011/01/10/the-toppling.

MacAskill, Ewan, and Richard Norton-Taylor. 2003. "10 Ways to Sex Up a Dossier." *The Guardian,* 27 March. Retrieved 23 November 2021. https://www.theguardian.com/politics/2003/sep/27/uk.davidkelly1.

Mazzei, Patricia, and Alan Feuer. 2022. "How the Proud Boys Gripped the Miami-Dade Republican Party." *The New York Times.* 2 June. Retrieved 8 June 2022. https://www.nytimes.com/2022/06/02/us/miami-republicans-proud-boys.html.

Mol, Annemarie. 2002. *The Body Multiple: Ontology in Medical Practice.* Durham: Duke University Press.

Mora Gámez, Fredy. 2016. "Reparation Beyond Statehood: Assembling Rights Restitution in Post-Conflict Colombia." PhD dissertation, University of Leicester.

Mulkay, Michael. 1985. *The Word and the World: Explorations in the Form of Sociological Analysis.* London: George Allen & Unwin.

NBC News. 2017. "Meet the Press." 21 January. https://www.nbcnews.com/meet-the-press/video/conway-press-secretary-gave-alternative-facts-860142147643.

Osborne, Thomas, and Nikolas Rose. 1999. "Do the Social Sciences Create Phenomena? The Example of Public Opinion Research." *British Journal of Sociology* 50(3): 367-396.

Oxford Dictionaries. 2016. "Word of the Year 2016." Retrieved 24 January 2023. https://languages.oup.com/word-of-the-year/2016/.

Restrepo Forero, Olga, and Malcolm Ashmore. 2017. "*For* Post-Truth, *Against* Alternative Facts." Presented at the Second Colombian National Colloquium of Social Studies of Science and Technology, "Making Peace in A Post-Truth World", Bogotá, September 2017.

The Righting. N.d. An A-Z Guide To Right Wing Media. Retrieved 24 November 2021. https://www.therighting.com/media-guide.

Rossi, Paolo. [1986] 1990. "Hechos científicos y estilos de pensamiento: apuntes sobre una revolución imaginaria." In *Las arañas y las hormigas; una apología de la historia de la ciencia*, 56-88. Translated by Juana Bignozzi. Barcelona: Crítica.

Schaible, Eric. 2010. "9/11 Iwo Jima Relation." May 4. Retrieved 7 December 2021. https://townhouse5visualarts.blogspot.com/2010/05/911-iwo-jima-relation.html

Segundo Coloquio Nacional de Estudios Sociales de las Ciencas y las Tecnologías. September 2017. "Hacer la Paz en el Mundo Posverdad". http://gesctm.unal.edu.co/historico-coloquio-escyt/.

Shapin, Steven. 1982. "History of Science and Its Sociological Reconstructions." *History of Science; an Annual Review of Literature, Research and Teaching* 20(3):157–211. doi: 10.1177/007327538202000301.

Shapin, Steven. 1995. "Here and Everywhere: Sociology of Scientific Knowledge." *Annual Review of Sociology* 21: 289-321.

Siddique, Haroon, and Clea Skopeliti. 2020. "BLM Protesters Topple Statue of Bristol Slave Trader Edward Colston." *The Guardian,* 7 June. Retrieved 25 November 2021. https://www.theguardian.com/uk-news/2020/jun/07/blm-protesters-topple-statue-of-bristol-slave-trader-edward-colston.

Sismondo, Sergio. 2017. "Post-truth?" (editorial) *Social Studies of Science* 47: 3–6.

Sokal, Alan. 1996. "A Physicist Experiments with Cultural Studies." *Lingua Franca,* May/June: 62–64.

Song, Runjie and Lu Gao. 2021. Daiwie Fu 傅大為, STS de yuan qi yu duo zhong jian gou - heng kan jin dai ke xue de yi zhong bian zhi yu da zao STS的緣起與多重建構—橫看近代科學的一種編織與打造 [A Genealogical History of STS and Its Multiple Constructions: To Weave an Extensive Network for Gazing upon the Modern Sciences] Taiwan: National Taiwan University Press, 2019.

607 pp. NT$850. *East Asian Science, Technology and Society: An International Journal* 15: 111–118. DOI 10.1080/18752160.2021.1877035

Strong, Jeremy. 7 July 2017. "Long read: Public opinion, legitimacy and Tony Blair's war in Iraq". *Democratic Audit.* Retrieved 23 November 2021. https://www.democraticaudit.com/2017/07/07/long-read-public-opinion-legitimacy-and-tony-blairs-war-in-iraq/

Taylor, Peter. 2016. "Chilcot Report: The Flawed Chain of Intelligence that Led to War." *BBC News,* 7 July. Retrieved 24 November 2021. https://www.bbc.com/news/uk-36737922.

The 2nd January Group. 1986. *After Truth: A Post-Modern Manifesto.* London: Inventions.

The Editors of *Lingua Franca,* eds. 2000. *The Sokal Hoax: The Sham that Shook the Academy.* Lincoln: Nebraska University Press.

Vattimo, Gianni. 2011. *A Farewell to Truth.* New York: Colombia University Press.

von Tunzelman, Alex. 2021. "The Toppling of Saddam's Statue: How the US Military Made a Myth." *The Guardian,* 8 July. Retrieved 24 November 2021. https://www.theguardian.com/world/2021/jul/08/toppling-saddam-hussein-statue-iraq-us-victory-myth.

Wells, Matt. 2003. "Study Deals a Blow to Claims of Anti-War Bias in BBC News." *The Guardian,* 4 July. Retrieved 23 November 2021. https://www.theguardian.com/media/2003/jul/04/Iraqandthemedia.politicsandthemedia.

Woolgar, Steve, ed. 1988. *Knowledge and Reflexivity: New Frontiers in the Sociology of Knowledge.* London: Sage.

Woolgar, Steve. 2017. "It Could Be Otherwise, In Times of Trouble." Paper presented at the Second Colombian National Colloquium of Social Studies of Science and Technology, "Making Peace in A Post-Truth World", September 2017.

Chapter 2

'New' way of critique of STS in the post-truth era

Kyunghwan Lee

Abstract: Does post-truth simply mean the end and failure of the age of truth? The neologism "post-truth" implies an existential bankruptcy of the 'fact' and the possibility of plural facts, raising concerns and fears among many people. STS scholars are also among those who express such concerns. At the same time, however, STS scholars are suspected of being the midwife of the post-truth. By reviewing the current debates on the relationship between STS and post-truth, this chapter responds to the misapprehension of the relationship and argues that 'critical' STS is requested to deal with various problems in the post-truth era since STS could raise the most important questions we should address. To refine the 'critical,' this chapter also looks at the "critique" discussed by Foucault and Boltanski and scrutinizes how the innovation of critique for STS is possible in the era of Post-truth. This paper confirms that Foucault and Boltanski's critique has a resonating point that differs from the ideological critique, emphasizing the importance of debunking the structural logic of obfuscating reality. In addition, they are offering essential questions, which can help innovate the critique in (and from) STS, as follows: How did we get to the present now? How did we come to know what we believe we currently know? How does our epistemological community's way of knowing to exist? Why do we have this way of knowledge production? Finding the answers to the question will be the first step to reinvigorating the 'critique.'

Keywords: STS. Post-Truth, Critique, Foucault, Boltanski

***

Paraphrasing the famous phrase from "The Communist Manifesto" by Marx and Engels, the present era could be understood as an era in which 'the specter of

Post-truth[1] is haunting the world. The name of the era, 'Post-truth', represented by fake news/misinformation, is analyzed and discussed at diverse levels to answer various questions on where the specter came from, whether it really exists or is simply discovered or invented. Multiple reasons are pointed out as the causal factors for the transition to the Post-truth era. Among them are the diversification of media, the demise of experts (and/or expertise), postmodernism, and the spread of (social) constructionism, suggested by voices of those concerned about the Post-truth era (Snyder 2021; McIntyre 2018; Nichols 2017; Kakutani 2018).

One of the popular solutions, generally suggested, to overcome the detrimental effects of social turmoil led by misinformation and Post-truth is fact-checking (Graves et al. 2016; Walter et al. 2020). Fact-checking, which has already spread widely, refers to a process of confirming the factuality and factual value of a specific statement and claim. Politicians' remarks and general information are mainly targeted (Walter et al. 2020). For example, when the presidential debate begins in the United States, several media outlets work as fact-checkers in real-time, posting on whether the presidential candidates' remarks are true or false. Statistical figures mentioned in candidates' remarks are typical objects of checking. Fact-checking, performed by a few media outlets initially, is now being carried out in various ways through numerous news media outlets. However, it is still questionable whether the expansion of fact-checking can be a driver for the end (or overcoming) of the Post-truth era. As we can see in the case of the spread of conspiracy theories surrounding COVID-19 and vaccines (Islam et al. 2021; Puri et al. 2020), 'factism'-if such a term can be coined- does not lead to the end of Post-truth and a return to the era of '(The) Truth' (of course, more empirical research may be needed).

The era of Post-truth is perplexing, not just in journalism and among various intellectuals. In addition, the sociology of (scientific) knowledge and Science and Technology Studies (hereafter STS), which have emphasized the social feature of (scientific) knowledge, are also embarrassed by the situation where epistemological democracy of (scientific) knowledge is castigated as epistemological anarchism or epistemological 'populism' under the Post-truth context. This embarrassment can also be confirmed by the emergence of diverse controversies on how to understand the present from the perspective of STS, and various 'public' intellectuals, which will be mentioned in the following section (Jasanoff and Simmet 2017; Sismondo 2017a; 2017b;

---

[1] The term Post-truth appears for the first time in a 1992 essay by Serbian-American writer Steve Tesich (1992). However, it can be said that his meaning of post-truth was 'after the truth was known'. In the present sense, which is used to mean "a time when truth is less important or irrelevant than one's opinion," the concept of Post-truth seems to have been first used in Ralph Keyes' book *The Post-Truth Era* published in 2004.

Fujimura and Holmes 2019). Such a conundrum leads STS to ask about its role in this turmoil. From the birth of its field, STS tried to reveal the social characteristics of technoscience, which is one of the essential critiques to go beyond the traditional understanding of science and knowledge (Sismondo 2008). However, the phenomenon of Post-truth paves the way to questioning STS's role in understanding the process of production and spread of knowledge. In the post-truth era, STS seems to be losing its power and legitimacy as a critique and/or critical ability to analyze the knowledge production process. In other words, the Post-truth era asks how STS can maintain its critical capacity to grasp the relationship between science (and knowledge) and the world.

In this context, this chapter will briefly examine the emergence of the Post-truth era and further review the position of social constructionism and STS in the Post-truth age. In other words, I will track the difficulties of the Post-truth period experienced by STS, which asks ontological questions on the knowledge while emphasizing the constructionist feature of expertise for a critical understanding of the world. In the last section, this chapter will examine how they can (re)solve the aporia they face. In particular, by looking at the "critique" discussed by Foucault and Boltanski, it will expand the discussion on how the innovation of critique is possible in the era of Post-truth.

The emergence of the post-truth era

The symptoms of Post-truth, represented by fake news/misinformation, are not only revealed by Trumpism and Brexit. As you can imagine from the words such as rumor and conspiracy, the phenomenon that can be understood as fake news has taken many forms throughout history.[2] However, the 2016 Oxford Dictionary's selection of words of the year (2016) and the political upheaval faced by the U.S., Britain, and Europe around that time raised 'Post-truth' to one of the most significant agendas of the times. Then, where does the era of Post-truth come from?

Psychological research introduces human cognitive bias as the answer to this question. Specifically, human cognitive limitations such as cognitive dissonance, group assimilation, and confirmation bias are the primary causes that impact Post-truth. Cognitive limitations mean that when we have the will and intention to believe in a (particular) fact, if the people around us believe in it, with only the slightest stimulus, we come to believe in it. In a nutshell, we believe in the fact as

---

[2] Of course, the situation in which the term 'Alternative Fact' is officially stated in the White House may be unfamiliar. However, it is still questionable whether it can be called a new phenomenon by the words such as a lie, an exaggeration, or a distortion are rare cases in politics and various other areas.

we want to believe in it as true via our limited cognitive capacity (McIntyre 2018). This is in line with the recent discussion of the echo effect, which is drawing more attention from scholarly works due to the spread of social media. It refers to a phenomenon in which beliefs and opinions are exchanged only with people of similar (social and political) disposition, strengthening specific tendencies and insulating from rebuttal. The spread of social media contributes to the situation where people of increasingly similar tendencies gather (online) with each other and share only identical opinions, strengthening self-confidence (Iyengar and Hahn 2009; Prior 2007; Kakutani 2018).

The spread of social media proceeds with the collapse of traditional legacy media. Current research on the crisis of journalism often points out web-based technology development, diversification of news channels, and changes in social and economic conditions as the leading causes of traditional journalism's crisis. In other words, the breaking down of the conventional news media production process -such as the absence of a gatekeeper – and changes in the (traditional) media news distribution process-such as heavy dependence on social media and news outlets' desperate strive for immediacy - contributed to the building of the social environment for Post-truth while bringing about a crisis in journalism (Kim 2018; Waisbord 2018).

Along with these factors, some scholars look at the demise of experts (expertise) as a primary driver for Post-truth (Nichols 2017). Tom Nichols argues that the development of the internet, including social media, deterioration of the quality of higher education (university education), and lack of exchange and communication between experts and the lay public led to the demise of expertise and an era of hatred of experts and expertise. Paradoxically, his concern is that the death of the expertise could be conducive to the age of anyone becoming an expert, contributing to the spread of disinformation and causing the collapse of traditional democracy symbolized by populism (Nichols 2017).

However, above all, it is postmodernism (and the spread of (social) constructionism based on it) that people concerned about Post-truth cast doubts on as a significant player in destabilizing the truth (and the facts). The central feature of postmodernism generally (mis)understood is the denial of Truth. For example, according to McIntyre (2018), "[T]here could be many answers, rather than just one, for any deconstruction. The postmodernist approach is one in which everything is questioned, and little is taken at face value. There is no right answer, only narrative (P.125)."

Ken Wilber (2017) offers a similar statement;

> If there were one line that summarizes the message of virtually all of the truly prominent postmodern writers (Jacques Derrida, Michel Foucault,

> Jean-François Lyotard, Pierre Bourdieu, Jacques Lacan, Paul de Man, Stanley Fish, etc.), it is that "there is no truth." Truth, rather, was a social construction, and what anybody actually called "truth" was simply what some culture somewhere had managed to convince its members was truth. Even science itself was held to be no truer than poetry. (Seriously.) There simply was no difference between fact and fiction, news and novels, data and fantasies. In short, there was "no truth" anywhere (P.17-18).

This critical remark also resonates with Kakutani's (2018) following explanation.

> [V]ery broadly speaking, postmodernist arguments deny an objective reality existing independently from human perception, contending that knowledge is filtered through the prisms of class, race, gender, and other variables. In rejecting the possibility of an objective reality and substituting the notions of perspective and positioning for the idea of truth, postmodernism enshrined the principle of subjectivity (P.37).

Academically speaking, the majority of the quotes mentioned above are too contentious to receive prompt approval. Still, they could be understood as snapshots reflecting on the general (mis)understanding of postmodernism[3] and (social) constructionism by people expressing apprehension over the concept of post-truth. Such a bitter attack on postmodernism naturally is reminiscent of the debacle named 'Science Wars' (Fischer 2021). As shown in the 'Science Wars,' rationalists seem to believe that attacks on 'objective' knowledge should pave the way for their cooptation where "the province of the left has been taken up by the right" (Bérubé 2011).

Although such a reproach on postmodernism seems to be a pervasive trend that reflects an accord of people concerned with the relationship between Post-truth and postmodernism, this diagnosis has at least two primary problems. The first one is related to the target of their criticism. Mostly, people worrying about Post-truth, including the authors mentioned above, group various French scholars under the umbrella of postmodernism. The philosophical concepts they produced did not assume a common intellectual ancestry or a

[3] How to understand 'postmodernism' is a highly controversial subject in many fields including sociology and STS, from the question of who the postmodernists are to the challenge of how to group such heterogeneous people in the same group. As François Cusset (2008) puts it, "[o]nly a gesture of rejection or direct opposition made it possible to assign them a unifying label (P.8)," although they expressed disagreement with each other in many ways. He provides a more detailed explanation of their discord (Cusset 2008).

common antithesis to counter together against, and their academic and political positions were always more discordant than congruent (Cusset 2008).

In addition, they sometimes distort or (at least) simplify their target theorists. For example, 'post-modern' theorists were blamed for denying (objective) reality and asserting that reality and truth could be constructed arbitrarily by language and power. However, this understanding is inconsistent with what theorists did in their work. For example, in his lectures at UC Berkeley, Foucault (2019) clearly explains what he studies under the name of problematization as follows;

> What I have tried to do from the beginning was to analyze "problematization," that is, how and why certain things, conducts, phenomena, and processes become a problem...... This kind of analysis, some people have interpreted it as a kind of "historical idealism." I think that this analysis of problematizations is completely different from any kind of historical idealism. When I say that I am studying the problematization of madness, or of sexuality, or of crime, **it's not a way to deny the reality of those things—crime, sexuality, or madness. On the contrary, I have tried to show, for instance, the real existence of social deviations which were the target of social regulation at a certain moment** (Emphasized by the author). And I have tried to show how real these social deviations were when they were returned to those institutions in which the doctors played an increasingly important role from the seventeenth century to the nineteenth century. I do not deny the existence of suffering, anxiety, deviant behaviors, and all those types of reactions and feelings due to brain damage and other [such causes]. The question I raise is this one: how and why were those different things brought together, and characterized, and analyzed, and treated as mental diseases? (P. 224-225)

As explained above, what Foucault is doing in his research projects is to place reality in the historical context. This project could be called 'historical ontology,' which is distinct from vulgar idealism. According to Paul Veyne, who is the great annotator of Foucault and an excellent archeologist and historian of the ancient world himself, Foucault's historical ontology is an effort to show that "what it could be otherwise" (Veyne 1997). The works of Foucault are not for the denial of reality but for digging into the context where reality exists. Namely, the opponents' blaming of Foucault seems to be attacking the straw man rather than finding the mark.[4]

[4] Besides reviewing Foucault's case, it is necessary to review how opponents misunderstand other French theorists' work in terms of reality (and fact). However, such

(Social) constructionism and the post-truth

Regardless of the misunderstanding of postmodernism by the opponents, French theory is considered to be an essential engine running the era of Post-truth. The other related target of accusation is (social) constructionism. Now that it is widely used in social science, the concept, which is not unfamiliar, has been spread after the publication of Thomas Luckmann and Peter Berger's book titled *The Social Construction of Reality,* which has been strongly influenced by Alfred Schultz's phenomenological sociology (Velody and Williams 1998; Hacking 1999). In particular, the sociology of scientific knowledge (and STS) strongly absorbed the influence of social constructionism and further developed it, even though they stem from different traditions[5]. The works of Edinburgh School's strong program and the Bath School of Harry Collins show significant achievements in how social constructionism can be related to the analysis of the (production of) scientific knowledge. Latour also borrows the term social constructionism for his first book's title, *Laboratory Life: (Social) Construction of Scientific Facts.* In the development trajectory of STS and the sociology of science (and scientific knowledge), constructionism has been used as an essential conceptual framework for various empirical research[6].

The power of constructionism comes from opening the black box of the production of (scientific) 'fact', focusing on various practices, technologies, and social interests surrounding the construction of reality rather than just leaving them as a priori or predefined. Through this approach, constructionism provides a new understanding of how scientific facts, understood as naturally given, are made. In other words, we have come to better understand the production of scientific facts through a constructionist approach, which leads us to delve into the complicated relationship between various scientific practices, which are more complex and multi-layered than commonly expected (Hong 1999).

---

a review is beyond the range of this chapter. François Cusset's eloquent work can be a good start to understand how French theory was (mis)understood and invented in the U.S. academia (Cusset 2008).

[5] Although social constructionism became popular jargon after publication of Luckmann and Berger's book, more influential work which contribute to wide application of the concept of social construction was Thomas Kuhn's seminal book, *The Structure of Scientific Revolutions.* For more detailed description of Kuhn's woks, please refer Hacking (2012).

[6] However, Latour erases the term 'social' from the second edition and uses the term constructionism without the modifier. For the debates between Latour and social constructionists in STS, see Bloor (1999a; 1999b) and Latour (1990; 1999).

The rise of the sociology of scientific knowledge can be considered a process of actively thinking about the characteristics of science as a social institution beyond the limitations of the functionalist sociology of science, such as Merton's approach. Of course, Merton's scientific sociology also mentions the relationship between science and society. However, it is understood that social influences in science should be avoided in functionalist sociology. When social factors influence science, it is regarded as false science or scientific errors rather than scientific advances. In other words, social influences in science are assumed to be an adverse effect that hinders scientists from establishing impartial facts. This approach has the limitation of mystifying and black-boxing science by distinguishing the realm of science from outside of science and stopping sociologists at the door of science (and laboratory).

Edinburgh School's Strong Program, which appears as an opponent of functionalism, opens the possibility of a sociological and (social) constructionist approach to scientific knowledge by presenting four methodological principles: Causality, Impartiality, Symmetry, and Reflexivity. These methodological principles require that when explaining beliefs about science, the same 'impartial' approach should be taken to analyze ideas about religion or political ideology. Furthermore, the form of explanation should be symmetrical when explaining scientific facts that are considered true and facts that are considered false. In other words, if the reason why scientific knowledge is considered false is related to social interests, truly successful knowledge should also be explained by social interests (Bloor 1976; Yearley 2005). Representative research under these tenets is Donald MacKenzie's (1981) discussion on statistical development and Shapin and Schaffer's (2017) work on the debate between Boyle and Hobbes.

The constructionist approach delves into how scientific facts were approved through various social practices by examining the historical process to produce scientific facts in more detail. By actively engaging with the constructionist approach, we come to the implicit assumption that "we could have other science (knowledge) that is not the science (knowledge) we have now," which implies that "our world could have been otherwise" (Hong 2004)[7]. These assumptions of social constructionism raise questions about the certainty of knowledge. Sometimes, however, when skepticism oscillates with the claim that anyone can produce knowledge and that knowledge has the same value regardless of its detailed production process, as Kakutani (2018) asserted, such claims can be converted to extreme relativism and scientific denialism. In other

---

[7] Of course, this assumption should be followed by the other critical assumption that 'however, it should be absolutely difficult to be otherwise.'

words, a constructionist understanding of knowledge can be (mis-)exploited and appropriated as the theoretical basis of Post-truth politics.

This cooptation has put aporia in the sociology of science (and knowledge) and STS, which are both based on a constructionist approach to knowledge-making. In other words, a question was raised about how the emphasis on the social and democratic characteristics of knowledge production, such as epistemological democracy, led by STS and sociology of science, should build a relationship with the Post-truth era (Jasanoff and Simmet 2017). STS scholars present various views on this issue. Collins himself was a scholar who led the social constructionist movement but argued that a new form of scientific (technical) approach is needed, providing a reflective assessment that "epistemological democracy" or "relativism," strongly impacted by social constructionism, eventually became the basis for analyzing Post-truth times. It is suggested that the boundaries for experts and non-experts should be established. Still, they can be drawn horizontally and flexibly according to problem setting rather than a hierarchical imposition. Collins and his colleagues add that we should study how this (differentiated) expertise can contribute to democracy and politics beyond technological limits (Collins and Evans 2002; Collins et al. 2017). In addition, Moore asserts the unavoidable need for 'expertise,' implying the relation of inequality or hierarchy even in the democratic regime. However, he also emphasizes that expertise should be legitimated democratically through diverse institutional arrangements (Moore 2017). Against these critiques, Sismondo emphasizes that the STS's detailed account of knowledge is not about the simplified epistemological democracy in which all knowledge has the same value but a complicated process requiring a variety of infrastructure, scientific practices, efforts, and structures to produce that fact (Simondo 2017a; 2017b). Pitts-Taylor also advocates the role of STS in the Post-truth era. Her main argument is that the task STS should achieve is to form ethical relations with scientific knowledge by revealing that (scientific) facts and knowledge are the result and reflection of various social relationships. This revelation can be made through transparent tracking of the process of knowledge production (Pitts-Taylor 2020). This controversy displays a critical attitude toward STS by revealing the social nature of knowledge production and asking how to establish an ethical relationship with knowledge.

However, this fierce controversy implies STS's difficulties in researching knowledge in the Post-truth era. One of the primary tasks of the STS has been to question the hierarchical relations of knowledge and its production process and emphasize the diversity of knowledge production through careful observation of the processes and practices of knowledge production. However, the 'post-truth' leads STS scholars to reflect on their academic practices concerning that STS contributes to flattening the relations between knowledge and confusing people with epistemic anarchy and democracy. It may have left room for misunderstanding epistemological democracy as synonymous with epistemological 'anarchy.'

What does having a 'critical' understanding of knowledge mean under these conditions? Social constructionism has offered a critical view by tracing and revealing the detailed knowledge-making process, which the functionalists did not address. Their approach enables sociologists to have a lens to understand (scientific) knowledge as a social product, not as a naturally given object. Can STS still have any 'critical stance' in the era of Post-truth? Can the research focusing on the social nature of knowledge (still) play a critical role by exposing the social prejudices and limitations of the various knowledge produced? Can we maintain a critical stance on STS by exposing the hybrid nature of knowledge production and emphasizing epistemological democratization through the work of blurring the boundaries between experts and non-experts that science and sociology have consistently questioned? For example, is there enough critique from STS to reveal the racist, sexist, and capital-oriented characteristics of the algorithm? Or is it a critical stance to understand algorithms not as an exclusive practice of programming engineers but as a result of a vast assemblage constituted by users, various technologies, and organizations?

Post-truth poses aporia to STS in asking how to understand knowledge and its production process and, simultaneously, how to renew the 'critique' that has been a significant feature. That means it seems to be asking how to renew the critique when it is widely spread or when the opponents criticize it. In other words, it can be understood as whether it is possible to restore its critique when the novelty and creativity of criticism are worn out through the process of social diffusion. Considering these questions, this chapter will examine the discussions on critique by Foucault and Boltanski and review how the critique of STS can be renewed.

Foucauldian critique

STS and the sociology of (scientific) knowledge have reviewed the formation, establishment, and utilization of knowledge from the standpoint of social constructionism. Still, Foucault and Foucauldian research was not considered a primary reference in the initial stages of STS. Despite limited exchange, there are two commonalities shared by the two camps. The first one is that both seek to apprehend knowledge rather than as a simple representation of reality (and nature). Secondly, they emphasize detailed tracking of the historical process of knowledge-making. These shared commonalities lead more STS scholars to pay attention to the Foucauldian approach to understanding knowledge and science, focusing on his concepts of power/knowledge and genealogy. Nowadays, it is hard for STS scholars to research the knowledge-production process without referencing the knowledge/power matrix, a primary Foucault's tenet.

However, how to understand the Foucauldian concept of 'critique' with the STS (and (social) constructionist) perspective remained unanswered. There are few works to bridge the two camps in terms of 'critique,' although it has received more attention from diverse fields (Fassin and Harcourt 2019; Fassin and Honneth 2022). To fill this gap, I would like to examine how Foucault's deliberation on 'critique' can contribute to STS in the era of Post-truth.

In a 1978 lecture organized by the Société Française de Philosophie, Foucault began his remarks by confessing the difficulty of entitling his lecture and mentioned that the question he wants to ask is 'What is the critique?' This question would later be confirmed as the title of the lecture, which is used as a document that raised important questions dealing with Foucault's central theme, 'subject'/'self.'

Foucault's understanding of critique begins with his interpretation of Kant's[8]. In his College de France lecture series in 1983, he suggests that Kant "have founded the two great traditions which have divided modern philosophy" (Foucault 2010: 20). The first tradition is "the question of the conditions of possibility of a true knowledge." This tradition developed as the 'analytic of truth' based on the scrutiny of "the conditions of possibility of a true knowledge," linked to Anglo-Saxon analytical philosophy. In the other tradition, critique could play a role as ethos, a genealogical inquiry for "an ontology of the present, of present reality, an ontology of modernity, an ontology of ourselves". (Foucault 2010:20-21) Similarly, Kant asked about how to understand the French Revolution, the critical thought concerned with "contemporary historical actuality" (Harcourt 2020). The former is an epistemological inquiry into truth, dealing with the conditions under which knowledge(/truth) can be established, with the structure and formation where perception or statements about objects can be claimed to be true (Harcourt 2020). In Foucault's understanding, critique should not be limited to recognition (of the truth). Instead, as the second tradition shows, the critical thoughts (of Foucault's more profound interest) should perform a historical-philosophical function. In this regard, Foucault explains as follows:

---

[8] In the lectures and discussions of late Foucault, Kant's critical philosophy and enlightenment (Aufklärung) emerged as critical topics. At first, Foucault emphasized the difference between the critique and Aufklärung, while later he highlights the close relationship between the two. A comparison of two lectures, Foucault (1996) and Foucault (2010), helps delve into Foucault's different understanding of the Kant's critique and Aufklärung.

> I believe that we can find, in the field of philosophical activities since the beginning of the 19th century, two poles which are related to each other, and which cannot be reduced to each other. At one of these poles, you find such questions as, "What is truth? How is it possible to know the truth?" It is the pole of philosophy as a formal ontology (ontologie formelle) of truth or as a critical analysis of knowledge. And at the other pole, you find such questions as: "What is our actuality? What are we as part of this actuality? What is the target of our activity of philosophizing insofar as we are part of our actuality?" Those questions deal with what I should call the historical ontology of ourselves or the critical history of thought. (Foucault 2015:84, translated by the citer)

Foucault's goal is to seek a critique that explores the changes in the 'subject,' the conditions of our own changes, and infinite possibilities while relating his critique to the latter. In this critical philosophy, critique is not a theoretical and abstract principle or doctrine but rather an ontological attitude and, further, a philosophical way of life, which aligns with Kant's concept of enlightenment that Foucault explicates. Kant's enlightenment, represented by the maxim "Sapere Aude," is related to immaturity in which humankind stays and is forced to stay, which is a particular state of incompetence "in which humanity would be maintained." Kant characterizes this incompetence through some interrelationship with the authority that keeps humankind immature. Foucault explains that Kant's understanding of this enlightenment has an appeal for courage (Foucault 2015).

Above all, this courage has deep shared points with the critique that Foucault understands. Foucault interprets critique in connection with governmentality/ governmentalization after the 16th century, saying there is something similar to virtue. In other words, the question of 'how to govern' is inseparable from "how not to be governed," considering Foucault's influential agenda of the inseparability of resistance and power. However, rather than meaning a statement of opposition as "we don't want to be governed at all," critique is a moral and political attitude with the question of "how not to be governed like that, by that, in the name of these principles, in view of such objectives and by means of such methods, not like that, not for that, not by them?" Foucault (1996) summarizes clearly as follows:

> [C]ritique is the movement through which the subject gives itself the right to question truth concerning its power effects and to question power about its discourses of truth. Critique will be the art of voluntary inservitude, of reflective indocility. The essential function of critique would be that of desubjectification in the game of what one could call, in a word, the politics of truth (P. 386).

After all, Foucault's critique can be understood as a process of questioning the device that allows me to exist in this way in order not to be ruled and a method of asking about the historical basis of this existence. The way he understands 'historical ontology' is to study the relationship we have with the truth and further analyze the way we have formed ourselves through the history of our thoughts (Foucault 1996:101). This means that we can question the beings we take for granted (including us), which is an exploration of the 'singularity' of existence. As Paul Veyne (2010) puts it, singularity is the ontological principle of Foucauldianism. In an understanding of Foucauldian ontology, "in every period, the historical universe has always been a chaos of singularities that stem from the previous period of chaos" (P.78). We are formed and characterized by specific institutions, practices, and technological devices through the combination of historical subjectivity in the chaos of specificities. Tracing this process can be a way to do a critical (historical) ontology, and it can be understood that clarification of the procedure is the way to carry out critique.

Boltanski's critique

If Foucault's understanding of critique begins with a conversation with Kant, Boltanski's understanding of critique stems from a tension with Bourdieu. Boltanski, who started his academic journey as Bourdieu's student and colleague, collaborated with him in publishing "La Production de l'Idéologie Dominante" in the academic journal *Actes de la Recherche en Science Sociales.* After their initial collaboration, Boltanski tried to distance himself from Bourdieu to overcome the limitations of Bourdieu's work.

Above all, one of the most significant limitations, Boltanski points out, is from his theory of habitus. According to Boltanski, Bourdieu's concept of habitus seems like a black box since in the theory, the actor is described as a mere being who cannot truly choose a strategy but is driven by the multiple causal forces operating behind him (Dosse 1999). Furthermore, what Boltanski criticizes most significantly regarding Bourdieu's discussion is the asymmetric relationship between sociologists and the lay public. In other words, in Bourdieu's sociology, a sociologist is an awake being armed with scientific knowledge and reveals the hidden truth to ordinary people who are ruled by social structure but do not understand it. In this landscape, the critical competencies of each actor are overlooked or underestimated (Celikates 2012).

Furthermore, this underestimation of critical competencies leads to excessive emphasis on dispositional attributes given by the structure, ignoring the attributes engraved on each actor. In addition, Boltanski explains that there is an explanatory system and language that actors already acquire, and based on this, they can perform their actions. However, these are ignored by Bourdieu's sociology. In other words, ordinary actors in society can also criticize and reflect by sufficiently absorbing and reinterpreting various (academic) knowledge and achievements created by sociology (Boltanski 2011).

Boltanski emphasizes the cyclical relationship between 'disposition' and structure castrated from the context, leading to overlook the uncertainty actors face in particular situations. Accordingly, in Bourdieusian sociology, it becomes impossible to understand the various choices and actions actors mobilize to solve problems through their capabilities in specific situations. In addition, Boltanski criticizes Bourdieu's critical sociology since it underestimates the significance of description as a primary weapon of sociology. The reason why description can act as a significant weapon of sociology is that it is a channel to understand actors' various social practices and the meanings of their self-understanding in specific social situations. In other words, for Boltanski, the mission of sociology is to reveal the meaning and importance of uncertainty in real social life and to understand how people behave, negotiate, and compromise in contradictions revealed through their capacity (Boltanski 2011).

With this approach, Boltanski adds a new understanding to the concept of critique. Unlike Bourdieusian sociology, where the critical capacity is given only to sociologists, Boltanski emphasizes the essential competence of the actor. According to him, in a society where various values coexist, an order is carried out through the consensus of actors. In particular, by securing more universality, values become hierarchized, and consensus can be reached based on them. The principle operating in this process may be called justification (Kim 2018). However, critique emerges when the agreed social order is shaken. In other words, critique comes from the judgment that it can no longer survive as it is now because the problematic order does not find the means to resolve contradictions within itself (Lee 2018).

An essential point in Boltanski's discussion on critique is that if it is the fundamental capability of actors rather than the privilege of researchers, the processes previously understood as revelations to Bourdieu would be carried out through justification in daily life, and reality should be (re)constructed through the actors who perform them. In other words, the critique always exists with those who have already been criticized (Susen 2014).

Toward the new way of critique

Above all, Foucault and Boltanski's critique has a resonating point that differs from the ideological critique, emphasizing the importance of debunking the structural logic of obfuscating reality. A 'critique' represented by Bourdieu is questioning the impossibility of free will and showing the magnitude of structural force. It is the habitus that uncovers the secret. In the Bourdieusian sense, 'truth' should be introduced from the outside to the inside. Liberation here can be awakening from illusion through the (outside intellectual's) critique (Fassin 2019).

On the other hand, a critique of Foucault and Boltanski is immanent and, therefore, performative. In other words, the capability of critique exists inside and is plural and, therefore, not directed in one direction. At the same time, the possibility of a change in internal order can be opened by performing critique. For Foucault, critique is a declaration that it will no longer exist like this by exploring its historical ontology with its genealogy. The core of Foucault's critique is to ask questions about how we came here. Based on this, you will be able to change the relationship between yourself, others, and the various technologies and devices surrounding it and carry out practices not to be ruled like this as an ethical subject. In addition, as Fassin (2019) pointed out in his essay on critique, critical thoughts were always profoundly intertwined with the locationality and temporality of the object. Foucault's illumination of the critique reminds us of the strength of STS as an opener of the black-box social arrangement, which is called for urgently to face the challenge posed by the Post-truth.

From Boltanski's point of view, critique can be understood as the power to reconstruct reality through actors. The only way to shake the justified social order is through the critical capabilities of the actors. Therefore, it is possible to declare that "justification is not necessary where critique does not exist" (Kim 2018). This implies that the request for new justification is only possible through critique from competent actors.

The discussion of critique between Foucault and Boltanski allows us to explore, among other things, the possibilities of a new analysis that STS and the sociology of science in the Post-truth era might envision. The renewed critique that we can update through Foucault and Boltanski will be related to the following questions; How did we get to the present now? How did we come to know what we believe we currently know? How does our epistemological community's way of knowing exist? Why do we have this way of knowledge production?

These questions resonate with the newly suggested research program, *Non-normative critique* (Hansen 2016). In an effort to bridge Foucault and pragmatic sociology, Hansen investigates new and innovative critique that problematizes the present and the assemblage structuring ourselves. Despite their differences in time frame, which means that Foucault's program is diachronic (more focusing on the historical) while pragmatic sociology is synchronic, each program questions the present not from the outside but from the inside. According to Hansen (2016), it is a new way to "repoliticize contemporary modes of governing (p.141)" This effort to reformulate the critique can be extended to STS.

Concluding remarks

Let's return to fact-checking. The fact-checking process can be understood as, above all, a strong critique of disinformation. For any reason, continuous verification of (fake) news may be necessary. However, as critics of Post-truth explain and argue, it does not seem easy to just wait for "truth" to save us. Such efforts matter and may be helpful, but to what end (Redfield 2019)?[9] In this sense, we need to reflect on where we stand, how we come to where we are, and a sincere tracing to capture the critical capabilities of actors in our community.

Sometimes, a deeper understanding of how people deal with information in a world of the diversified media environment and political polarization and capturing the (various) responses that people exhibit toward the Post-truth could be more critical than a normative critique of disinformation and laments on losing facts which should be restored[10]. In other words, STS and the sociology of science need to capture the various symptoms of Post-truth, which emerges as 'social facts,' and to explore the various actors, practices, and tools that consist of social arrangement, making this possible. By examining the social reactions and (dis)obedience exhibited by the arrangement, we could deal with the historical ontology of Post-truth accounts.

This is the way to overcome the limitations and legacy that the 'science war' left behind. As mentioned above, the discussion over 'Post-truth' politics reminds us of the science war in the 90s, which still bears the question of the legacy of such a fierce debate among diverse camps. In order to obtain more productive results (legacy), STS needs to refine (and/or revise) what role critique can play in the Post-truth era. Foucault and Boltanski (and the French pragmatic school) seem to be good resources to move the critique beyond traditional ideological analysis. Their novel way of critique resonates with a reflexive attitude to understand what we know. In this sense, the following statements of Jasanoff (2017) are worth citing.

> Only through sustained reflection on why we know what we think we know can we find ways to strengthen, even augment, our collective knowing—and so push back against those who would dismantle the

---

[9] In his essay on the 'critique', Redfield (2019) questions on how critique could be useful in the development/international aid field where the hyper-utilitarian ethos is dominant. He concludes his essay suggesting that "the real question always remains "useful to *what end?* (Emphasis in the original)"

[10] One of the serious limitations of fact-checking as a critical praxis is that it does not throw the doubt on the binary option between the true and false, which overlook the significance of the context and contingency in understanding the world. For more detailed explanation, please refer the Zerilli (2020).

human institutions we have entrusted with the hard task of making public knowledge.

This reflexive attitude has strong shared points with Foucault and Boltanski's critique. Genealogical critique on the historical ontology and tracing the inner capabilities of actors of our community could deepen our understanding of ourselves and where we are (and possibly what has to be done now). Despite the significance of having a reflexive attitude, there are still tasks that need to be accomplished. One of them is how to understand critique in practical terms. The modern social and political theoretical tradition has emphasized the importance of uncovering the tensions and relationships between critique and praxis (practice) as a central issue. It requires us to consider how to relate praxis with reflexive (and non-normative) critique. In this sense, Harcourt (2020) offers an important clue to develop our understanding of their relationship. According to him, one of the traditional tenets of the relationship could be summarized as the following question; "What is to be done?". The question is placed in Marxian tradition, emphasizing the importance of practical social change with critique. In this sense, critique has played the role of a lighthouse guiding social change to a desirable world. However, he asserts that we need to go beyond this question with the following question, "What more am I to do, and how does what I am doing work?" (Harcourt 2020: 24). These questions can be a good starting point to (re)build the relationship between the praxis and critique with reflexive attitude.

References

Bérubé, Michael. 2011. "The Science Wars Redux." *Democracy: A Journal of Ideas* 19: 64-74.

Bloor, David. 1976. *Knowledge and Social Imagery*. Chicago: University of Chicago Press.

Bloor, David. 1999a. "Anti-Latour." *Studies in History and Philosophy of Science* 30(1):81–112. doi: 10.1016/s0039-3681(98)00038-7.

Bloor, David. 1999b. "Reply to Bruno Latour." *Studies in History and Philosophy of Science* 30(1):131–36.

Boltanski, Luc. 2011. *On Critique: A Sociology of Emancipation*. Translated by Gregory Elliott. Cambridge: Polity.

Celikates, Robin. 2012. "Systematic Misrecognition and the Practice of Critique: Bourdieu, Boltanski and the Role of Critical Theory." In *Recognition Theory and Contemporary French Moral and Political Philosophy: Reopening the Dialogue*, edited by M. Bankovsky and A. Le Goff. London: Palgrave Macmillan.

Collins, Harry M., and Robert Evans. 2002. "The Third wave of science studies: Studies of expertise and experience." *Social Studies of Science* 32(2): 235-296.

Collins, Harry, Robert Evans, and Martin Weinel. 2017. "STS as Science or Politics?" *Social Studies of Science* 47(4): 580-586.

Cusset, François. 2008. *French Theory: How Foucault, Derrida, Deleuze, & Co. Transformed the Intellectual Life of the United States.* Translated by Jeff Fort. Minneapolis, MN: University of Minnesota Press.

Dosse, François. 1999. *Empire of Meaning - The Humanization of the Social Sciences.* Translated by Hassan Melehy. Minneapolis, MN: University of Minnesota Press.

Fassin, Didier. 2019. "How Is Critique" In Fassin and Harcout 2019, 13-35.

Fassin, Didier, and Axel Honneth, eds. 2022. *Crisis Under Critique: How People Assess, Transform, and Respond to Critical Situations.* New York, NY: Columbia University Press.

Fassin, Didier, and Bernard E. Harcourt, eds. 2019. *A Time for Critique.* New York, NY: Columbia University Press.

Fischer, Frank. 2021. *Truth and Post-Truth in Public Policy.* New York, NY: Cambridge University Press.

Foucault, Michel. 1996. "What Is Critique?" Pp. 382-98 in *What Is Enlightenment?: Eighteenth-Century Answers and Twentieth-Century Questions,* edited by James Schmidt. Berkeley, CA: University of California Press.

Foucault, Michel. 2010. *The Government of Self and Others: Lectures at the Collège de France 1982–1983.* Translated by Graham Burchell. New York, NY: Palgrave Macmillan.

Foucault, Michel. 2015. *Qu'est-Ce Que la Critique?: Suivie de La Culture de Soi,* edited by Henri-Paul Fruchaud and Daniele Lorenzini. Paris: Librairie Philosophique J. Vrin.

Foucault, Michel. 2019. *"Discourse and Truth" and "Parresia",* edited by Henri-Paul Fruchaud and Daniele Lorenzini. Chicago, IL: University of Chicago Press.

Fujimura, Joan H., and Christopher J. Holmes. 2019. "Staying the Course: On the Value of Social Studies of Science in Resistance to the 'Post-truth' Movement." *Sociological Forum* doi: 10.1111/socf.12545.

Graves, Lucas, Brendan Nyhan, and Jason Reifler. 2016. "Understanding Innovations in Journalistic Practice: A field Experiment Examining Motivations for Fact-Checking." *Journal of Communication* 66(1): 102-138.

Hacking, Ian. 1999. *The Social Construction of What?* Cambridge, MA: Harvard University Press.

Hacking, Ian. 2012. "Introductory Essay." In *The Structure of Scientific Revolutions,* edited by Thomas S. Kuhn, vii-xxxvii. Chicago, IL: University of Chicago Press.

Hansen, Magnus Paulsen. 2016. "Non-Normative Critique: Foucault and Pragmatic Sociology as Tactical Re-Politicization." *European Journal of Social Theory* 19(1):127–45. doi: 10.1177/1368431014562705.

Harcourt, Bernard E. 2020. *Critique and Praxis.* New York, NY: Columbia University Press.

Hong,Sungook. 1999. *Science and Technology as Productivity and Culture.* Paju: Moonji Publishing Co. (In Korean).

Hong,Sungook. 2004. *Science and Science Studies.* Seoul: Seoul National University Press. (In Korean).

Islam, Md Saiful, Abu-Hena Mostofa Kamal, Alamgir Kabir, Dorothy L. Southern, Sazzad Hossain Khan, SM Murshid Hasan, Tonmoy Sarkar. 2021. "COVID-19 Vaccine Rumors and Conspiracy Theories: The Need for Cognitive Inoculation Against Misinformation to Improve Vaccine Adherence." *PloS one 16(5)*: e0251605. https://doi.org/10.1371/journal.pone.0251605.

Iyengar, Shanto, and Kyu S. Hahn. 2009. "Red Media, Blue Media: Evidence of Ideological Selectivity in Media Use." *Journal of Communication* 59(1): 19–39. doi:10.1111/j.1460-2466 .2008.01402.x

Jasanoff, Sheila, 2017 "What Should Democracies Know?" First 100 Days. Available at: https://first100days.stsprogram.org/2017/03/28/what-should-democracies-know/

Jasanoff, Sheila, and Hilton R. Simmet. 2017. "No Funeral Bells: Public Reason in a 'Post-Truth' Age." *Social Studies of Science* 47(5): 751–70.

Kakutani, Michiko. 2018. *The Death of Truth: Notes on Falsehood in the Age of Trump*. New York, NY: Tim Duggan Books.

Keyes, Ralph. 2004. *The Post-Truth Era: Dishonesty and Deception in Contemporary life*. London: Macmillan.

Kim, Ju-Ho. 2018. "The Spirit of the 68 Movement, from Critique of Capitalism to its Justification: Focusing on Boltanski and Chiapello's 'New Spirit of Capitalism'." *Journal of Western History* 138(0): 66-96. (In Korean)

Kim, Sumi. 2019. "Regarding the Truth In/About the 'Post-Truth' Era : A View on Journalism and the Politics of Truth." *Media & Society* 27(4): 49-103. (In Korean)

Latour, Bruno. 1990. "Technology Is Society Made Durable." *The Sociological Review* 38(1_suppl):103–31. doi: 10.1111/j.1467-954x.1990.tb03350.x.

Latour, Bruno. 1999. "For David Bloor... And beyond: A Reply to David Bloor's 'Anti-Latour.'" *Wordpress.Com*. Retrieved July 29, 2023 (https://reclus.files.wordpress.com/2009/03/latour-bloor-and-beyond.pdf).

Lee, Sang-Gil. 2018. *Atlas' Feet: Reading Bourdieu in Postcolonial Context*. Paju: Moonji Publishing Co. (In Korean)

MacKenzie, Donald. 1981. *Statistics in Britain: 1865-1930: the Social Construction of Scientific Knowledge*. Edinburgh: Edinburgh University Press.

McIntyre, Lee. 2018. *Post-Truth*. Cambridge, MA: MIT Press.

Moore, Alfred. 2017. *Critical Elitism: Deliberation, Democracy, and the Problem of Expertise*. Cambridge, England: Cambridge University Press.

Nichols, Tom. 2017. *The Death of Expertise: the Campaign Against Established Knowledge and Why It Matters*. New York, NY: Oxford University Press.

Oxford Dictionaries. 2016. "Word of the Year 2016." https://languages.oup.com /word-of-the-year/2016/.

Pitts-Taylor, Victoria. 2020. "Science, Critique, Authority, and Accountability." *Sociological Forum*, 35(1): 229-233.

Prior, Markus. 2007. *Post-Broadcast Democracy: How Media Choice Increases Inequality in Political Involvement and Polarizes Elections*. New York, NY: Cambridge University Press.

Puri, Neha, Eric A. Coomes, Hourmazd Haghbayan, and Keith Gunaratne. 2020. "Social Media and Vaccine Hesitancy: New Updates for the Era of COVID-19 and Globalized Infectious Diseases." *Human Vaccines & Immunotherapeutics* 16(11): 2586-2593.

Redfield, Peter. 2019 "The Usefulness of Uncertain Critique." In Fassin and Harcout, 71-90.

Shapin, Steven, and Simon Schaffer. 2017. *Leviathan and the Air-Pump: Hobbes, Boyle, and the Experimental Life.* Princeton, NJ: Princeton University Press.

Sismondo, Sergio. 2008 "Science and Technology Studies and an Engaged Program" In The Handbook of Science and Technology Studies (Third Edition), edited by Edward J. Hackett, Olga Amsterdamska, Michael Lynch, Judy Wajcman, 63-67. Cambridge, MA: MIT Press.

Sismondo, Sergio. 2017a. "Post-Truth?" *Social Studies of Science* 47(1): 3–6. https://doi.org/10.1177/0306312717692076.

Sismondo, Sergio. 2017b. "Casting a Wider Net: A Reply to Collins, Evans and Weinel." *Social Studies of Science* 47(4): 587–92. https://doi.org/10.1177/0306312717721410.

Snyder, Timothy, 2021. "The American Abyss: A Historian of Fascism and Political Atrocity on Trump, the Mob and What Comes Next." *The New York Times*, January 9. https://www.nytimes.com/2021/01/09/magazine/trump-coup.html

Susen, Simon. 2014. "Is There Such a Thing as a 'Pragmatic Sociology of Critique'? Reflections on Luc Boltanski's *On Critique*" Pp. 173-210 in *The Spirit of Luc Boltanski: Essays on the 'Pragmatic Sociology of Critique',* edited by S. Susen and B. S. Turner. London: Anthem Press.

Tesich, Steve. 1992. "A Government of Lies." *The Nation*, January 6.

Velody, Irving, and Robin Williams. 1998. *The Politics of Constructionism.* London: SAGE Publications.

Veyne, Paul. 1997. "Foucault Revolutionizes History." Pp. 146-82 in *Foucault and His Interlocutors* edited by Arnold Ira Davidson. Chicago, IL: University of Chicago Press.

Veyne, Paul. 2010. *Foucault: His Thought, His Character.* Translated by Janet Lloyd. Cambridge: Polity.

Waisbord, Silvio. 2018. "Truth Is What Happens to News: On Journalism, Fake News, and Post-truth." *Journalism Studies* 19(13): 1866-1878.

Walter, Nathan, Jonathan Cohen, R. Lance Holbert, and Yasmin Morag. 2020. "Fact-Checking: A Meta-Analysis of What Works and for Whom." *Political Communication* 37(3):350–75. doi: 10.1080/10584609.2019.1668894.

Wilber, Ken. 2017. *Trump and a Post-Truth World.* Boulder, CO: Shambhala Publications.

Yearley, Steven. 2005. *Making Sense of Science Understanding the Social Study of Science.* London: SAGE.

Zerilli, Linda M. G. 2020. "Fact-Checking and Truth-Telling in An Age of Alternative Facts." *Le Foucaldien* 6(1): 1–22.

Chapter 3

Social reflexivity as an approach to critical life science research on human diversity

Tino Plümecke and Laura Schnieder

Abstract: Analyzing empirical data about researchers in human genetics and the neurosciences who are concerned with human diversity, this paper investigates research approaches that understand themselves to be reflexive in the sense of a critical and transformative approach to dominant paradigms and to society. We are interested in *how* this reflexivity is constructed, considered, and enacted. Therefore, we take reflexivity 1) as an *empirical object* of inquiry through which scientific actors describe their actions and 2) as an *analytical concept* that can be used to capture the distinct characteristics of the respective research projects. Using examples from our research on critical life science, we analyze in what respects the researchers present themselves as diverging from the mainstream of their discipline—from the concepts and theories related to their work—and what kind of conflicts this causes within the scientific community. To analyze these forms of reflexive research, we use a preliminary concept which we call "social reflexivity." We elucidate this concept in relation to "varieties of reflexivity" (Woolgar 1991), briefly differentiate it from approaches such as "strong reflexivity" (Harding 1993) and "diffraction" (Barad 2007; Haraway 1997) and discuss its possibilities and limitations.

Keywords: Reflexive Research, Human Diversity in Genetics and Neuroscience, Social Reflexivity, Critical Life Science

***

Introduction

"Reflexivity" today is, first and foremost, a buzzword. While reflexivity, in the sense of a practice of thinking about the possibilities of knowledge, is as old as the written records of reasoning itself (Sandywell 1996; Schällibaum 2001), the widespread sense of the word as "turning one's thoughts back" comes from the Renaissance and modern European thought. Currently, the term appears in a number of academic contexts and with a variety of semantic connotations. Its different meanings can be outlined as follows: the term "reflexivity" describes the need for self-consciousness or self-awareness in the research process and qualitative social sciences, especially in ethnographic research (Davies 2008; Dean 2017; Flick and Hoppe 2021; Hammersley and Atkinson 2019; May and Perry 2011; Steinert 1999); it refers to what thinking can and cannot do in philosophy (Grim and Rescher 2012); it can denote, in the sociology of science, the analysis of the social nature of scientific knowledge (Bloor [1976] 1991; Lynch 2000); and it can be used as a heuristic principle for social scientific research (Bourdieu and Wacquant 1992; Wacquant 1992) or to analyze specific societal modes such as "reflexive modernity" (Beck 1997; Beck, Giddens, and Lash 1994). The term is also used in everyday language, which adds to its wealth of meanings. Attempts to systematize the various approaches are so numerous that they could become the subject of a separate metastudy (e.g., Ashmore 1989; Barla 2021; Finlay 2002; Lynch 2000; Subramani 2019; Whitaker and Atkinson 2021; Zienkowski 2017). Moreover, because of these many accounts and the widespread presence of the concept, a large part of contemporary social science research could be subsumed under the paradigm of a "reflexive turn" (Forster 2014:591; Mauthner and Doucet 2003).

Seeking to reconstruct the meaning of the term in a field of specific use, in the following study, we take an empirical look at research projects in the life sciences, which, in one way or another, claim to mobilize a kind of reflexive scientific approach. Researchers in these fields often characterize such approaches by describing the science they practice with adjectives such as "critical," "progressive," "transformative," "responsible," "reflexive," or "alternative," or as "feminist," "queer," "radical," or "good" (e.g. Allen 2019; Bryant, Grossi, and Kaiser 2019; Harding 2008; Mayberry, Subramaniam, and Weasel 2001; Rose and Rose 1976a; 1976b). What these approaches have in common is that their proponents all understand themselves to be critical vis-à-vis the disciplinary mainstream. We take up the term "critical," also used by these authors, to describe research approaches that aim to be transformative—intended to intervene in their own discipline and possibly into society.

In investigating the specific characteristics of such approaches and research projects, we are interested in how these propositions are generated and enacted. As "reflexivity" and "being reflexive" are key terms in some of these approaches, we specifically analyze how this entity of reflexivity is performed.

The two research areas we look at are human genetics, with its classifications based on ethnicity, ancestry, and population, and the neurosciences, with their conceptualizations of sex and gender. Both are particularly suited to looking at the practice of reflexivity in knowledge production because they recognize that their objects of inquiry are heavily affected by sociocultural and political influences. We take our empirical data from document analyses, interviews, and participatory observation. Focusing on reflexivity as one of the critical elements for understanding the specific practice in both fields, we explore the term in two ways: firstly, we investigate reflexivity as an *empirical object* that scientific actors use to describe their activity; and secondly, we use the term as an *analytical concept* that we hope can capture specific characteristics of the research projects we are examining.

While most work on reflexivity focuses either on conceptual questions or on descriptions of ways in which the term is used, we are interested in a recursive connection between the two approaches. We thereby seek to understand the specifics in the fields of "critical science" through an interactive approach linking conceptual considerations with empirical findings and vice versa. We believe such an approach can strengthen both the empirical analysis and the theoretical content of the concept of reflexivity. Using abductive logic, we outline the essentials of an analytically sophisticated concept that we programmatically term "social reflexivity." We begin by presenting some empirical observations, followed by selected considerations about the analytical power of reflexivity, to then summarize our findings and pose questions for further discussion.

Empirical observations: "I want my experiments! They just have to be different ..."

To empirically illustrate how some life scientists consider reflexivity to be a valuable concept in practice, we begin by highlighting insights taken from accounts given by four scientists about their own work. We glean these insights by analyzing these researchers' publications, examining their projects through participatory observation, and directly interviewing them. Each of these four researchers represents different approaches to reflexive practices, and their characterization thereby underlines the richness of reflexivity in the form of critical, interventional approaches in the life sciences.

We have given the first researcher the pseudonym of Catharina A. She is a scientist in the field of neurosciences and is currently using fMRI technology in a research project on the mental representation of sex/gender in the human brain. She describes herself as a "critical" "gender and brain researcher" who "is searching for alternatives" (Catharina A. 2019; 2022). In her publications, C.A. repeatedly uses terms such as "reflect," "reflexive," and "reflexivity" to explain what she is trying to implement in her research, and in particular, she uses approaches grounded in feminist and queer theory. She writes, for example, that "a queer-neuroscientific approach has to reflexively address how neuroscience deals with the categories of sex/gender and sexuality" and that "neuroscience should use reflexivity as a main inside tool for working towards a better neuroscience." She also criticizes the "unreflective generalization of results," which she understands to be typical in mainstream neuroscience. In our interview, she delineates her task as a researcher along the themes of implementing critique and facing the challenges in the very practice of *doing* research in a critical way: "And I think you can stay a critic; you simply have to. And at the same time, you have to do your experiments to bring some intervention into the world." So, while C.A. decidedly continues to work in the field of the neurosciences, she wants her experiments to be "different."

In the two in-depth interviews, we conducted with C.A. and an ethnography of research meetings among her team, two possible hindrances to such an endeavor *in practice* become quite visible. Firstly, C.A. considers a lack of interdisciplinarity in her research field to be a barrier to her reflexive project. She criticizes that inter- and transdisciplinarity is mainly reduced to the notion of multidisciplinarity, so instead of working together, it remains a juxtaposition of the respective disciplinary paradigms. From her point of view, a close collaboration between different—social scientific, philosophical, and neuroscientific—approaches are needed to correctly investigate questions of how sex/gender are to be represented in the brains that she investigates. Secondly, in both the interviews and our ethnography, methodological challenges to implementing feminist and queer approaches in neuroimaging experiments in humans become apparent. C.A. explains that well-established statistical methods (e.g., cluster analysis) are hard to employ in a reflexive manner because categories of sex/gender that should, in fact, be the subject of investigation are often taken for granted. Entanglements between society and science are obscured in the procedures, concepts, and methods of evaluation used.

C.A. strongly emphasizes the importance of reflexivity in her publications and in our interview. At the same time, the specific strategy she might take to implement this stance remains largely unclear to her, as such an approach is not yet sufficiently developed in her field. The difficulties that C.A. highlights were also brought to our attention in a discussion in which we were participant observers, where C.A.'s working group struggled over the possibilities of implementing research methods that deal as critically as possible with categories of sex/gender while still claiming to produce scientifically sound and reproducible results. Thus, in addition to reflecting on how sex/gender and sexuality are categorized in neurosciences, C.A. uses the term "reflexivity" primarily to describe choices about methodology and general research practice and about implementing feminist and queer approaches in the research process.

The second researcher, whose work and interview we would like to examine here, expresses her relation to the concept of reflexivity in a similar but more explicit way. Katja L., a life scientist by training who nowadays describes herself as a feminist STS scholar, explains: "Reflexive means for me firstly ... that I consider it essential to transparently reveal any decisions that are made during the research process" (Katja L. 2022). Here, K.L., like C.A., goes beyond the question of how to deal with categories of sex/gender in the research process. She further posits reflexivity as a necessity to make visible multiple influences on her research. In this regard, she notes that this stance is a matter of reflecting on "the social and cultural framework" in which one conducts research, on "how it is shaped, [on] what historical background" has an effect. She ultimately frames her practice with the concept of "situated knowledges," pointing out with reference to Harding's idea (see further details below) that "strong objectivity always includes strong reflexivity."

Reflexivity thus functions in the accounts of the two scientists presented so far as a term that refers to at least two coupled variants of meaning: firstly, as a designation of a *research practice* through which both scientists incorporate feminist and queer-theoretical approaches into their research (in order to "transparently reveal any decisions that are made in the research process" and to "address how neuroscience deals with the categories of sex/gender and sexuality"); and secondly as a reference to a *normative claim* that both researchers address as a future commitment to improvement in the sciences (as a "tool for working towards a better neuroscience"). We will follow up on these two variants of meaning in our synthesis section after giving two further examples of the manifold references to reflexivity we have collected so far.

The third scientist we interviewed was Hannah R. She was trained in molecular biology and science and technology studies. Unlike C.A., she left "the lab" and today defines herself as working across a broad spectrum: "I am clearly an STS scholar, a scholar in science and technology studies, I would say ... that's what I am. I consider myself a critical race theorist as well" (Hannah R. 2018). In our interview with her, she described how, for some time during her career, reflexivity became an overarching principle for the collective scientific efforts she was involved in and a stance that she sees as strongly connected to questions of power relations:

> We really looked at what scientists do [... and wondered] how might that change how science actually gets done? We'll have to then be accountable for power, we can't just pretend it doesn't exist. So when we stop to pretend it doesn't exist, do we then start creating methods for sharing the power, for coming with a different way of organizing a laboratory? So this is it. I guess that was what we were thinking about when we were thinking about reflexivity. (H.R.)

Reflexivity in H.R.'s account can be described as a tool that enabled researchers from different areas in the project to come together and do science differently, even if it has now been replaced by a more comprehensible term: for H.R. today, the fundamental questions related to doing a different kind of science are more adequately grasped by the term "justice."

The fourth scientist we would like to introduce works in the areas of biology and gender studies and is an eminent figure in the field of critical life sciences. We have given her the pseudonym Rebecca M. In several of her books and research papers, R.M. has written extensively about sex/gender and race in biosciences from a critical perspective. In her work, she uses models that assume an intrinsic entanglement of biology and culture, thus presupposing a reflexive understanding of the relationship between nature and nurture. She also refers to several proponents of critical or feminist approaches. But surprisingly (from the perspective of the authors), in her books and some 100 papers, R.M. refers only very rarely and not very explicitly to the term and the concept of reflexivity. It can be said that she does not reflect on reflexivity—at least in her writings; and she did not use this term in her answers to our questions. This is remarkable because overall, R.M. calls for a more accurate consideration of how we produce scientific knowledge, extensively contextualizes her own research, and argues that bioscientific research must acknowledge and respond to the far-reaching responsibilities that follow from its social implications. She describes her work as "political to the core" and depicts herself as a firm believer in careful argument based on full and mature consideration of the evidence. She calls for scientists to examine science's own myths about how science works and about its self-image as a practice that is always objective and apolitical. Based on her statements, it is not surprising that for many researchers in the field of "critical" sciences, and especially in research on sex/gender, R.M. stands for a research practice in which the societal framings of gender and racial categories are also reflected in biological research.

The last example shows that a simple empirical determination based on an in vivo code—which names practices as reflexive only if the term "reflexivity" is used—would be insufficient for our purposes. We should also note that this would not be a satisfactory way to systematize the diversity of conceptualizations of the term reflexivity. For us, the need to analytically clarify the concept of reflexivity explicitly arises from the observation of the diversity of meanings it has and the realization that practices of reflexivity are, at times, not conceptualized as such.

On the analytical capabilities of reflexivity

There are numerous fields where concepts of reflection and reflexivity have been elaborated and further developed, including philosophy, cultural studies, anthropology, and sociology, as well as psychology and the educational sciences. The breadth of its application in various disciplines and the abundance of its conceptual intentions and goals are also expressed in a multitude of understandings that arise from its theoretical and methodological functions and from the qualities attributed to reflexivity in each particular context. In this regard, Steve Woolgar aptly speaks of "varieties of reflexivity" ([1988]1991:21), and Michael Lynch complains about a "confusing variety of ways" (2000:26). Some scholars also distinguish between reflection and reflexivity (Stirling 2006). However, synonymous usage of the two terms is much more common in the literature on reflexivity than attempts to distinguish them conceptually. Despite or precisely because of this multitude of versions of reflexivity, confusion and misconceptions persist in the debate about reflexivity as it pertains to the semantic space under consideration here.

At the beginning of the twentieth century, the notion of reflexivity was eventually extended in the direction we are interested in, first in new developments in the study and sociology of knowledge and science and later in science and technology studies. One of the founders of the sociology of science, the German sociologist Karl Mannheim, for instance, used conceptual variants of reflection to grasp the epistemological issues of his research approach (Mannheim [1929] 1997; 1964); and one representative of the "newer" science studies, David Bloor, determined in the four principles of his outline of a sociology of scientific knowledge that this endeavor "would be reflexive" (Bloor [1976] 1991).

A further significant development took place in the field of feminist science studies and criticism of science from the late 1970s onwards (Rose and Rose 1976a; 1976b; Roberts 1981; Bleier 1986; Harding 1986; Haraway 1988; Schiebinger 1999). These approaches are of particular interest here because they imply considerations that go beyond the preoccupation with merely egological—i.e., cognitive, methodological, and logical—questions about the possibilities of knowledge and seek to bring the conditions and effects of the social and the societal into focus.

In 1970, the US sociologist Alvin W. Gouldner formulated an early program for a "reflexive sociology," which, as he argued, would have to be a "radical sociology": "Radical because it would recognize that knowledge of the world cannot be advanced apart from the sociologist's knowledge of himself and his position in the social world" (Gouldner 1970: 489). He argued further that reflexive sociology seeks not "an insulation but a transformation of the sociologist's self, and hence of his praxis in the world" (490) (see also additional elaborations by O'Neill 1972 and Friedrichs 1970). Gouldner's draft thus opposes what he considers to be limited academic sociology. It advocates a transformation of the researching subject and research practice, emphasizing above all the social conditions of each specific society—including the social conditions for knowledge production and for the possibility of engaging in the world through scientific practice. The most appealing thing here is Gouldner's version of reflection with a focus on the social conditionality of knowledge production and its strong point of a double radicalization—in the form of the necessity of reflexive knowledge on the part of the researching subject and the transformation of practice in the world by this subject (note: this means not only the practice of knowledge production but also practice in the world) (Gouldner 1970: 489).

Two decades later, the French sociologist Pierre Bourdieu took up several aspects of Gouldner's proposals (albeit with some explicit criticism; see, for example, Bourdieu and Wacquant 1992). Bourdieu argues that epistemic practices are inevitably socially shaped, and he strongly advocates for the embedding of a reflexive worldview as an integral part of, and necessary precondition for, a proper recognition of the world. For this purpose, it is not the individual researcher that Bourdieu sees as the primary object but rather the social and intellectual unconscious that has been merged into the scientific tools and operations. For him, such an endeavor can only be a collective undertaking, and it is not something that can be individually attributed to the scientist (see also Wacquant 1992: 36). In concrete terms, Bourdieu stresses that to "avoid becoming the object of the problems that you take as your object; you must retrace the history of the emergence of these problems" (Bourdieu and Wacquant 1992: 238). The researcher, in Bourdieu's view, has to "objectify the objectifying distance and the social conditions that make it possible" (Bourdieu 1992:14). On this basis, he argues that the most vital task for a scientist (and especially for the social sciences—though we can take this as a general statement for the sciences) is to establish "the conversion of thought, the revolution of the gaze, the rupture with the preconstructed and with everything that buttresses it in the social order—and in the scientific order" (Bourdieu and Wacquant 1992: 251).

Bourdieu is thus committed to an understanding of reflexivity that looks beyond the individual research subject to the conditions and structures of knowledge production, to its social foundations and limitations, and to its categorial axioms. He distances himself from forms of narcissistic introspection and mere examination of the subject and seeks instead a consideration of the "object constructed by science for the social conditions of possibility of the 'subject'" (Bourdieu and Wacquant 1992:214). In this sense, he clearly speaks in favor of a collective enterprise that would use reflexivity as a tool to possibly bring about "more responsible politics" (Bourdieu and Wacquant 1992:194).

In a quite similar understanding based on Bloor's ideas about "strong objectivity," the US feminist philosopher and postcolonial theorist Sandra

Harding has proposed some conceptual outlines for what she calls "strong reflexivity." Harding uses this concept to implement what she regards as a central insight provided by feminist standpoint theories and a symmetrical understanding of the entanglements of science and society. "Strong objectivity," she writes, "requires that the subject of knowledge be placed on the same critical, causal plane as the objects of knowledge" (Harding 1993:69). Just as she understands the standpoint as something that "take[s] science and politics to achieve," which means that standpoints are "socially mediated," she also conceptualizes the notion of strong reflexivity (Harding 1991: 276). By "being mediated," she means that science and knowledge are always "deeply permeated by the social relations through which they come into existence" (Harding 1991:ix). Although Harding is still very much concerned with the individual subject, who, for her, must be understood as "part of the object of knowledge" (69), she also makes clear that strong reflexivity captures the "dominant beliefs of an age" that are "collectively" and not only individually held (52). As an example of how culture-wide beliefs function as evidence, she mentions, among other things, how sexist and androcentric assumptions are collectively shared and not (sufficiently) recognized as such in mainstream accounts of objectivity. Harding sees strong reflexivity as a mode of feminist standpoint epistemology and as something that can be a resource to maximize objectivity.

Harding's elaborations on reflexivity have been widely adopted and have been particularly influential in feminist science studies and feminist critiques of science. It is on this basis that the feminist STS researcher Donna Haraway refers to Harding's elaborations. In her work, however, and especially in her book *Modest_Witness@Second_Millennium* (1997), Haraway in fact, fundamentally rejects the concept of reflexivity since she suspects that the idea of reflection leads to a problem of representationalism inasmuch as it tends to presuppose what is to be reflected as an objective fact and does not look at the productive site of the mediation of what has to be reflected with the objects of investigation. Far from turning away from the goal of reflexivity, Haraway proposes another physical metaphor and introduces the term "diffraction" as a way of escaping from the illusion of simply mirroring an empirical entity. Diffraction creates a pattern of a new image by means of a grid. Following the concept of interference, which refers to the superposition of two or more waves and the formation of characteristic patterns, diffraction tries to make explicit that knowledge production is about making a difference in the world. Haraway declares: "Diffraction is the production of difference patterns in the world, not just of the same reflected – displaced - elsewhere" (268).

The feminist theorist and STS scholar Karen Barad takes this metaphor from Haraway and uses it as an overarching trope in her seminal book *Meeting the*

Universe Halfway (2007; see also Barad 2003). Building on Niels Bohr's treatment of material phenomena, Barad attempts to analytically apprehend the entanglement of "the natural" with "the social." But far from merely highlighting their inextricability, she is interested in the productive relations between the natural and the social or the object and the subject. For this purpose, Barad uses the term "diffraction" as a heuristic to address a number of critical questions, such as the possibility of objectivity, the nature of measurement and meaning-making, and the relationship between discursive practices and the material world (ibid.: 94). Certainly, it is worth exploring further the notion of diffraction or a diffractive approach in (social scientific) empirical practice (Barla 2021; Serra Undurraga 2021). For our purposes, however, it is essential to emphasize that reflexivity—while arguably staying tied to a humanist understanding of the entanglements of science and "the social"—goes, in the views of the researchers we interviewed, far beyond the representationalist framework targeted by Barad, Haraway, and others.

While Barad's early representations of reflexivity are of little help in our endeavor because they remain far too reductive, ignoring essential work in feminist research by social scientists and related disciplines (Lumsden 2019; Serra Undurraga 2021), it is very well worth taking up the aspects of the concept of diffraction. Barad explains that "reflection and diffraction are not opposites, not mutually exclusive, but rather different optical intra-actions highlighting different patterns, optics, geometries that often overlap" (2014:184). However, Haraway and Barad repeatedly make clear that they are concerned with a performative understanding of scientific practices and with strengthening the perspective of different kinds of knowledge-making practice. Haraway elaborates: "Diffraction does not produce 'the same' displaced, as reflection and refraction do. Diffraction is a mapping of interference, not of replication, reflection, or reproduction. A diffraction pattern does not map where differences appear, but rather maps where the *effects* of difference appear" (Haraway 1992: 300, emphasis in original). Haraway thus points out that she is not concerned with a pure representation but with making "a difference in the world" (1997: 16). What makes the concept of diffraction so interesting here is precisely the program of focusing on the interaction between researchers and the researched objects via a supposedly technically improved reproduction of reality, which thus generates an attempt to make a change in the field and the world. It is precisely the latter that is at the heart of many of the statements made by the scientists we interviewed.

Synthesis and conclusion

Reflexivity is clearly one of the core concepts employed by the scientific endeavors examined in this paper, which are often described as critical, responsible, feminist, etc. In our analysis of interviews with such scientists and their publications, we used the concept of reflexivity to analytically describe the core elements of these approaches. Reflexivity can be found in these approaches both as a *programmatic stance of a normative requirement* to make a difference, i.e., to improve science and the world, and as *specific research practices* in the form of actions that aim to go beyond the mere presentation of identity-biographical reflexivity to consider the social conditions and prerequisites of science within the research process (Flick and Hoppe 2021). The researchers mobilized an understanding of reflexivity that encompasses the research methods they apply in their work, their own entanglement with their objects of study, and political-ethical claims of how to do science in practice. This is what C.A. means by "reflexivity as a main inside tool for working" in the neurosciences. It is also what K.L. understands as a core content of reflexivity, namely to "transparently reveal all decisions that are made during the research process." Since such practices are by no means always carried out under the name of reflexivity, it is all the more necessary to connect empirical observations with analytical understandings of reflexivity, which goes beyond an in vivo code of qualitative research.

Furthermore, the frequent use of the term in some of the interviews and texts is also marked by a reference to a broad understanding of reflexivity that goes beyond a narrow focus on the relationship between a researcher and what they are researching or on mere epistemological questions. The strong reference to social conditions and the attempt to understand both social-cultural conditions and their effect on research, as well as the possible impact of knowledge production on society, appear to us as the central characteristics here. When K.L. speaks of "critical reflexivity," of a "social and cultural framework" and the effects of the "historical background," or C.A. reports on the question of methods for "sharing the power," their comprehension of reflexivity includes a dimension that we have tentatively named "social reflexivity."

However, the abundance of meanings attached to this term and the widely divergent understandings of what it means in practice constrain any simple descriptive analysis. Furthermore, the normative charge of reflexivity that is invoked by a number of actors to present their approach as having an epistemological and methodological advantage tends to be more of an obstacle for the analytical version of reflexivity (Lynch 2000). Therefore, in order to analyze reflexivity in its current usage as a buzzword and, more specifically, to understand its use and application in practice, we saw it as necessary to investigate the conception of reflexivity in the social sciences and feminist science studies.

We have gladly taken up Gouldner's formulation of the necessity of reflexive knowledge and the transformation of practice in the world. Equally helpful for us were Bourdieu's elaborations on reflexivity about the conditions and structures of knowledge production and its social foundations and limitations. Harding's remarks on placing the subject of knowledge on the same critical, causal level as the objects of knowledge moreover resonate with Gouldner's outline of a "radical sociology" and the relatively similar conceptualizations by Bourdieu. Above all, Haraway's and Barad's focus on the performativity of reflexivity (or diffraction) captures central aspects of the practices of reflexivity in the field under study. What Harding has called "strong reflexivity" can be understood with Bourdieu's (and also Gouldner's) concept of "reflexive objectification" as a collective task to grasp the social dimensions and conditions of science. From this perspective, the entanglements of science and society apply to all scientific knowledge production, but they should be even more valid for the life sciences that deal with human beings as objects of knowledge. As we have shown, in the scientific endeavors we examined here, reflexivity has been employed as a tool or an action in a decidedly collective transformative process that aims to make a difference—in the double sense of productivity and intervention.

References

Allen, Ben. 2019. "Radical Science Returns." *SftP Magazine* 22(1). Accessed April 2, 2020. https://magazine.scienceforthepeople.org/vol22-1/radical-science-returns/.

Ashmore, Malcolm. 1989. *The Reflexive Thesis: Wrighting Sociology of Scientific Knowledge.* Chicago: University of Chicago Press.

Barad, Karen. 2003. "Posthumanist Performativity: Toward an Understanding of How Matter Comes to Matter." *Signs: Journal of Women in Culture and Society* 28(3): 801–31.

Barad, Karen. 2007. *Meeting the Universe Halfway: Quantum Physics and the Entanglement of Matter and Meaning.* Durham: Duke University Press.

Barad, Karen. 2014. "Diffracting Diffraction: Cutting Together-Apart." *Parallax* 20, 3: 168–87. https://doi.org/10.1080/13534645.2014.927623.

Barla, Josef. 2021. "Beyond Reflexivity and Representation: Diffraction as a Methodological Sensitivity in Science Studies." *Distinktion: Journal of Social Theory*: 1–23. https://doi.org/10.1080/1600910X.2021.1934506.

Beck, Ulrich. 1997. *The Reinvention of Politics: Rethinking Modernity in the Global Social Order.* Cambridge: Polity Press.

Beck, Ulrich, Anthony Giddens, and Scott Lash. 1994. *Reflexive Modernization: Politics, Tradition and Aesthetics in the Modern Social Order.* Stanford: Stanford University Press.

Bleier, Ruth, ed. 1986. *Feminist Approaches to Science.* New York: Pergamon.

Bloor, David. [1976]1991. *Knowledge and Social Imagery.* 2nd ed. Chicago: University of Chicago Press.

Bourdieu, Pierre. 1992. *The Logic of Practice.* Stanford: Stanford University Press.

Bourdieu, Pierre, and Loïc J. D. Wacquant, eds. 1992. *An Invitation to Reflexive Sociology.* Chicago: University of Chicago Press.

Bryant, Katherine, Giordana Grossi, and Anelis Kaiser. 2019. "Feminist Interventions on the Sex/Gender Question in Neuroimaging Research." *S&F Online* 15(2). Accessed September 9, 2020.http://sfonline.barnard.edu/neurogenderings/feminist-interventions-on-the-sex-gender-question-in-neuroimaging-research/.

Catharina A. 2019. Unpublished interview and ethnography with a neuroscientist.

Catharina A. 2022. Unpublished interview and ethnography with a neuroscientist.

Davies, Charlotte Aull. 2008. *Reflexive Ethnography: A Guide to Researching Selves and Others.* 2nd ed. ASA Research Methods in Social Anthropology. London: Routledge. http://www.loc.gov/catdir/enhancements/fy0649/98029216-d.html.

Dean, Jon. 2017. *Doing Reflexivity: An Introduction.* Bristol: Policy Press.

Finlay, Linda. 2002. "Negotiating the Swamp: The Opportunity and Challenge of Reflexivity in Research Practice." *Qualitative Research* 2, no. 2: 209–30. https://doi.org/10.1177/146879410200200205.

Flick, Sabine, and Katharina Hoppe. 2021. "Reflexivität als Mantra? Voraussetzungen und Grenzen partizipativer Forschung." In *Zur Kritik der partizipativen Forschung. Forschungspraxis im Spiegel der Kritischen Theorie,* edited by Sabine Flick and Alexander Herold, 18–40. Weinheim: Beltz.

Forster, Edgar. 2014. "Reflexivität." Pp. 589–98. in *Handbuch Pädagogische Anthropologie,* edited by Christoph Wulf and Jörg Zirfas. Handbuch. Wiesbaden: Springer VS.

Friedrichs, Robert W. 1970. *A Sociology of Sociology.* New York: The Free Press.

Gouldner, Alvin W. 1970. *The Coming Crisis of Western Sociology.* London: Heinemann.

Grim, Patrick, and Nicholas Rescher. 2012. *Reflexivity: From Paradox to Consciousness.* Frankfurt am Main: Ontos.

Hammersley, Martyn, and Paul Atkinson. 2019. *Ethnography: Principles in Practice.* 4th ed. New York: Routledge.

Hannah R. 2018. Unpublished interview with a biologist and STS scholar.

Haraway, Donna. 1988. "Situated Knowledges: The Science Question in Feminism and the Privilege of Partial Perspective." *Feminist Studies* 14(3): 575. https://doi.org/10.2307/3178066.

Haraway, Donna. 1992. "The Promises of Monsters: A Regenerative Politics for Inappropriate/d Others." Pp. 295–337 in *Cultural Studies* edited by L. Grossberg, C. Nelson, and P. A. Treichler. New York: Routledge.

Haraway, Donna. 1997. *Modest_Witness@Second_Millennium. FemaleMan_Meets_OncoMouseTM: Feminism and Technoscience.* New York: Routledge.

Harding, Sandra. 1986. *The Science Question in Feminism.* Ithaca: Cornell University Press.

Harding, Sandra. 1991. *Whose Science? Whose Knowledge? Thinking from Women's Lives.* Ithaca: Cornell University Press.

Harding, Sandra. 1993. "Rethinking Standpoint Epistemology: What Is 'Strong Objectivity'?" Pp. 49-82 in *Feminist Epistemologies* edited by L. Alcoff and E. Potter. New York: Routledge.

Harding, Sandra. 2008. *Sciences from Below: Feminisms, Postcolonialisms, and Modernities.* Durham: Duke University Press.

Katja L. 2022. Unpublished interview with a neuroscientist.

Lumsden, Karen. 2019. *Reflexivity: Theory, Method and Practice.* London: Routledge.

Lynch, Michael. 2000. "Against Reflexivity as an Academic Virtue and Source of Privileged Knowledge." *Theory, Culture & Society* 17(3): 26–54. https://doi.org/10.1177/02632760022051202.

Mannheim, Karl, 1964. *Wissenssoziologie: Auswahl aus dem Werk.* Edited by K. H. Wolf. Berlin: Luchterhand. Accessed August 16, 2019.

Mannheim, Karl. 1964. *Wissenssoziologie: Auswahl aus dem Werk.* Edited by K. H. Wolf. Berlin: Luchterhand.

Mannheim, Karl. 1997 [dt. 1929]. Ideology and Utopia. London, New York: Routledge.Mauthner, Natasha S., and Andrea Doucet. 2003. "Reflexive Accounts and Accounts of Reflexivity in Qualitative Data Analysis." *Sociology* 37(3):413–31. https://doi.org/10.1177/00380385030373002.

Mauthner, Natasha S., and Andrea Doucet. 2003. "Reflexive Accounts and Accounts of Reflexivity in Qualitative Data Analysis." *Sociology* 37(3):413–31. https://doi.org/10.1177/00380385030373002.

May, Tim, and Beth Perry. 2011. *Social Research & Reflexivity: Content, Consequences and Context.* London, Thousand Oaks: Sage.

Mayberry, Maralee, Banu Subramaniam, and Lisa H. Weasel, eds. 2001. *Feminist Science Studies: A New Generation.* New York: Routledge.

O'Neill, John. 1972. *Sociology as a Skin Trade: Essays Towards a Reflexive Sociology.* New York: Harper and Row.

Roberts, Helen, ed. 1981. *Doing Feminist Research.* London: Routledge & Kegan Paul.

Rose, Hilary, and Steven Rose, eds. 1976a. *The Political Economy of Science: Ideology of/in the Natural Sciences.* London: Macmillan.

Rose, Hilary, and Steven Rose, eds. 1976b. *The Radicalisation of Science.* London: Macmillan.

Sandywell, Barry. 1996. *Logological Investigations.* London: Routledge.

Schällibaum, Urs. 2001. *Reflexivität und Verschiebung.* Vienna: Passagen Verlag.

Schiebinger, Londa L. 1999. *Has Feminism Changed Science?* Cambridge, MA: Harvard University Press.

Serra Undurraga, Jacqueline Karen Andrea. 2021. "What If Reflexivity and Diffraction Intra-Act?" *International Journal of Qualitative Studies in Education,* QSE 1–14. https://doi.org/10.1080/09518398.2021.1900622.

Steinert, Heinz. 1999. "Reflexivität: Zur Bestimmung des Gegenstandsbereichs der Sozialwissenschaften." In *Ansichten der Gesellschaft: Frankfurter Beiträge aus Soziologie und Politikwissenschaft,* edited by Wolfgang Glatzer, 59–71. Opladen: Leske + Budrich.

Stirling, Andy. 2006. "Precaution, Foresight, and Sustainability: Reflection and Reflexivity in the Governance of Science and Technology." Pp. 225–272 in *Reflexive Governance for Sustainable Development,* edited by Jan-Peter Voss, Dierk Bauknecht, and René Kemp. Cheltenham: Edward Elgar.

Subramani, Supriya. 2019. "Practising Reflexivity: Ethics, Methodology and Theory Construction." *Methodological Innovations* 12(2): 205979911986327. https://doi.org/10.1177/2059799119863276.

Wacquant, Loïc J. D. 1992. "Toward a Social Praxeology: The Structure and Logic of Bourdieu's Sociology." In *An Invitation to Reflexive Sociology*, edited by Pierre Bourdieu and Loïc J. D. Wacquant, 1–60. Chicago: University of Chicago Press.

Whitaker, Emilie Morwenna, and Paul Atkinson. 2021. *Reflexivity in Social Research*. Cham: Palgrave Macmillan.

Woolgar, Steve. [1988] 1991. *Knowledge and Reflexivity: New Frontiers in the Sociology of Knowledge*. London: Sage.

Zienkowski, Jan. 2017. "Reflexivity in the Transdisciplinary Field of Critical Discourse Studies." *Palgrave Commun* 3(1). https://doi.org/10.1057/palcomms.2017.7.

PART 2.
Post-factual politics

Chapter 4

The spread of the 'gender ideology' and the peace referendum in Colombia

Javier Enrique Guerrero Castro

Abstract: Gender Ideology has been an overarching topic used by the political right during elections, and it refers to the apparent effort of the left or, in some cases, the political center homosexualizes children to end traditional family and traditional sexual relationships, meaning between heterosexual men and women, and to allow individuals to swap sex/gender as they please. Gender ideology is defined by activists and academics as a political tool, a form of propaganda, and a rhetorical distortion aimed to create moral panic toward the advances in policies of gender equality. Gender Ideology was deployed by right-wing politicians and promoted by Christian evangelical leaders and catholic ones to reject the peace agreement between the Colombian government and the FARC, proposing that such an attempt was 'encoded' within the peace agreement between the Colombian government and FARC. The use of gender ideology is closely associated with conspirational ideas regarding the existence of a 'lobby gay' and 'castrochavismo'. In this chapter, I consider Gender Ideology as part of the discussion on post-truth politics, as different camps accused each other of spreading fake news to win the election. In this chapter, I present an analysis of 3234 tweets containing the term "ideología de género" (gender ideology) and related to the peace referendum for 2016. STS provides both analytical and methodological tools to engage in situations where actors disagree but cannot avoid each other, such as in controversies. Gender Ideology is analyzed as a controversy on the ways sexual politics entangle with social change.

Keywords: Gender Ideology, Political Right, Post-Truth Politics, Conspirational Ideas, Social Change

***

On September 15, 2016, a couple of weeks before the Colombian peace agreement referendum, Colombian former president Uribe published a tweet with a link to a video of a conference given by Alberto Bárcena, a professor of history at Universidad CEU San Pablo, a private Catholic in Madrid, and a well-known defender of Franco's legacy. The description of the video argued for the existence of a global project sponsored by the United Nations and aimed at eroding the traditional family and imposing a 'gender ideology.' It also argued that there is a process of antichristian social engineering aimed at imposing a new family model in which parents will lose the right to educate their offspring and changing gender would be promoted. In the video, Bárcena considers the Catholic church the last bulwark against gender ideology. On the 2nd of October, 2016, Colombians voted 'No' in the Colombian peace agreement referendum to end the war between FARC and the Colombian State. Together with castrochavismo and the fear of socialism and communism, the imposition of the so-called 'gender ideology' was promoted by the No camp during the campaign toward the vote as arguments for rejecting the peace agreement. The No camp proposed that such an attempt was 'encoded' within the peace agreement between the Colombian government and FARC (Semana 2016b).

Defining gender ideology is a difficult task. While for conservative movements, it is a well-defined group of ideas that takes away the fundamental content of the complementary and natural (biological) relationship between man and woman and aims to destroy traditional values promulgated by the global gay lobby. The most extreme version of gender ideology states that the aim of the progressive movements is to 'homosexualize' children, to end traditional family and traditional sexual relationships between heterosexual men and women, and to allow individuals to swap sex/gender as they please. Activists and academics define gender ideology as a political tool, a form of propaganda, and a rhetorical distortion aimed at creating moral panic toward the advances in policies of gender equality (Beltrán and Creely 2018). As expressed by Rondón (2017), the term 'gender ideology' constitutes a set of classification practices between the 'normal order' and those attempting to subvert the 'natural order' (Rondón 2017). The gender ideology debate can be thought of as a controversy, where there exists not only a multitude of positions, actors, and so on but also different realities (Law 2004).

In this chapter, the term 'gender ideology' is considered part of the discussion of post-truth politics, as actors framing the discussion mutually accuse one another of utilizing fake news or alternative facts to convince voters or to impose the global homosexual agenda, depending on which camp. The gender ideology controversy is related to science and technology: facts, expertise, and infrastructures are contested in a very public way, including biological definitions of gender and sex, childhood psychology, the organization of the school system, and ultimately, like many other controversies in the 'post-truth era' are deployed in social platforms. Post-truth has been studied from several disciplines: sociology, anthropology, political science, psychology, philosophy, history, and journalism. The methods of studying post-truth have moved from purely conceptual or theoretical to utilizing quantitative and qualitative methodologies (Gorrell et al., 2019). However, most studies of such phenomena remain firmly rooted in considering post-truth as an anomaly, a pathology, or a malady of the post-modern world. The diffusion of fake news and conspiracy theories, together with fake news, is considered a threat to democratic institutions (Rosenfeld 2018; Rosenblum and Muirhead 2020;) and to the epistemic authority of science (Vernon 2017; Collins et al. 2017; Collins et al. 2020), among others. There is a trend to imply that lying, untruthful behavior is more prevalent now than in the past and that we are currently living in a post-truth era (Farkas and Schou 2019; Brotherton 2020), while other authors remind us that "post-truth" has a long political and intellectual history, and goes hand in hand with modern liberal democracy (Rosenfeld, 2018; Finchelstein, 2020). On the other hand, the rise of social media has been considered as providing a new impetus for fake news and a forum for the spread of conspiracy theories, as well as a key for political developments such as Brexit and the election of Donald Trump (Journell 2017; Wilber 2017; Bufacchi 2021) and other political events.

Söderberg (2021) argues that STS has discussed post-truth from three different perspectives. Collins represents a minority, accepting that STS is in some way complicit in the development of science denial, and urges a move back to the security of facts and expertise. Steve Fuller embraces the takeover of STS and postmodern thought by right-wingers while finally, and according to Söderberg, the most prominent camp, denies the link between STS and post-truth and challenges the very idea that post-truth is a novelty. In this chapter, I do not attempt to deepen the already ample and interesting discussion within STS, involving some of the most important scholars in the field (Collins et al. 2017; Fuller 2017; Jasanoff and Simmet 2017; Lynch 2020; Sismondo 2017), but to argue that at its core STS suggest for an empirical study of post-truth.

I suggest that the issue of gender ideology in Colombia during the peace referendum is an opportunity to understand political mobilization online around a 'post-truth' topic. I propose that STS's emphasis on the production of socio-technical realities from a symmetrical standpoint and a commitment to studying the co-production of technology and social realities may further our understanding of an issue such as gender ideology as they unfold in social media. Therefore, I do not start from judgments regarding the truth of such ideas but rather attest to their consequences on social action. Put it in terms of the Thomas Theorem, "when people perceive things to be real, they will be real in their consequences." Therefore, it is possible to apply perspectives capturing the dynamics of controversies in social media, such as issue mapping (Marres

2015). Adopting a methodological approximation from controversial studies avoids considering the different positions in the discussion as marred by irrationality and instead focuses the attention on the dynamics of the issue. Lynch has reminded us that despite the possibilities of the politicization of symmetry, the application of the symmetry principle does not mean accepting the sameness of the contending arguments (Lynch 2020). Therefore, utilizing a controversial perspective allows us to consider the several registers that make up the gender ideology discourse, the circulation of digital objects, and to describe the repertories used to advocate or dismiss post-truth discourses, all of this while considering that "Post-truth politics is more a matter of duplicating the appearances, rather than erasing the relevance of objective truth." (Lynch 2020:54)

This chapter contributes to the study of Post-truth in Latin America, especially in Colombia. I argue that STS and issue mapping are fruitful tools for studying the construction of issues, such as gender ideology, through an empirical analysis of the dynamics of gender ideology in Twitter in 2016.

Colombian politics, the context of the 'Gender Ideology'

The uses of gender ideology in the Colombian public sphere must be interpreted in the framework of this country's recent political history and geopolitical relations. Since those uses are commonly associated with the social groups congregated around former president Álvaro Uribe Vélez (in office 2002-2010), his rise to power could be a starting point for this account. During his 8-year term, Uribe dealt big blows against Farc, to the point that the end of the guerrilla movement seemed so near, as Uribe and their officials explicitly and repeatedly advertised. Despite this never happening, and the fact that by then, the allegations of abuses had become public scandals of great resonance and profound institutional consequences, Uribe was still able to justify the need to continue with his policy of military confrontation to avoid a resurgence of the guerrillas and to achieve their definitive defeat (Galindo Hernández 2005; Cárdenas and Villa 2012). Since, according to the constitutional reform that had approved re-election, no president could run for a third presidential term, Juan Manuel Santos, Minister of Defense in Uribe's second presidential term and a member, unlike Uribe, of Colombia's liberal political elite, was designated as the candidate to continue his program.

The arrival of Santos to power in Colombia would bring about a change of direction in this respect, with the reestablishment of relations between Bogotá and Caracas, the first act of public resonance with which Santos distanced himself from what a large part of the national political class and his electorate considered his role as faithful heir of the policies defined during Uribe's rule (Sierra 2016). The definitive breakup between Santos and Uribe occurred in

2012, when the beginning of peace negotiations between the leaders of the FARC and representatives of the Colombian government in Havana, Cuba, was announced. Uribe would rise as the main contradictor of this process, managing to deal with its most significant setback in 2016 by pushing for a negative vote in the referendum through which Santos sought to grant popular legitimacy to the agreements reached with the FARC. One of the main slogans of the No campaign was the affirmation that the peace agreement also represented the imposition of gender ideology in Colombia. Uribe and his political followers took the concessions made by the government to the guerrillas as signs of Juan Manuel Santos' betrayal of the right-wing agenda outlined by his predecessor.

The first draft of the peace agreement had a substantial section on gender, emphasizing women's and LGBTI rights, and was met with disapproval by the right and religious leaders. This section of the peace agreement was presented one year later than the controversy about the Sentence by the High Court in relation to the need for schools to teach about gender and sexuality. The High Court ordered the Ministry of Education to produce a booklet with the purpose of training schoolteachers to prevent discrimination and homophobia in school environments and (to prevent discrimination and homophobia in schools). The far-right sectors of the country disseminated images of the pornographic comic strip In Bed with David and Jonathan and passed them off as the Ministry of Education booklet. The dissemination of these pornographic illustrations, with which, according to the Colombian anti-gender groups, the Government wanted to homosexualize children.

While the agreement's purpose was to recognize and amend the disproportionate effects of violence on women and LGBTI communities, the same political forces that rallied against the Ministry of Education organized protests against what they considered the codification of the gender ideology in the peace agreement. With the use of the term 'gender ideology,' the conservative religious right was able to secularize their discourse and uphold their values (Beltrán and Creely 2018).

Several terms allowed expressing eloquently and synthetically the opposition to the peace agreement and accounted for the political opposition to Uribe's policies, which Santos would be leading Colombia through the agreement with the FARC. Castrochavismo (Guerrero-C and Jaraba-Barrios 2022) and gender ideology. Although analysts tend to point out other factors as more relevant in explaining the triumph of the negative vote in the referendum, it is also known that the campaign promoted by Uribe, aimed at provoking voter indignation and anger through strategic communication via social networks, was instrumental in outnumbering the voters who supported the peace process advanced by the Santos government (Botero 2017; Matanock and García-

Sánchez 2017). On October 2, 2016, Colombians voted in a referendum on a peace agreement between the government and the FARC, the strongest and oldest left-wing guerrilla force. The peace accord was narrowly rejected by 50.2%. The 'No' campaign managed to mobilize historic hatred for the guerrillas, fear of the left ("Colombia will become another Venezuela") along with a now-familiar mixture of right-wing cultural themes, including homophobia, anti-feminism, and resistance to gender equality, encompassed in the term 'gender ideology' which the right used as a motto.

Gender ideology as an 'issue'

This article aims to utilize the studies of controversies and issue mapping developed within STS as a suitable strategy to understand issues of post-truth, especially in providing a way to understand how those are played out in social media (Marres 2015). I see a homology between the production and study of controversies online and the formation of post-truth issues in social media, one that produces forms of engagement using platforms' affordances, circulation of memes, gifs, URLs, and other digital objects.

I argue that the issue of 'gender ideology' has come about due to the interaction of the many statements/registers, but we are also preoccupied with digital platforms' interactive dimensions that have long been recognized as influencing the public and the construction of issues (Bruns and Burgess 2015). Digital approaches to studying controversies have recognized how platforms preconfigure specific practices, shaping the actions of those involved in controversies (Weltevrede and Borra 2016). The platforms, therefore, influence the development of online controversies (Koed Madsen 2012). As in any controversy, those discussing the arguments endorsing or rejecting the issue of gender ideology utilize and negotiate with the platform affordances, using similar or the same digital objects, such as hashtags sharing URLs and images, to convince, mock or debunk the counterpart. Controversy is when actors disagree and cannot ignore each other (Venturini 2010).

The confluence of controversy studies and matters of concern or issues is demonstrated by Moats (2019), who uses the terms controversies and issues interchangeably. Adopting controversy as a point of departure allows us to understand both the substantial issue and the interactive relation with the digital platforms, assuming a symmetric approximation that avoids normative impositions, pathologizing and epistemological quandaries. Such approximation draws from the principle of symmetry, one of the central tenets of the Strong Programme of the Sociology of Scientific Knowledge (SSK). The principal urges the analyst to adopt a skeptical view concerning 'truth.' In a dispute, all positions are considered viable (Fuller 1995). Such principles and the empirical approach to studying controversies, suggested by STS, signal the possibilities of studying controversies or issues online (Marres and Moats 2015).

Mapping issues inherit an empirical approximation of studying scientific controversies. Controversy analyses of scientific disputes have been a central methodology for Science and Technology Studies (Marres and Moats 2015; Vertesi and Ribes 2019). Since the 1970s, STS has studied the production of knowledge and provided an alternative understanding of the politics of knowledge. Bruno Latour, Michel Callon, and others developed techniques for mapping technoscientific controversies to observe and describe socio-technical controversies during the late 1980s and early 1990s. Controversies afford the opportunity for everyone to argue on their account and to question the justifications of others (Callon et al. 2009:33). Digital technologies provide the tools for visualizing public controversies and making the networks visible (Munk et al. 2019). Although the controversies in STS focus on socio-technical issues, mapping issues offers techniques to study all kinds of public disputes (Venturini 2012).

Inspired by media studies and digital sociology, researchers have utilized automated techniques to follow the traces left by actors and deployed digital techniques (Marres 2017), mainly network analysis, to capture, analyze and visualize public discussion. Nevertheless, as Marres has shown, 'In digital controversy analysis, however, the question of bias touches on a serious methodological problem concerning the viability of digital media as settings for the enactment of controversy and their analysis' (2015:664). Digital technologies configure public debates in the form of issues. The formation of issues online is mediated by how platforms are structured. Taking this consideration seriously forces the researcher to move from analyzing controversies to mapping issues. Issue mapping, beginning with a topic, allows us to understand the emergence of issue formation and matters of concern, tracing the relations among digital objects, stakeholders, and statements (Burgess and Matamoros-Fernández 2016). The mapping of issues implies the representation of previous knowledge, often in the form of networks, to reduce the complexity of the data.

Methodology

I utilize issue mapping (Marres 2015) as the primary methodological orientation. Issue mapping is the implementation of controversy analysis to trace public discussions and issues utilizing digital methods (Rogers 2013). Issue mapping is influenced by the project cartography of controversies developed by the MediaLab of SciencePo in France. MediaLab turns to the empirical tradition of STS in analyzing controversies and provides both methodological and theoretical reflections when dealing with digital data

(Venturini et al. 2017) as well as situating the discussion within the digital ecosystem where they occur. The perspective seeks to understand the interactive relationships between the platforms, the public issues, and the methods deployed by researchers to understand them. Mapping issues is in debt with the dictum to 'follow the actors' (Muniesa 2015), and therefore, it is significantly influenced by the Actor-Network-Theory (Moats 2019), that is to say, an approach that aims to study the social in the making (Latour 2005). As such, this interpretation proposes a new encounter between the micro and the macro and moves away from structural perspectives (Latour et al. 2012), and produces new methodological approximations between the quantitative and qualitative (Blok and Pedersen 2014; Munk 2019).

To put this into practice, I used Twitter as the primary source of information and combined it with an extensive news search concerning gender ideology. Álvaro Uribe Vélez, the Colombian ex-president and main opponent of the peace agreement, had more than five million followers, and his tweets are often reproduced in traditional media, therefore reaching populations beyond Twitter. The analysis consisted of an approach that combined SNA visualizing tools with a hermeneutic approach to individual tweets. I collected data from Twitter using the web scraper parsehub. The final data collection consisted of 3234 individual tweets containing the term "ideología de género" (gender ideology) and related to the peace referendum for 2016. I am aware of the many issues surrounding the co-production of matters of interest in digital infrastructures, such as the economy, metrification of engagement and tracking infrastructures, and the monetization of attention (Gray et al. 2018), that might introduce noise into the data collection. However, I engage such possible occurrences as part of constructing the issue under study, purposely adopting an empiricist stand in the research (Marres 2015; Marres and Moats 2015). I did not attempt to delete forms of digital participation, such as bots and paid users, or to make a difference between 'real' vs. 'fake' content.

To analyze the data, I used Gephi. These strategies help visualize the emerging thematic clusters and the characterization of discursive repertoires of the gender ideology. As pointed out by different works within the STS (Breslin et al. 2020; Marres and Gerlitz 2016), Twitter allows us to explore not only the production of narratives based on measures focused on the popularity of the actors and their statements about themselves but also to detect co-occurrences between keywords as signs of thematic and social convergences. We follow not only the links between actors but also digital objects such as hashtags and URLs. Tweets are statements, and as controversy studies demonstrate, statements are never isolated (Venturini et al. 2015). To detect the various registers and provide a qualitative reading of the data, I codified the data using MAXQDA (VERBI Software 2019) and ANTconc (Anthony 2020) to read the uses of the term 'gender ideology' within the data. We complemented Twitter data with an extensive news search, allowing us to compare the dynamics of the term on Twitter with the political cycle, specifically in relation to the referendum of 2016. As a way to characterize the many ways in which gender ideology emerges, I use Potter's (1996) characterization to identify two main repertories within the conversation. The empiricist repertory defends the facticity and the existence of gender ideology, while the ironizing repertory undermines the trust, validity, and credibility of gender ideology as a representation of reality.

The mobilization of 'gender ideology' on Twitter

To analyze the gender ideology, I captured several of the specific markers collected from social media (Bruns and Stieglitz 2014; Marres and Weltevrede 2013), such as hashtags (Omena et al. 2020) and URLs. Hashtags are a valuable instrument for following conversations and identifying important actors in a controversy (Bruns and Burgess 2015; 2011). Including a hashtag in a tweet is a statement of participation in a discussion. I call to attention both the ambiguous uses of the same hashtag, the same hashtag in different contexts (Procter et al. 2015), and the modularity use of the hashtag. I see in this latter a demonstration of 'gender ideology' as a controversy.

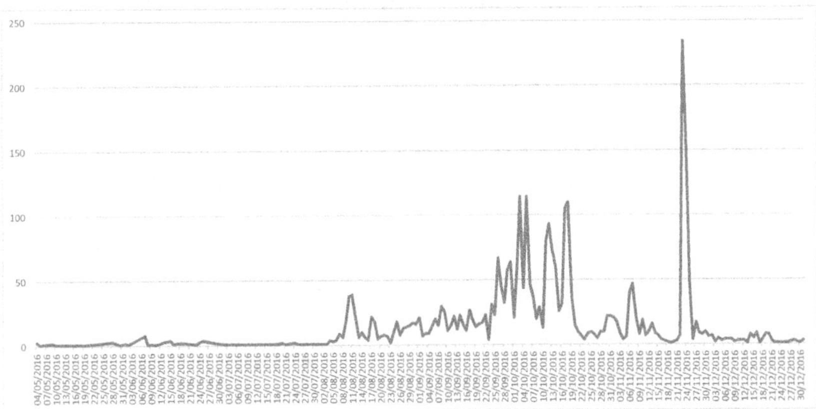

Figure 4.1. Gender ideology on Twitter (2016)

Mentions of 'gender ideology' (ideología de género, in Spanish) on Twitter appeared in 2008 and mostly referred to denunciations made by a Spanish archbishop. The following year, the term gender ideology became a motto to call for protest in Medellín against the project of creating a women's hospital in the city. The first URL directed to an op-ed in an alternative news outlet appeared and was shared (now truncated). Overall, the temporal distribution of the uses of the term 'gender ideology' indicates a close relationship with the electoral cycle of Latin American politics, especially the Colombian. Although the term was more or less constant during the years before the referendum, it was in 2016 that it gained prominence and is closely related to the peace agreement. During the campaign before the referendum, the No campaign argued that the peace agreement had either a clause or encoded the gender ideology, and, bringing back the statement made by Archbishop Cañizares in 2008, compared 'gender ideology' with the Nazi Empire (Semana 2016a). The use of gender ideology is closely associated with conspirational ideas regarding the existence of a 'gay lobby' and 'castrochavismo.'

The top five most mentioned hashtags are of ambiguous use: #colombia, #Paz, #ColombiaDecide, #FARC, and #DelAcuerdoNoNosSacan. Those hashtags allow us to identify the creation of communities. The use of hashtags grew during the protests organized by Uribe in the days leading to the peace referendum. The use of hashtags allows us to identify how users decided to participate in a conversation and, with whom to engage, and either to agree or disagree. The visual exploration of the co-occurrence of hashtags allows for identifying topics under discussion concerning gender ideology, providing a clearer picture of the issue under discussion than a solely frequency-based interpretation (Marres 2015). The co-occurrence analysis shows (Figure 4.2) how connected one hashtag is to the other (Marres and Gerlitz 2016).

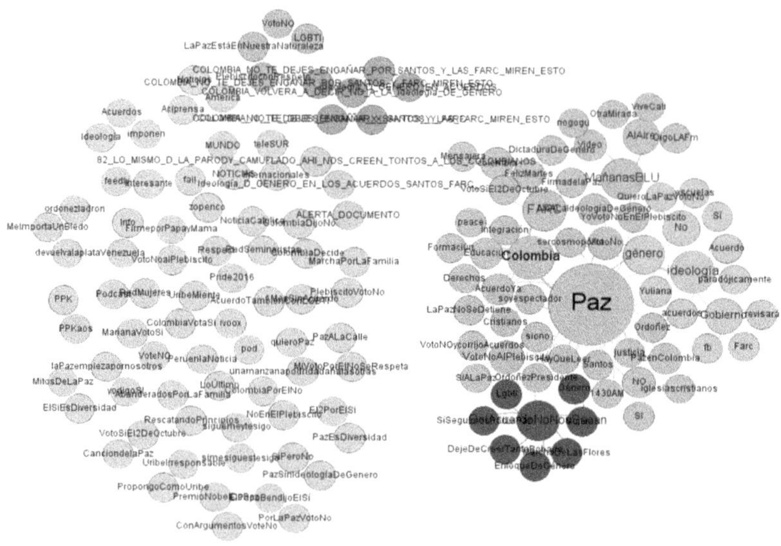

Figure 4.2. Co-occurrence of hashtags related to gender ideology (2016)

Figure 4.2 is a visualization of the frequency in which two hashtags appear together in a tweet. Several significant clusters are identified. Not surprisingly, #paz is the biggest of the clusters. However, more interestingly, following the previous assertion of gender ideology as associated with local politics, the clusters revolve around the peace process and its main actors. The cluster #FARC shows how the idea of gender ideology in Colombia is built around key ideas, such as gender dictatorship and LGBTI. There are two clear groups, one engaging with each other and the other lacking discernible cohesion.

The visualization also allows identifying the main cluster revolving around the central mottos of the No camp, #VotoNoAlPlebiscito, #votono. They are associated with the political candidates of Centro Democrático, the political party founded by ex-president Uribe and opposed to the peace agreement with FARC. Zooming in allows us to visualize the relations on the other side of the discussion. Again, most of the hashtags are associated with the electoral process or supporting the peace agreement. There are unconnected hashtags in which it is possible to distinguish attempts to mobilize against or in favor of the peace agreements in relation to gender ideology.

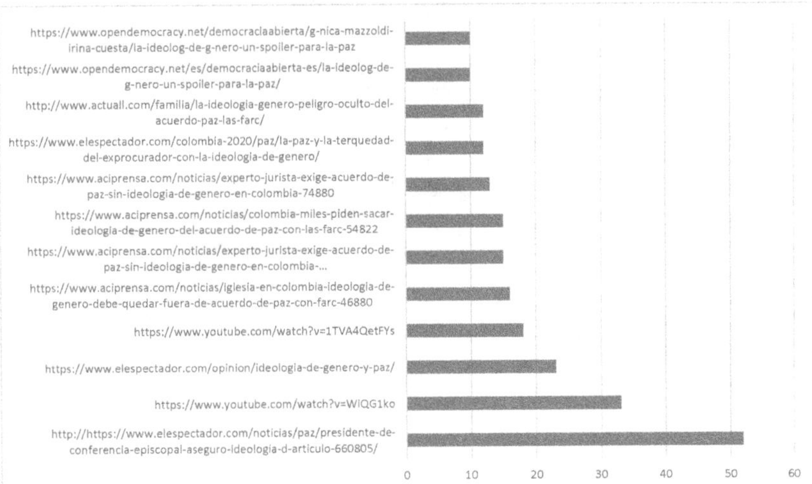

Figure 4.3. Most shared URLs in the data set (2016)

To investigate the role of media in the unfolding of gender ideology, I selected the domain of shared links. This provides a sense of the platform ecology in which gender ideology is shared and assembled. Links to Twitter occupy a higher position. I identified 353 instances of sharing Twitter posts, while a heuristic facilitated our understanding between the platform and other spaces.

Another prominent of the links shared corresponded to an online campaign to remove any trace of gender talk in the agreement after the No vote won; the link was retweeted 371 times[1].

The above description identifies key digital objects in the issue of gender ideology, such as URLs (Figure 4.3) and hashtags. The issue of gender ideology is discussed in traditional outlets, where journalists and pundits mainly discuss gender ideology and alternative outlets. To further demonstrate the lack of engagement between the two main sides, I found it helpful to construct a graph based on the relationship between the links shared and the sources used (Figure 4.4).

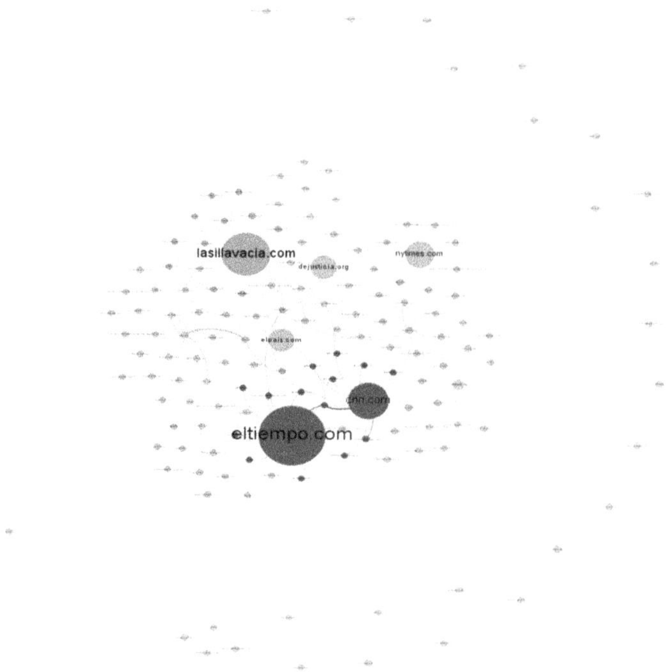

Figure 4.4. Sources and links shared during the gender ideology discussion in 2016

What the graphic demonstrates is the lack of cohesion in the discussion. Sources on each side of the debate rarely cite each other. I extracted the dataset of accounts co-mentioned in each tweet (Figure 4.5). The network was found

[1] The campaign was signed by some 27000 persons of the 50000 pledged.

using Gephi (Bastian et al. 2009). This network helps to analyze the issue formation of gender ideology further.

A large cluster containing accounts is identified. The cluster contains the names of prominent political figures, AlvaroUribeVel, petrogustavo, and Juan Manuel Santos. Within this cluster, we identify different accounts in both fields, empiricists and ironizing engaged in controversy. The network allows us to argue for what we claim in the introduction: Gender ideology has been a critical issue in the Colombian political discourse, and it was one of the main topics when discussing the peace agreement. Contrary to what happens with sources and hashtags, the network of accounts provides a different view of the discussion. By analyzing hashtags, links, and sources, it seems that the different camps do not engage with one another. Nevertheless, the analysis of accounts/persons mentioned demonstrates that users are well aware of who is promoting what.

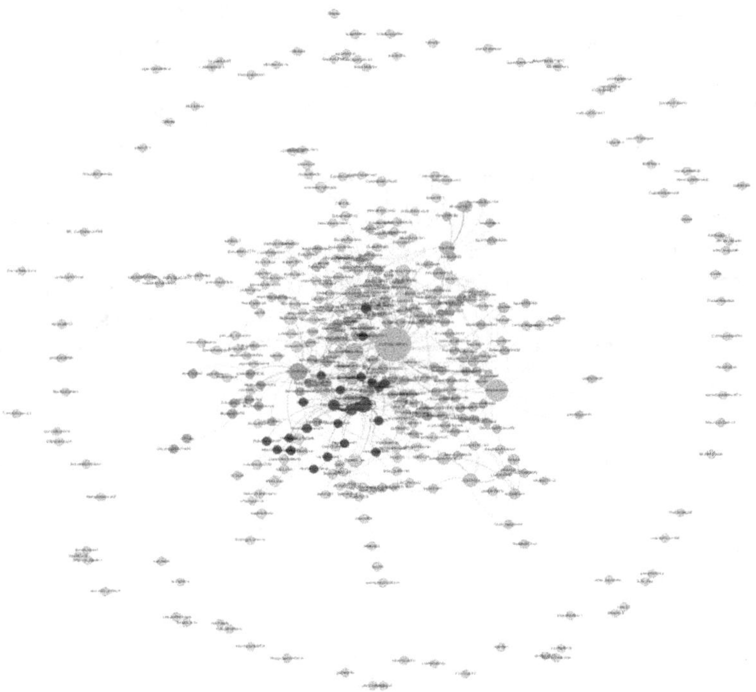

Figure 4.5. Network of accounts mentioned in the dataset (2016)

As highlighted, one of the main characteristics of gender ideology is its fluidity and modularity. Gender ideology is adapted to the local political context, but it is also connected with global networks of conservative activism (Rondón 2017). Besides visually exploring the controversies, I adopted an approach that allows us to move to the granular level. To do so, I analyzed tweets and hashtags in their context. Following Potter (1996), I provide a distinction between empiricist and ironizing discourses; I explored the digital objects and individual tweets, categorizing the gender ideology in various tones in both versions. The cost of participation is low, allowing for the creation of opposing and participant communities.

The exploration of tweets demonstrates that gender ideology is not homogenous. There are perceivable differences between various kinds of realities portrayed by the actors using the term gender ideology. In doing so, I recognize the difficulties of classifying the statements in a particular version. Within the empiricist repertoire, I classify three primary tones, soft, intermediate, and hard. In the ironizing repertoire, I identify two distinctive tones.

The soft version of the empiricist repertoire refers to alleged sympathy or tolerance towards gender ideology or any other social actor considered as aligned with any of the previous. In such a version, there is a warning of falling to take the necessary measures to avoid falling into the trap of gender ideology. The role of those opposing gender ideology would be to reveal the encoded messages or omissions within the peace agreement.

The second version, which I have labeled as intermediate, creates a situation in which there is a strategy already in place. This would be achieved by alliances with local politicians of any political inclination. Members of the elite, such as President Juan Manuel Santos or left-wing presidential candidates, could be either separate or together, and the FARC would attempt to implement the gender ideology. The strategy of the second version is to ask the government to remove any trace of 'gender ideology.'

The third version would be closest to the traditional definition of conspiracy theory and portrays the existence of machinations to challenge the traditional order or society. In this case, gender ideology results from forces outside Colombia, the so-called 'gay lobby,' and is portrayed as an attempt to transform society. Here, the idea is that behind gender ideology, it is possible to find the governments of Cuba and Venezuela who would also be plotting to suspend democracy in Colombia in order to create a dictatorial regime and abolish all current rules and values in which traditional family values are one the keystones.

On the other hand, the ironizing repertories reveal two significant versions. One such attempt is to undermine empiricists by accusing them of irreflexively accepting a fictitious plot and blindly believing in gender ideology. The second version on the ironizing side seeks to contrast the idea of gender ideology, utilizing facts to demonstrate that it was not part of the peace agreement or that such an issue was not considered part of the peace talks with the FARC. As part of such a strategy, the Colombian president, Juan Manuel Santos, and some other high-ranking officials of his government gave interviews to major news outlets in an attempt to convince voters.

These different versions of gender ideology are part of a deeply rooted rhetoric in the political discourses of the Colombian right in general and Uribism in particular: the mention of the traditional family and Catholic values. Those discourses speak of the close connection between traditional conservative values regarding sexuality and Colombian institutions. According to Curiel (2013), heterosexuality is consecrated within the Colombian constitution, in which ideas such as the heterosexual nuclear family represent the nation.

Conclusions

Gender ideology has been a permanent fixture of Colombian politics. In this article, I have argued that the flexibility of the term gender ideology, together with the affordances of digital objects, are entangled in producing the effects of gender ideology. Using the issue mapping methodology, I have identified gender ideology as a controversy and revealed the various tones or variations of the field in dispute.

I have approached gender ideology using the tools provided by the STS. Specifically, I aimed to understand gender ideology not in truth or falsity but as an issue (Marres 2015; 2017). The approach proposed here engages with the phenomenon of 'gender ideology' following its many iterations and argues that approaching post-truth as a controversy allows us to overcome many of the theoretical and methodological dead ends of the current research on post-truth. The issues of post-truth are inherently political, but not because they challenge officially sanctioned understanding or reality. Regarding gender ideology, the power distribution is closest to the political party in power. It should also be understood as resulting in the performative properties of the digital (Marres 2017). An important characteristic of the discussion as it was held on Twitter is that, while the different camps produce their hashtags and circulate their own sources, and in that way conform to ideas such as filter bubbles, people discussing gender ideology on Twitter direct their comments to political figures associated with the opposing camp.

Issue mapping recognizes that when using digital tools to analyze controversies or issues online, there is the danger of being overdetermined by the platform ontologies or the methods of data analysis. To overcome this issue, I used a research design that combines the analytical capacities of social network visualization with the granularity of qualitative analysis. We propose an approach that challenges the binary distinction of most social sciences' interpretation of post-truth between what is considered false or true and rational or irrational in relation to the social. The categories of STS have allowed us to overcome such binaries while simultaneously proposing a description of what the actors consider to be true or false. I have shown that in the contemporary political arena in Colombia, values such as objectivity and validity are relational and are subjected to negotiation. The actors involved in the gender ideology issue negotiate with platforms, engage in creative uses of digital objectives, and make statements. To paraphrase one of the tweets analyzed, the reader, now equipped with the analysis presented here, should decide about the existence or not of gender ideology.

References

Anthony, L. 2020. *AntConc (Version 3.5.9) [Computer Software]*. Waseda University. https://www.laurenceanthony.net/software

Bastian, M., Heymann, S., and Jacomy, M. 2009. Gephi: An Open Source Software for Exploring and Manipulating Networks. *International AAAI Conference on Web and Social Media; Third International AAAI Conference on Weblogs and Social Media*. https://www.aaai.org/ocs/index.php/ICWSM/09/paper/view/154

Beltrán, William Mauricio, and Sian Creely. 2018. "Pentecostals, Gender Ideology and the Peace Plebiscite: Colombia 2016." *Religions* 9(12):418. doi: 10.3390/rel91 20418.

Blok, Anders, and Morten Axel Pedersen. 2014. "Complementary Social Science? Quali-Quantitative Experiments in a Big Data World." *Big Data & Society* 1(2):205395171454390. doi: 10.1177/2053951714543908.

Botero, Sandra. 2017. "El Plebiscito y Los Desafíos Políticos de Consolidar La Paz Negociada En Colombiac" *Revista de Ciencia Política (Santiago)* 37(2):369–88. doi: 10.4067/s0718-090x2017000200369.

Breslin, S. D., Enggaard, T. R., Blok, A., Gårdhus, T., & Pedersen, M. A. 2020. "How We Tweet About Coronavirus, and Why: A Computational Anthropological Mapping of Political Attention on Danish Twitter during the COVID-19 Pandemic." *The COVID-19 Forum III. University of St. Andrews. Somatosphere.* http://somatosphere.net/forumpost/covid19-danish-twitter-computational-map/

Brotherton, Rob. 2020. *Bad News: Why We Fall for Fake News*. London, England: Bloomsbury Sigma.

Bruns, Axel and Jean E. Burgess 2011. "The use of Twitter hashtags in the formation of ad hoc publics." Pp. 1-9 in *Proceedings of the 6th European Consortium for Political Research (ECPR) General Conference 2011 edited by* A. Bruns and P. De Wilde.

Bruns, Axel and and Jean E. Burgess. 2015. "Twitter hashtags from ad hoc to calculated publics." Pp. 13-27 in *Hashtag Publics: The Power and Politics of Discursive Networks* edited by N. Rambukkana. United States of America: Peter Lang Publishing.

Bruns, Axel, and Stefan Stieglitz. 2014. "Twitter Data: What Do They Represent?" *It - Information Technology* 56(5):240–45. doi: 10.1515/itit-2014-1049.

Bufacchi, Vittorio. 2021. "Truth, Lies and Tweets: A Consensus Theory of Post-Truth." *Philosophy & Social Criticism* 47(3):347–61. doi: 10.1177/0191453719 896382.

Burgess, Jean, and Ariadna Matamoros-Fernández. 2016. "Mapping Sociocultural Controversies across Digital Media Platforms: One Week of #gamergate on Twitter, YouTube, and Tumblr." *Communication Research and Practice* 2(1):79–96. doi: 10.1080/22041451.2016.1155338.

Callon, Michel, Yannick Barthe, and Pierre Lascoumes. 2011. *Acting in an Uncertain World: An Essay on Technical Democracy.* London, England: MIT Press.

Cárdenas, Ernesto, and Edgar Villa. 2013. "La política de seguridad democrática y las ejecuciones extrajudiciales." (*Democratic security policy and extra-judicial execution) Ensayos sobre Politica Economica 31*(71):64-72. http://www.scielo.org.co/scielo.php?script=sci_arttext&pid=S0120-44832013000200004&lng=en&tlng=

Collins, Harry, Robert Evans, and Martin Weinel. 2017. "STS as Science or Politics?" *Social Studies of Science* 47(4):580–86. doi: 10.1177/0306312717710 131.

Collins, Harry, Robert Evans, Darrin Durant, and Martin Weinel. 2020. *Experts and the Will of the People: Society, Populism and Science.* Cham, Switzerland: Palgrave Macmillan.

Cosentino, Gabriele. 2020. *Social Media and the Post-Truth World Order: The Global Dynamics of Disinformation.* 1st ed. Cham, Switzerland: Springer Nature.

Curiel Pichardo, R. Y. 2013. "La nación heterosexual. Análisis del discurso jurídico y el régimen heterossexual desde la antropologia de la dominación" *La Frontera (GLEFAS).*

D'Ancona, Matthew. 2017. Post-Truth: The New War on Truth and How to Fight Back. London: Random House.

Farkas, Johan, and Jannick Schou. 2019. *Post-Truth, Fake News and Democracy: Mapping the Politics of Falsehood.* London, England: Routledge.

Finchelstein, Federico. 2022. *A Brief History of Fascist Lies A Brief History of Fascist Lies.* Berkeley, CA: University of California Press.

Fuller, Steve. 1995. "Is There Life for Sociological Theory after the Sociology of Scientific Knowledge?" *Sociology* 29(1):159–66. doi: 10.1177/00380385950290 01011.

Fuller, S. 2017. "Is STS all talk and no walk. " *EASST Review, 36*(1): 21–22.

Galindo Hernández, C. 2005. "De la "seguridad nacional" a la "seguridad democrática": nuevos problemas, viejos esquemas." *Revista Estudios Socio-Jurídicos, 7*(1): 496–546.

Gorrell, G., Bakir, M. E., Roberts, I., Greenwood, M. A., Iavarone, B., & Bontcheva, K. 2019. "Partisanship, propaganda and post-truth politics: Quantifying impact in online debate." *ArXiv Preprint ArXiv:1902.01752.*

Gray, Jonathan, Carolin Gerlitz, and Liliana Bounegru. 2018. "Data Infrastructure Literacy." *Big Data & Society* 5(2):205395171878631. doi: 10.1177/2053951718786316.

Guerrero-C, Javier, and Bruno Jaraba-Barrios. 2022. "You the Readers Will Complete the List. The Castrochavismo Conspiracy Theory." *Convergence The International Journal of Research into New Media Technologies* 28(4):1083–1102. doi: 10.1177/13548565221112377.

Jasanoff, Sheila, and Hilton R. Simmet. 2017. "No Funeral Bells: Public Reason in a 'post-Truth' Age." *Social Studies of Science* 47(5):751–70. doi: 10.1177/0306312717731936.

Journell, Wayne. 2017. "Fake news, alternative facts, and Trump: Teaching social studies in a post-truth era. " *Social Studies Journal 37*(1): 8–21.

Koed Madsen, Anders. 2012. "Web-Visions as Controversy-Lenses." *Interdisciplinary Science Reviews: ISR* 37(1):51–68. doi: 10.1179/0308018812z.0000000004.

Latour, Bruno. 2010. *Reassembling the Social: An Introduction to Actor-Network-Theory.* Cary, NC: Oxford University Press.

Latour, Bruno, Pablo Jensen, Tommaso Venturini, Sébastian Grauwin, and Dominique Boullier. 2012. "'The Whole Is Always Smaller than Its Parts': A Digital Test of Gabriel Tardes' Monads: 'The Whole Is Always Smaller than Its Parts.'" *The British Journal of Sociology* 63(4):590–615. doi: 10.1111/j.1468-4446.2012.01428.x.

Law, John. 2004. "After method: Mess in social science research." In *After Method: Mess in Social Science Research.* Routledge. https://doi.org/10.4324/9780203481141

Lynch, Michael. 2020. "We Have Never Been Anti-Science: Reflections on Science Wars and Post-Truth." *Engaging Science, Technology, and Society* 6:49–57. doi: 10.17351/ests2020.309.

Marres, Noortje. 2015. "Why Map Issues? On Controversy Analysis as a Digital Method." *Science, Technology & Human Values* 40(5):655–86. doi: 10.1177/0162243915574602.

Marres, Noortje. 2017. *Digital Sociology: The Reinvention of Social Research.* 1st ed. Oxford, England: Polity Press.

Marres, Noortje, and Carolin Gerlitz. 2016. "Interface Methods: Renegotiating Relations between Digital Social Research, STS and Sociology." *The Sociological Review* 64(1):21–46. doi: 10.1111/1467-954x.12314.

Marres, Noortje, and David Moats. 2015. "Mapping Controversies with Social Media: The Case for Symmetry." *Social Media + Society* 1(2):205630511560417. doi: 10.1177/2056305115604176.

Marres, Noortje, and Esther Weltevrede. 2013. "SCRAPING THE SOCIAL?: Issues in Live Social Research." *Journal of Cultural Economy* 6(3):313–35. doi: 10.1080/17530350.2013.772070.

Matanock, Aila M., and Miguel García-Sánchez. 2017. "The Colombian Paradox: Peace Processes, Elite Divisions & Popular Plebiscites." *Daedalus* 146(4):152–66. doi: 10.1162/daed_a_00466.

Moats, David. 2019. "Following the Fukushima Disaster on (and against) Wikipedia: A Methodological Note about STS Research and Online Platforms."

Science, Technology & Human Values 44(6):938–64. doi: 10.1177/016224391881 5234.

Muniesa, Fabian. 2015. "Actor-Network Theory." Pp. 80–84 in *International Encyclopedia of the Social & Behavioral Sciences*. Elsevier. https://doi.org/ https://doi.org/10.1016/B978-0-08-097086-8.85001-1

Munk, Anders Kristian. 2019. "Four Styles of Quali-Quantitative Analysis: Making Sense of the New Nordic Food Movement on the Web: Making Sense of the New Nordic Food Movement on the Web." *The NORDICOM Review of Nordic Research on Media & Communication* 40(s1):159–76. doi: 10.2478/nor-2019-0020.

Munk, Anders Kristian, Axel Meunier, and Tommaso Venturini. 2019. "Data Sprints: A Collaborative Format in Digital Controversy Mapping." Pp. 472–96 in *digitalSTS*. Princeton University Press.

Omena, Janna Joceli, Elaine Teixeira Rabello, and André Goes Mintz. 2020. "Digital Methods for Hashtag Engagement Research." *Social Media + Society* 6(3). doi: 10.1177/2056305120940697.

Potter, Jonathan. 1996. *Representing reality: Discourse, Rhetoric and Social Construction*. Thousand Oaks CA: Sage.

Procter, R., Voss, A., and I. Lvov. 2015. "Audience research and social media data: Opportunities and challenges." *Participations: Journal of Audience Reception Studies*.

Rogers, Richard. 2015. *Digital Methods*. London, England: MIT Press.

Rondón, Manuel Alejandro Rodríguez. 2017. "La Ideología de Género Como Exceso: Pánico Moral y Decisión Ética En La Política Colombiana." *Sexualidad, Salud y Sociedad: Revista Latinoamericana* (27):128–48. doi: 10.1590/1984-6487.sess.2017.27.08.a.

Rosenblum, Nancy L., and Russell Muirhead. 2020. *A Lot of People Are Saying: The New Conspiracism and the Assault on Democracy*. Princeton: Princeton University Press.

Rosenfeld, Sophia. 2018. *Democracy and Truth: A Short History*. Baltimore, MD: University of Pennsylvania Press.

Semana. 2016a. Ideología de género: una estrategia para ganar adeptos por el "No" al plebiscito. *Semana*. August. https://www.semana.com/nacion/articulo/ideologia-de-genero-una-estrategia-para-ganar-adeptos-por-el-no-al-plebiscito/488260/

Semana. 2016b. Ideología de género, el caballo de batalla del No al plebiscito. *Semana*. September. https://www.semana.com/nacion/articulo/ideologia-de-genero-el-caballo-de-batalla-del-no-al-plebiscito/493093/

Sierra, Jerónimo Ríos. 2016. The narcotisation of FARC and ELN guerilla activities, 1998-2012. *UNISCI Discussion Papers* 2016(41): 205–234. https://doi.org/10.5209/rev_RUNI.2016.n41.52680

Sismondo, Sergio. 2017. "Post-Truth?" *Social Studies of Science* 47(1):3–6. doi: 10.1177/0306312717692076.

Söderberg, Johan. 2021. *Post-truth Imaginations*. Routledge

Venturini, Tommaso. 2010. "Diving in Magma: How to Explore Controversies with Actor-Network Theory." *Public Understanding of Science* 19(3):258–73. doi: 10.1177/0963662509102694.

Venturini, Tommaso. 2012. "Building on faults: how to represent controversies with digital methods." *Public Understanding of Science 21*(7):796–812.

Venturini, Tommaso, Mathieu Jacomy, Axel Meunier, and Bruno Latour. 2017. "An Unexpected Journey: A Few Lessons from Sciences Po Médialab's Experience." *Big Data & Society* 4(2):205395171772094. doi: 10.1177/2053951717720949.

Venturini, Tommaso, Donato Ricci, Michele Mauri, Lucy Kimbell, and Axel Meunier. 2015. "Designing Controversies and Their Publics." *Design Issues* 31(3):74–87. doi: 10.1162/desi_a_00340.

VERBI Software. 2019. *MAXQDA 2020 [computer software]*. maxqda.com

Vernon, Jamie. 2017. "Science in the Post-Truth Era." *American Scientist* 105(1):2. doi: 10.1511/2017.124.2.

Vertesi, Janet, and David Ribes. 2019. "Introduction". In *digitalSTS*. edited by J. Vertesi, D. Ribes, L. Forlano, A. Camus, D. Vinck, D. Ribes, N. Calvillo, D. K. Rosner, C. Dunbar-Hester, X. (Charalampia) Kerasidou, L. Stark, S. Couture, S. J. Jackson, A. S. Chan, C. A. Hawthorne, C. Ilten, P.-B. McInerney, D. Nemer, P. Chirumamilla, M. Jacomy. Princeton University Press. https://doi.org/10.1515/9780691190600

Weltevrede, Esther, and Erik Borra. 2016. "Platform Affordances and Data Practices: The Value of Dispute on Wikipedia." *Big Data & Society* 3(1):205395171665341. doi: 10.1177/2053951716653418.

Wilber, Ken. 2019. *Trump and a Post-Truth World*. Boston, MA: Shambhala Publications.

Chapter 5

Living in post-truth: anti-politics and the power of the powerless in the twenty-first century

Michael Kilburn

Abstract: A post-truth war on reality poses serious challenges to policy, governance, historical understanding, and accountability. Considering the proto-post-truth regime in late socialist Czechoslovakia may provide insights, best practices, and a warning for navigating our current ontological and political crisis. During the period of so-called "Normalization" that followed the 1968 Soviet invasion, Václav Havel's philosophical injunction to "live in truth" was central to the praxis of his dissident activities that helped instigate the collapse of socialism in Czechoslovakia during the 1989 Velvet Revolution. As the first post-communist president of that country, he attempted to leverage his position to promote a new "anti-political politics" based on "practical morality… service to the truth, and care for our fellow humans...the phenomenon of human conscience." (Havel 1991:269, 271). This paper examines the context and strategy of the dissident move to counter ideological manipulation with one of "living in truth," the multidimensional nature of Havel's truth claims, and the devolution of their currency in the post-communist period. Ironically, Havel's uncompromising moral positioning in the dissident struggle against Communism may have inadvertently contributed to the very distrust of authority, anti-elitism, and right-wing populism that undermines a commitment to shared truth as the basis of social cohesion and progressive politics. The humanist skepticism of a dissident playwright philosopher against the overarching ideological claims of a technocratic regime may have lessons, both salutary and cautionary, for the role of STS in a post-truth age.

Keywords: Truth; post-truth; living in truth; Havel; Czechoslovakia; Normalization; post-totalitarianism; post-communism; antipolitical politics; ideology; Heidegger

***

> "Seek the truth, hear the truth, learn the truth, love the truth, speak the truth, hold the truth, and defend the truth until death."
>
> –Jan Hus, 1415

> "Truth isn't truth."
>
> –Rudy Giuliani, 2018

Introduction

In his New Year's address as the first post-Communist president of Czechoslovakia, former absurdist playwright and dissident activist Václav Havel mocked the tropes and platitudes of the former regime and laid out a new compact with the nation.

> For forty years, you heard from my predecessors on this day different variations on the same theme: how our country was flourishing, how many million tons of steel we produced, how happy we all were, how we trusted our government, and what bright perspectives were unfolding in front of us. I assume you did not propose me for this office so that I, too, would lie to you. (Havel 1997: 3)

This was a tipping point, not only in Havel's transition from dissident to president but in the life of the nation. It was appropriate that Havel staked his presidency on the fundamental distinction between truth and lies. He himself had leveraged his personal moral philosophy and his dissident practice on an ethic of "living in truth."[1] (Havel 2018:367). Havel's alethiological praxis was that, in a regime based on lies and hypocrisy, small-scale, quotidian practices of honesty and decency would have a cumulative, corrosive effect on the pillars of the Communist system, eventually leading to individual moral rehabilitation, existential revolution, the reconstitution of civil society, and the possibility of a rejuvenated liberal democratic politics. As Orwell once said, "In a time of universal deceit, telling the truth is a revolutionary act" (cited in McIntyre 2018:1).

This chapter critically interrogates Havel's theory and its implications for Czechoslovakia's post-Communist transition. Specifically, it describes the historical, political, and cultural context of Havel's theory of "living in truth," parses the philosophical sources and operant modes of his truth claims, and explores his unique understanding of naturalism as a check on ideology. The conclusion briefly examines the unintended consequences of his dissident praxis, which may have inadvertently contributed to the growth of illiberal

[1] Or "living within the truth." See Paul Wilson's translator's introduction, note 3 in Havel (2018:253).

post-truth politics in central Europe. The humanist skepticism of a dissident playwright philosopher against the overarching ideological claims of a technocratic regime may have lessons, both salutary and cautionary, for the role of STS in a post-truth age.

Post-totalitarianism: the cultural logic of late socialism

Havel's life's work was to critically analyze the political conditions in which he found himself; to understand their broader philosophical, social, and moral implications; and to develop a "dissident" strategy to salvage one's dignity, authenticity, and soul from the dehumanizing "aims of the system" (Havel 2018:360). The political conditions in which he found himself were some of the most absurd and trying of the twentieth century. Born in 1939 to an haute bourgeois family in Prague, only a year into the Nazi annexation and occupation of Czechoslovakia, Havel's childhood coincided with World War II. He was just nine years old when the Communists came to power in a parliamentary coup. Denied admission to college because of his class background, he trained as an engineer and worked as a laboratory assistant but was always at heart a humanist, philosopher, and man of letters. He eventually found his way into the theater, first as a stagehand, then a writer in residence, and gained international recognition for his acerbic plays skewering the absurdities of life under socialism. After the 1968 Soviet invasion put an end to the democratic reforms of the Prague Spring and imposed a reactionary regime of so-called "Normalization" - a set of policies and disciplinary practices designed to quash the spirit of democratic reform and stuff the genie of "socialism with a human face" back in the bottle - Havel was banned from the theater and publication and found himself thrust into the role of a dissident political activist.

As Popescu (2011), Danaher (2015), and Williams (2023) document, Havel's project to make sense of the nonsense of late socialism was a comprehensive endeavor written across his philosophical treatises, open and prison letters, plays, interviews, and political speeches. Despite the variety of writing styles, disciplines, and contexts, Havel's message was remarkably consistent: that human dignity, integrity, and identity were under duress from a crisis of modernity -manifest on both sides of the iron curtain- that threatened to alienate the essence of being and subordinate it to a technocratic limbo. Resistance or "dissent" to this creeping entropy was urgent and essential to avoid a spiritual and existential crisis.

In 1975, as the process of post-invasion "Normalization"[2] was more or less consolidated, he brashly laid out these concerns in an open letter pointedly

[2] There is no evidence that anyone in Husak's government was aware of "Discipline and

addressed to "Dr." (rather than "Comrade," "General Secretary," or "President") Husák, warning that this creeping inauthenticity, apathy, fear, and humiliation would have dire consequences for Czechoslovak culture and society.[3]

> It needs little imagination to see that such a situation can only lead toward the gradual erosion of all moral standards, the breakdown of all criteria of decency, and the widespread destruction of confidence in the meaning of values such as truth, adherence to principles, sincerity, altruism, dignity, and honor...the loss of hope and the loss of the belief that life has meaning. ...Order has been established. At the price of a paralysis of the spirit, a deadening of the heart, and devastation of life. Surface "consolidation" has been achieved. At the price of a spiritual and moral crisis in society. (Havel 1991:62)

The most sustained analysis of the philosophical and political implications of this crisis, however, is found in Havel's famous 1978 essay, "The power of the powerless." Here he developed a praxis of "living within the truth" as a response to the spiritually debilitating conditions of late socialism in Czechoslovakia. In practice, Normalization entailed a wave of purges of the Communist party and the professions, a ratcheting down of individual and social possibility and expression in accordance with the Party line, and a ritual consensus and complicity enforced through personal and professional microaggressions and institutionalized humiliations. In spirit, it required a moral capitulation to the demands of the regime, a daily ritual sacrifice of the "aims of life" to the "aims of the system" (Havel 2018:360-361). Though not yet operationalized, this dichotomy was foreshadowed in his earlier open letter.

> Life rebels against all uniformity and leveling; its aim is not sameness, but variety, the restlessness of transcendence, the adventure of novelty and rebellion against the status quo. .... On the other hand, the essence of authority (whose aim is reduced to protecting its own permanence by forcibly imposing the uniformity of perpetual consent) consists basically in a distrust of all variety, uniqueness, and transcendence; in an aversion to everything unknown, impalpable, and currently

Punish," published that year, though Foucault's analysis of social normalization through the institutionalization of disciplinary practices does bear some resemblance to the ambitions of the post-invasion regime.

[3] Husák did not reply, at least not until Dec 10 1989 (Human Rights Day!), when he resigned.

> obscure; in a proclivity for the uniform, the identical, and the inert; in deep affection for the status quo. (Havel 1989:23-25)

The distinction between the two is crucial not only to Havel's analysis of the existential crisis wrought by the post-totalitarian system but to his counterstrategy of civil disobedience by "living in truth." Life, for Havel, is characterized by "plurality, diversity, independent self-constitution and self-organization…striv(ing) to create new and improbable structures." The system, by contrast, sought "conformity, uniformity, and discipline…contriv(ing) to force life into its most probable states." The "yawning abyss" between the aims of the system and the aims of life is bridged by ideology, which, while disingenuous, provides excuses to rationalize the cognitive dissonance engendered by acceding to the demands of the system (Havel 2018:360-361).

The key theoretical insight of the "Power of the powerless" essay is in diagnosing the root causes of social dysfunction and personal alienation in a new tectonic of power Havel termed called "post-totalitarian." Like other "posts," –postmodernism, postcolonialism, post-traumatic stress disorder, etc. – post-totalitarianism denotes not the historical conclusion and overcoming of the totalitarian project but rather its evolutionary recapitulation in a subtler and more insidious form, a self-conscious farce to the tragedy of the original iteration.[4] If, for Orwell, the image of totalitarianism was a boot stamping on a human face forever, post-totalitarianism was more like a boot stepping on a human toe in a crowded tram forever. At the same time, while the boot-wearer loudly sucks a toothpick and breathes cigarette smoke and bad breath into your face, forever.

This kinder, gentler, more cynical version of totalitarianism, instituted by the reactionary post-invasion regime to effect a "restoration of order" (Šimečka 1984), did not entail the kind of blunt, brutal tactics of the Stalinist 1950s but relied instead on Taylorist gestures of obedience like signing loyalty oaths or hanging flags in your flat or propaganda posters in the greengrocers, nagging daily rituals of subservience and complicity that Havel compared to radiation poisoning or asthma, gradually wearing down the populace into a kind of drab compliance. Gone were the spectacles of show trials, prison camps, and executions, replaced by social networks of discipline, surveillance, and bureaucratic inconveniences. Big Brother was replaced by faceless bureaucrats and terror by anxiety, humiliation, and a cloying sense of guilt and shame. Obedience was replaced by abeyance. The system was maintained by thousands of tiny, seemingly insignificant but cumulative compromises and

---

[4] McIntyre describes post-truth in similar terms: "The prefix post is meant to indicate not so much the idea that we are "past" truth in a temporal sense as in "postwar" but in the sense that truth has been eclipsed; that it is irrelevant." (2018:5)

minor acts of civil obedience, each rationalized by readymade ideological excuses.

Havel illustrates this point with his famous parable of the greengrocer, who is compelled to put up a propaganda poster that arrives along with a box of tomatoes. The gesture itself is meaningless; the greengrocer's consent is automatic, as ritualized as stocking the produce itself. His acquiescence is as semantically shorn as the slogan on the poster. But the cynical and insidious calculation of post-totalitarianism, as opposed to the revolutionary idealism of its antecedent, is that participation in the ritual, however desultory, is enough to sustain the trappings of the system. Post-totalitarianism did not demand a heartfelt expression of love for Big Brother but only perfunctory daily rituals of acquiescence and debasement in a loveless, arranged marriage. Of course, it was possible to live, even to live fairly comfortably, in these conditions, but it required a kind of moral surrender.

> In a way, we are all being publicly bribed. If you accept this or that official position at work not, of course, as a means of serving your colleagues, but of serving the management you will be rewarded with such and such privileges. If you join the Youth League, you will be given the right and access to such and such forms of entertainment. If, as a creative artist, you take part in such-and-such official functions, you will be rewarded with such-and-such genuine creative opportunities. Think what you like in private; as long as you agree in public, refrain from making difficulties, suppress your interest in truth, and silence your conscience, the doors will be wide open to you. (Havel 1989:9)

Thus, post-totalitarianism begat post-truth. Because the system was knitted together in bad faith, a self-contradictory ideological facade, a Potemkin village of appearances trying to pass itself off as reality, people were forced to lie to themselves or at least become alienated from the truth and acquiesce to the demands of the system in order to survive. They had to go along to get along, suspending their disbelief in a mutual game of let's pretend or what Miroslav Kusy (1985:164) described as an ideology of als ob.[5] Nonetheless, despite the lack of consent, millions of individuals absently going through the motions

---

[5] Ash (1985) defines this principle as "a subjective, idealistic self-definition in which the idea takes absolute precedence over reality and consciousness determines being. In the condition of unfreedom, it proclaims the principle of As If. Try to live *as if* you live in a free country."

created what might be called a spartakiada effect[6]: a social spectacle of unity, orderliness, and common purpose. But this society of the spectacle, like the ideology that bound it together, was based on a ruse. Living in a state of desultory acquiescence had a debilitating effect on morale and morality, on truth itself. As Havel writes,

> This is why life in the system is so thoroughly permeated with hypocrisy and lies. Because the regime is captive to its own lies, it must falsify everything...Individuals need not believe all these mystifications, but they must behave as though they did, or at least tolerate them in silence… For this reason, however, they must live within a lie. They need not accept the lie. It is enough for them to have accepted their life with it and in it. For by this very fact, individuals confirm the system, fulfill the system, make the system, and are the system (Havel 2018: 361).

Havel also realized that this disingenuous legitimacy through false consciousness, this principle of *als ob*/As If, could work both ways. If the regime could sustain itself by arbitrarily conjuring the conditions of its own existence and convincing a critical mass of citizens to praise the emperor's new clothes, perhaps this subjunctive mode could also be deployed in self-defense. If a suspension of belief had created the conditions of servitude, perhaps a suspension of this operant disbelief could undo it. Havel's antidote to the moral, social, and cultural debasement, degradation, and degeneration was simply to refuse to play one's part in the farce, to boycott the ideological industrial complex, and to live in truth.

The Czechs are better known for their irony and skepticism than grand gestures of principle, but this mode of resistance need not entail abnormal moral courage or heroic sacrifice. Just as post-totalitarianism did not demand total submission, Havel suggested that the principle of living in truth could likewise find expression in simple, modest gestures. Refusing to speak in the euphemisms of a systematically distorted language; turning off the self-censor in the activities of daily life; the rediscovery of banned and distorted histories and traditions; a re-centering of the body and the every day in resistance to excessive idealism; and the creation of alternative communities and social networks which functioned as ideology-free zones could act as a kind of asymmetric warfare against the (lack of) integrity of the system.

---

6 *Spartakiada* was the archetypical socialist spectacle; an immense synchronized gymnastic display performed in huge sports arenas, remnants of which may still be seen in North Korea and the opening ceremonies for Olympic games.

To live in truth is to simply live in accordance with the aims of life; to live within the lie is to submit/subvert one's humanity to the aims of the system. But, given the Foucauldian subtleties of post-totalitarianism, it is not always clear when the line has been crossed. There is no baptism anymore — theology has long since devolved into doctrine and faith subsumed by ritual— just weekly confessions and a thousand, dispirited Hail Marys. There are no initiation rites, just daily hazing. The Faustian bargain is not struck by signing a blood pact with the devil at a midnight crossroads but by absently humming along to a mediocre hymn night after night. The gray zone of moral corruption is particularly foggy under post-totalitarianism and made even hazier by the preference falsification (Kuran 1997) and gaslighting that Havel documents so cringingly well in his plays, with readymade excuses and rationalizations for accepting the many enticements of collusion.

There would be a cost to this defiant, albeit humble, moral posture, of course, as Havel details in his parable of the greengrocer cum dissident, a loss of privileges and position (though still better than losing one's fingernails as under pre-post-totalitarianism); perhaps a demotion or inconvenient transfer. One's children might not be able to get into school or Pioneer camp. There would be social isolation and disapproval, disapproval, surveillance, and interrogations, perhaps even prison. But the fruits of this strategy were moral and spiritual recovery, a recognition and reassertion of one's authentic humanity.

Truth in the Czech context

Havel's moral philosophical praxis and political strategy of "living in truth" might have had a particular resonance in Bohemia. With its national motto of *Pravda vítězí* (the truth shall prevail) and a historical legacy of honoring scholars and writers over politicians and generals, Czechs should have been receptive to Havel's appeal to transcendent truth as a measure of personal and social integrity. The country's founder and first president, Tomáš Masaryk, had been a philosophy professor and statesman who resisted political expediency in deference to the truth[7]. It was he who chose Jan Hus' invictus *Pravda vítězí* as the motto of the new nation. His own counsel to his new countrymen, *nebát se a nekrást* (don't be afraid and don't steal), was somewhat more prosaic but evinced a similar, if more pragmatic, appeal to civic integrity. Havel's position represented a synthesis of these two founding fathers, combining quasi-theological submission to moral absolutes with quotidian practice in the *drobná práce* (small-scale work) of principled everyday conduct.

---

[7] In his exposure of the Dvůr Králové and Zelená Hora manuscripts as forgeries and his defence of Leopold Hilsner, a Jew accused of ritual murder.

But the Czechs are also a skeptical lot, with a sometimes ambivalent relationship with the truth. It is the world's most atheist country and delights in irony and ambiguity in its culture and humor. *Pravda vítězí!* may be inscribed on the national seal and currency, but the unofficial motto of the nation is *všechno je jinak*, which roughly translates to "everything is indeterminate" or "nothing is what it seems."

Here in central Europe, truth claims are nuanced by magical realism, alchemy, and the theater of the absurd. Tycho Brahe and Johannes Kepler were in the court of Rudolf II, but so were Edward Kelley, John Dee, and a bevy of alchemists and hucksters. Bohemia is the land of Charles IV, who founded the second oldest university in Europe, and Jan Amos Komensky (Comenius), the father of modern education theory and teacher of nations. But it is also the land of Kafka and Freud and Surrealism. Einstein developed his theory of relativity and gravitational lensing while living in Prague. Even Hus' famous aphorism of *Pravda vítězí* is an abbreviation of a somewhat more ambiguous quote: "Love the truth. *Let others have their truth*, and the truth will prevail" (Wikipedia 2022). "Truth prevails," said Jan Masaryk as he struggled to balance the tradition of his father's legacy of liberal democracy with the insurgent Communist ideology in postwar Czechoslovakia, "but it's work" (*Pravda vítězí, ale dá to fušku*). This complicates but does not necessarily contradict Havel's truth claims. After all, Havel himself philosophized about living in truth while also writing absurdist theater plays that seemed to orbit endlessly around meaning without ever landing. Relativity does not relativize truth but points to a deeper, more complex, multidimensional reality.

What is Havel's truth?

Given this ambiguity and the various, sundry, and tawdry disincentives to realizing and acting upon the better angels of our nature within the post-totalitarian system, how might one recognize the aims of life? How does one, theoretically and practically, "live in truth?" While expansively analytical on the moral and social implications of a system built on living within the lie, forensic in his analysis of the cascading effects of dissent, and quixotically optimistic on the prospects for an existential revolution, the essay is curiously mum on the topic. What, then, are the roots of Havel's alethiology?

As Williams (2023:1) outlines, it is clear that Havel's notion of truth cannot be reduced to simply "'correct' or 'accurate' facts and information." Rejecting the "Byzantine theocracy" that the "center of power is identical with the center of truth" (Havel 2018:357), Havel implies that such a narrow, positivist definition belies a "technocratic fallacy" (Williams 2023:1), whether practiced by the scientific socialism of Marxist truthers or by western technocrats, that represents a symptom of the very crisis of spirit and meaning that he was fighting against. Truth for Havel was much deeper and richer than the mere abstractions of modernity and more primal than the convolutions of theology. He sought a holistic expression of integral truth that incorporated context and respected complexity, truth as a dynamic and self-referential ecosystem.

Of course, defining "living in truth" as not "living within the lie" begs the question: where is the line beyond which the human spirit feels constrained by its straitened circumstances? At what point does human personality rebel against its degradation, particularly in a system that incentivizes false consciousness?

Truth and lies

Without an operant definition of truth itself or any ideological or theological reference point, Havelian truth is mainly defined in negative terms, as a refusal to live within the Lie demanded by the system and illuminated in the breach. Like Alan Dershowitz's (2004) embodied theory of human rights in "Rights from wrongs," truth may be revealed in the breach. Havel implies that the existence of truth can be deduced from the light bent by the mass of the surrounding system of lies. "Living the truth," he writes in the Power of the Powerless, "...is thus woven directly into the texture of living a lie. It is the repressed alternative, the authentic aim to which living a lie is an inauthentic response" (Havel 2018: 369).

Living in Truth entails conscientious objection to ideological coercion, to the manipulations and ritual humiliations of the system. This is first a call for clear and independent thought, but also for its practical expression, such as Masaryk once advocated for building democracy in the First Republic. In the context of the post-totalitarian regime of normalization, Havel (2018:370) advocates principled everyday behaviors such as opting out of the farcical ritual of elections, speaking or writing one's mind; participating in strikes, protests, or demonstrations, or even independent initiatives like rock concerts or even walks in the woods. Truth is mundane, antiheroic like his characters, standing in the gutter but looking at the stars.

To guide behavior and make ethical choices, he speaks of being true to oneself, to one's conscience, to the "obscure arena of being itself" (Havel 2018:370). Like a Socratic gadfly (or a Jiminy Cricket), he counseled his fellow citizens to "always let your conscience be your guide." If one is not deluded by ideology, he suggests, conscience and inherent human dignity instinctively recoil from debasement. He describes "dissent," which he always puts in scare quotes, not as a function of abnormal courage or surfeit of the spine but more of a gag reflex. For Havel, dissidents are defined not by any vested interest, professional capacity, or unique provenance. They are not some rare "protected species" of "exotic insects" but only defined by the "intrinsic moral aspect of their activity" (Havel 2018:381-382). Thus, anyone is potentially a dissident, waiting only for environmental stresses to activate this common recessive gene.

The illuminating contrast of truth and lies helps explain and integrate the moral function of Havel's dramaturgy, which might otherwise be considered a somewhat nihilistic anomaly in the larger context of his oeuvre. Havel's main translator Paul Wilson, for example, describes the plays as exploring the "dark side" of human longing for meaning, a dialectical confrontation with meaninglessness and absurdity, while his political and philosophical writings map out a sunnier terrain.

> There's a deep irony here. Many of Havel's plays, regardless of their comic aspects, are essentially tragedies in which the protagonists begin by trying to resist the relentless and impersonal pressure of the system but then, despite their best intentions, end up serving it…His dramatic protagonists end up choosing to "live within the lie." In other words, they end up perpetuating the system that has enslaved and humiliated them. On the other hand, Havel's essays—and particularly *The Power of the Powerless*—radiate his belief that positive change, in harmony with the genuine aims of life, is not only possible but almost inevitable, and that the truth that will set people free is best revealed in a clash of opposites. What Tom Stoppard has called Havel's "moral imagination" has two poles, a yin and yang, a deeply felt awareness of the contradictions that give both his dramatic and his non-dramatic writing its powerful creative tension. In his drama, for aesthetic reasons it is the dark side—the system, entropy, falsehood, death—that triumphs. In his essays, on the contrary, it is life, truth, moral integrity, and human authenticity that have the upper hand. (Wilson 2018:234)

Williams (2023:10) appears to concur with this assessment, writing, "Havel's essays put the case for avouching truth and taking responsibility positively, while his plays did so mostly negatively by showing the consequence of their absence." For Popescu, Havel's drama represents an effort to "deconstruct, rather than create, meaning," though she notes that, following his philosophical mentor Jan Patočka's method of negative Platonism (see below), deconstructing meanings has the same truth-seeking values in Havel's plays that it does in Socratic dialogues (2011:12-13).

Danaher affirms this constructivist spirit, characterizing Havelian drama as not just meaninglessly absurdist but as a kind of appeal, soliciting the audience in a co-creation of meaning or resistance to the bleak *huis clos* of the characters. Havel's absurdist plays draw not so much from the French existentialist tradition of abject contemplation of meaninglessness but on the Czech tradition of the *apelativní divadlo* [the theater of appeal] (Wilson 2018:232; Danaher 2015:74), which entreats the audience to fill the void of the theatrical spectacle with their own participation, to try to make sense of the senselessness on the stage and practice audience, and by extension, civic engagement. This is the effect of Havel's theater of appeal, presenting dark matter that exerts a gravitational pull on the moral universe, a photographic negative outlining the contours of truth for the audience to color in.

But this negative outline of truth is too subjective, positivist, and contextual to suffice as an ethical guide to behavior, much less the basis of a future democratic society. Havel's uncompromising philosophical stance and his consistent advocacy of absolute and universal human rights imply a deeper faith in their self-evident and inalienable truth that there must be a universal constant to measure it against. "Something in the greengrocer snaps," writes Havel (2018:367), but what? Without a clear referent for truth, how does one recognize and resist the Lie? A negative definition of truth as the opposite of lies may reveal its shadow but does not substantively illustrate its character and makeup. What, then, is the foundation of Havel's truth claims?

Positive sources of truth

Philosophy

Lacking religious faith, Havel sought the consolation of philosophy, believing that "all human beings ultimately have access to a repository of moral anchors, which they can recognize epistemologically through self-reflection and apply in their lived experience." (Popescu 2018:n. 11). Though denied admission to the philosophical faculty at Charles University because of his class background, he was a lifelong thinker and autodidact, cobbling together a coherent personal philosophy and worldview from his own idiosyncratic reading of Heidegger and a long mentorship and collaboration with the phenomenologist Jan Patočka.

In Heidegger, Havel found a philosophical framework that helped make sense of his personal alienation and the general malaise that seemed to afflict Czechoslovak society in the 1960s and 70s. Heidegger had postulated human existence as a condition of Dasein [Being], a sort of philosophical locked-in syndrome combining partial self-awareness with a complete lack of control and agency. The human condition, he said, is one of "thrownness" [*Geworfenheit*]: we are all arbitrarily cast into a situation over which we have no control and must struggle to find meaning and reconnect with our authentic selves, which have been shaken loose by the rough landing and are constantly corrupted and coopted by others. Heidegger's vision of modernity was like a vast Ponzi scheme whose promises of progress, freedom, happiness, and fulfillment could only be kept solvent by the constant reinvestment of fresh capital by glassy-eyed, soul-benumbed gambling addicts tethered to their slot machines.

For Havel, this bleak vision of hapless beings scammed or seduced into serving a system of repression seemed to resonate with the post-totalitarian Communist regime of arbitrary power, social control, and technocracy. Finding chinks in the auto-totality of socialist repressive desublimation, much less organizing any collective civil disobedience, seemed impossible, and subscribing to ideological rationalizations of complicit behavior only masked one's alienation or "fallenness" from authentic identity and responsibility.

Unlike Heidegger, whose prognosis for modernity's alienation was progressive and terminal, Havel managed to salvage hope for overcoming the conditions of post-totalitarianism. Whereas Heidegger's conceptualization of human authenticity was mainly negative, with an absence of inauthenticity or alienation, Havel attempted to infuse it with something positive; an expression of substantive being. Heidegger, Patočka, and Havel all saw Plato and especially Aristotle as a fatal turning point in the privileging and blind pursuit of human rationality that would ultimately lead to scientism, technocracy, bureaucracy, the concentration camp, and the atomic bomb, but, inspired by Patočka, Havel salvaged a pre-Socratic sense of the soul and sought a practical and constructive response to the crisis of modernity, rather than capitulation to technologic nihilism (Tucker 2020:16,155,162-64).

In a similar vein, Havel saw the primal condition of thrownness as not only a shattering or disorientation but also an opportunity for transcendence (Brennan 2013:23). Existential phenomenology's deductive notion of authenticity as a clearing away of false consciousness through a deep skepticism of appearances and reliance on direct intuition seemed to provide a path to navigate around the socialist diktat and spectacle of ideology. The measure of one's authentic self, Havel gleaned, was to be found in honest and critical self-reflection and consideration of the extent of one's subjection to 'others,' be they externalities or the inauthentic parts of the self. Trying to understand and escape this trap required constant vigilance and discipline to disaggregate one's authentic self from the surrounding distractions, but it seemed possible.

Moreover, it might be possible to build a support network of like-minded truth-seekers. While Heidegger seemed to think that any social interaction with "not-I/Others" (Tucker 2000:138) was a distraction from the subjective work of philosophical self-care, Havel also saw expressions of authenticity in collective action, such as the Czech musical underground or the dissident community. Even before his affiliation with Charter 77, Patočka had theorized a "solidarity of the shaken" (Brennan 2013:111), the mutual recognition of those who question the status quo.

A key aspect of Havel's approach to truth as authenticity was that it had a practical as well as theoretical dimension, and Heidegger's framework also seemed to confirm Havel's native Czech pragmatism; that the truth of *Dasein* [Being-in-the-world] would be tested in the *drobná práce* of everydayness (Tucker 2000:136). Whereas Patočka tended to define "life in truth" narrowly as an authentic orientation of the self toward the universe, Havel saw it as having multiple interrelated dimensions: existential, noetic, moral, and political (Tucker 2000:142).

"Living in truth" was not, for Havel, a mere theoretical exercise or philosophical discipline but entailed a practical responsibility to live out the true meaning of one's creed. Patočka's concept of "understanding," following Arendt, also conjoined knowledge and action (Brennan 2013:112), though it was not until his affiliation with Havel in the dissident NGO Charter 77 that he actively engaged in civil disobedience. With eerie foreshadowing, he contended, "a life not willing to sacrifice itself to what makes it meaningful is not living" (Brennan 2013:6). Shortly after accepting the position of spokesman for the group, Patočka was subjected to a lengthy police interrogation and died of a brain hemorrhage, a martyr to the cause of living in truth.

Adapting Heidegger's existential philosophy to his own situation, recasting some key terms for his own purposes, and enlisting Patočka's concepts of solidarity and engagement -both theoretically and literally- Havel developed a unique constructive phenomenological praxis that construed Dasein [Being] as a chrysalis of becoming rather than a state of purgatory; "others" as potential allies rather than alienated selves; and the modern world as redeemable through existential revolution. Living in truth had found its legs. It still, however, needed to learn how to walk.

Modes of truth

In a 2023 collection on Havel's keywords, Kieran Williams contributes a forensic chapter on the central concept of "Truth: *Pravda*," genealogically and archaeologically parsing Havel's history and body of work to suggest three crucial modes of truth in action: avouching, the Absolute; and dialectics. Each provides a different angle on how Havel enacted his philosophy of truth in practice.

Avouching truth

The first mode, which Williams translates as "avouching," involves three integrated moves connecting the private and the public: a) consulting one's "God and/or conscience" in an inner dialogue to determine the truth; b) publicly endorsing the position and offering oneself as guarantor; and c) enacting, or living out this truth in public engagement. In his memoir *Summer Meditations,* Havel clarifies the social responsibility of this mode. Morally autonomous individuals can build what he calls a "spiritual (as distinct from an ideological) state by freely "extricating a human being from the carapace of all ideological interpretations and rehabilitating it as an autonomous subject of individual conscience, of one's own thinking avouched by one's own existence, of individual responsibility and an altogether non-abstract love for one's neighbor" (Havel 1993:128).

In his 1993 inaugural address as President of the Czech Republic (following the dissolution of Czechoslovakia), Havel spoke of the civic virtues necessary for founding and building the new state, first among them, "faith in a truth for which it is necessary to vouch personally" (as cited in Williams 2023:n.8). Individual faith or conviction must be publicly affirmed to be meaningful: freedom of conscience and speech must be publicized by also exercising the right/responsibility of petition.

While philosophical rather than religious in orientation, Havel's resort to a personal conscientious avouching of truth against official dogma recalls the legacy of the Protestant Reformation in Bohemia, which countered Catholic hierarchy, hegemony, and Propaganda (Fide) with a radically democratic theology of equality and recognition before God. That "truth would prevail" was not for the Bohemian brethren an abstract theoretical claim, but was screamed by Hus at his Catholic inquisitors from the heretic's pyre at the Council of Constance in 1415. The truth was for Havel, immanent and specific, realized through a personal encounter with the divine, an embodied practice against a vicious and soul-killing ideological regime.

> Havel asserted that there was no single, universal guide to living life correctly, so each person must turn to his or her God and/or conscience, conduct an inner dialogue and then personally vouch (*ručit*) for (not simply believe in) the resultant standards of conduct. To avoid that process by relying instead on a ready-to-hand collection of maxims and commandments would only erode human integrity and make the world worse. (Williams 2023:10)

Absolute truth

The first mode of personal avouchment of truth is not entirely subjective, however, but derived its imprimatur of authenticity from resonance with a transcendent, metaphysical truth, a "shared reservoir of fundamental, universal human norms (which Havel calls 'absolute')" (Popescu 2018:140).

It's not a perfect analogy, but there is sympathy or resonance in Havel's truth-seeking with Martin Luther King's justification of civil disobedience to resist unjust laws. In his 1963 "Letter from Birmingham jail," King outlines his moral justification for direct action in the face of legalized oppression and systemic racism. Responding to the concerns of his fellow clergy that his strategy of nonviolent direct action was too radical and disruptive and that refusing to abide by laws and customs threatened anarchy, King responded that there were two kinds of laws: just and unjust, and that contravening unjust laws was not only justified, but a moral obligation.

> A just law is a man-made code that squares with the moral law or the law of God. An unjust law is a code that is out of harmony with the moral law. To put it in the terms of St. Thomas Aquinas: An unjust law is a human law that is not rooted in eternal law and natural law.... I submit that an individual who breaks a law that conscience tells him is unjust and who willingly accepts the penalty of imprisonment in order to arouse the conscience of the community over its injustice, is in reality expressing the highest respect for law. (King 2019:1140)

King's guidance on how to recognize unjust laws in the context of systemic discrimination is analogous to Havel's parsing of the aims of life in everyday practice. But lacking King's, much less Aquinas', Christian faith, Havel's referent for the flash of moral recognition is unclear. King's second line of defense in the essay in recognizing the dispiriting and dehumanizing effects of unjust laws in practice is closer to Havel's phenomenological approach.

> Any law that uplifts human personality is just. Any law that degrades human is unjust. All segregation statutes are unjust because segregation distorts the soul and damages the personality. It gives the segregator a false sense of superiority and the segregated a false sense of inferiority. Segregation, to use the terminology of the Jewish philosopher Martin Buber, substitutes an "I-it" relationship for an "I-thou" relationship and ends up relegating persons to the status of things. Hence segregation is not only politically, economically and sociologically unsound, it is morally wrong and awful. (King 2019:1141)

Despite similarities to King's invocation of a higher "moral law, or law of God," Havel was not conventionally religious. His version of the Absolute was closer to Platonic forms, an "unseen, cosmic order" (Williams 2023:10) animated by the fecund and exuberant "aims of life." Like Platonic ideals, which he had salvaged and reimported into his remix of Heidegger, or Jefferson's Deist invocation of the "Laws of Nature, or Nature's God" (Declaration of Independence), the Absolute was, however imperfectly, manifest in reality, filtered through the limited scope of human comprehension, and constrained and deformed by the bureaucratic and hegemonic "aims of the system."

Nonetheless, Havel believed that the apparent chaos and randomness of the world were not symptoms of man's fallenness but rather fragmentary glimpses into the Absolute, filtered by a cognitive and temporal limitation rather than a philosophical hoax. Careful attention to moments of transcendence, estrangement, and the uncanny, especially in art and theater, shamanistically "opens the doors to a truly fresh, sharp, and penetrating vision…that places us face to face with the truth" (Havel cited in Williams 2023:10).

Williams also cites an early letter demonstrating Havel's youthful discretion:

> Truth is, first of all the Absolute. […] The Absolute is everything, the Relative is how the human consciousness perceives everything, how things appear to it, how it judges. The Relative here is the reflection of the Absolute, as in Plato's cave, where phenomena are reflections, shadows, ideas. … There will be truth, even if there will not be matter or man. (cited in Williams 2023:11)

While the mature Havel had let go of some of his youthful idealism, the existence of a common, universal, absolute truth that could be mined for ethical guidance remained an important touchstone for his conception of truth.

Dialectical truth

For Havel, the Absolute existed metaphysically but could not be directly grasped, only intuitively avouched in a leap of faith or glimpsed in rare moments of estrangement and inspiration. As a lightning bolt instantaneously illuminates and momentarily resolves the contradictions between atmospheric ionicity and the earth, seeking concordance with the Absolute in a world of mundane approximations and distractions involved a constant critical engagement of opposed theses and antitheses: truth/lies, aims of life/aims of the system; antipolitics/politics; etc., to yield flashes of insight.

The third mode of truth for Havel, according to Williams, was this energizing dialectical clash of opposing ideas and forces, producing heat and light and allowing the dynamic and collective construction of truth. This was, however, a continual work in progress; the goal was not to finally arrive at a static and universal truth but to engage in a communal process of truth creation, a reasonable facsimile of the Absolute.

> Truth, for Patočka, is not the universal attainment of Being, but rather the continual strife between what is concealed and an openness to Being – in other words, it is becoming. … The task of philosophy is not, as Plato thought, to supply a metaphysical certainty (the world of ideas) to the problem of the natural world. It is to continually struggle, as Socrates demonstrated in his ceaseless questioning of Athenians. (Brennan 2013:115)

Thus, Patočka described his approach as "negative Platonism," employing the means, but not the ends of the Socratic method. For Havel, truth is not a static state but is revealed dialectically in the tension between the aims of the system and the aims of life, which constantly stress and rupture the artificially closed and static ideological world of appearances. In "Anatomy of a Reticence," for example, he describes truth-seeking as a struggle and contrasts the "mental short circuit" of ideology with the "demanding unending unpredictable dialogue with conscience or with God" (Havel 1991:302).

According to Havel's dialectic, a "lasting residue of his adolescent Hegelianism" (Williams 2023, 16), truth emerges in the course of the struggle to grasp the Absolute. The aim of life is one of "restlessness of transcendence" (Havel 1991:71), an eternal process of self-realization. Truth is not being but becoming; the aims of life are processual, tending towards "new and improbable structures" (Havel 2018:360), while the aims of the system slouch toward entropy, evermore stilted, soul-confining forms.

This isometric, dialectical approach can also be seen in Havel's theater pieces, which explore the absurdity of a world out of sync, the impossibility of communication, and the unrequited longing for meaning and connection, where tension is generated between a clash, however, muted, of the irreconcilable worldviews of the characters. Reconciliation/synthesis is never achieved onstage, but the awkward, lingering energy appeals to audience members to participate in the resolution, to "engage with the world around them, and through this civic engagement, to weaken the system's grip and restore their moral integrity and their capacity to act responsibly as citizens" (Wilson 2018:233).

This required participatory democracy and mutual responsibility, a kind of community theater. Through the collective co-construction of truth in the public sphere, consensus could be achieved. But it had to grow organically through a dialectical – or, better, a dialogical – process of ethical communicative action. Truth can and should never be dictated, much less mandated, by the mouthpieces of any ideological system.

"The Absolute = Nature"[8]

In addition to philosophically anchoring his positive truth claims in phenomenology, avouchment, appeals to the Absolute, and the alchemy of dialectics, Havel also sought to ground truth in a literal naturalism: in the ground itself. The *grunt*[9], or traditional Czech family farm familiar to Havel from his boyhood, symbolized sustainability and harmony within a natural order that was the concrete expression of the order of being. With its intuitive and symbiotic relationship with nature, "rooted in the nature of its place, appropriate, harmonious, personally tested by generations of farmers and certified by the results of their husbandry," the *grunt* was Havel's archetype for "living in truth," a halcyon ideal that had been uprooted and destroyed by modern agribusiness and brutal collectivization, an "arrogant, megalomaniac, and brutal invasion by an impersonally objective science, represented by a newly graduated agronomist or bureaucrat in the service of the 'scientific world view'" (Havel 1991:253-254).

> Havel longed for a return to an authentic rooted existence, to the "natural" order of things. He missed the rootedness of the peasant or the boy in Patočka's "natural world," which he conflated with Husserl's *Lebenswelt*, life-world, as the person's genuine pre-technological-scientific subjective experience. For Havel, children, peasants, and medieval European humanity were not yet alienated from their experiences, from the natural order of things in a world that starts in the morning and ends in the evening, a world that has its up (the heavens) and down (the earth), a simpler world with a clear difference between home and abroad, good and evil, beauty and ugliness, near and far, duty and work. The modern age, in a fit of hubris, seeks to transcend the boundaries of the natural world of experiences, but pays the price for this act of pride. (Tucker 2000:142)

Havel often used ecological metaphors to illustrate this distinction: the aims of life are depicted as organic, while the system is technocratic and tends toward entropy, monoculture, and ecocide. His most sustained treatment of ecology -the natural world or *Lebenswelt*- as a metonym for truth and integrity in social and political life occurs in the 1984 essay, "Politics and Conscience." In it, he describes a childhood memory of walking through the Bohemian countryside and encountering the foul spectacle of a factory noisily belching

---

[8] Used by Havel (cited in Williams 2023:11)

[9] Derived etymologically from the German *Grund*, meaning ground, foundation, or groundedness. (Havel 1991:253).

out smoke. "Each time I saw it," he writes, "I had a sense of something profoundly wrong; of humans soiling the heavens…It seemed to me that, in it, humans are guilty of something, that they destroy something important, arbitrarily disrupting the natural order of things, and that such things cannot go unpunished." (Havel 2010:440)

The degradation of the environment was a tangible measure of civilization's alienation from truth, the natural order of things, and from itself. That Czechoslovakia was at the time one of the most polluted countries on earth reflected for Havel the depth of its spiritual degradation as well and the structural corruption of the post-totalitarian system.

> All of us, east and west, face one fundamental task…resisting vigilantly, thoughtfully and attentively…at every step and everywhere the irrational momentum of anonymous, impersonal and inhuman power-the power of ideologies, systems, apparat, bureaucracy, artificial languages and political slogans. …we must draw our standards from the natural world, heedless of ridicule, and reaffirm its denied validity. We must honor with humility the limits of the natural world and the mystery which lies beyond them, admitting that there is something in the order of being which evidently exceeds all our competence. (Havel 1991:267)

Not only were the policies of Normalization crushing the spirit of the country, Havel warned, but sooner or later, such arrogance would have lethal consequences for the regime itself. Havel uses metaphors of geomorphology to stress the folly of trying to contain such elemental forces. "Life cannot be destroyed for good," he writes in his letter to Dr. Husák, "a secret streamlet trickles on beneath the heavy crust of inertia and pseudo-events, slowly and inconspicuously undermining it. It may be a long process, but one day it has to happen, …the crust cracks and the lava of life rolls out…life and history - demand their due." Elsewhere, he compares the situation to pus straining to escape a putrefying wound, carbon monoxide poisoning produced by incomplete combustion, and a "tornado whirl[ing] through the musty edifice of petrified power structures." (Havel 1989:28-34)

Such biological, geomorphological, and ecological metaphors are common in his writing, functioning to counterpose the organically emergent "aims of life" with the alienating, destructive, and unsustainable "aims of the system." It is no coincidence that he chose a greengrocer, rather than a bus driver or factory worker, as the epitome of the ordinary Czech citizen driven by conscience and inherent connection to the natural world to stand up to the system and assert his individuality and dignity. In his first presidential New

Year's address, cited above, Havel made clear this correlation, describing the view from his presidential plane of a ruined environmental and cultural landscape, an analogue to the "contaminated moral atmosphere" (Havel 1991:390-391) bequeathed by decades of misguided Communist technocracy.

This intuitive understanding of the natural world as an expression of the absolute truth of the order of being was bolstered by his reading of Heidegger, who also evinced a deep skepticism toward technoscience as a monstrous Golem that had slipped its master's control and threatened to wreak havoc on the world as punishment for our hubris.

> Our attention, therefore, inevitably turns to the most essential matter: the crisis of contemporary technological society as a whole, the crisis that Heidegger describes as the ineptitude of humanity face to face with the planetary power of technology. Technology—that child of modern science, which in turn is a child of modern metaphysics—is out of humanity's control, has ceased to serve us, has enslaved us and compelled us to participate in the preparation of our own destruction. And humanity can find no way out: we have no idea and no faith, and even less do we have a political conception to help us bring things back under human control. We look on helplessly as that coldly functioning machine we have created inevitably engulfs us, tearing us away from our natural affiliations (for instance, from our habitat in the widest sense of that word, including our habitat in the biosphere) just as it removes us from the experience of Being and casts us into the world of "existences." (Havel 1978: section XX)

Thus, politics, like science, had become simply a technology of power, devoid of conscience, morality, humility, and any connection to the *Lebenswelt.* "This impersonal power has achieved what is its most complete expression so far in the totalitarian systems," writes Havel in "Politics and Conscience" (2010: 447) "...natural science, rationalism, scientism, the industrial revolution and also revolution as such, as a fanatical abstraction, through the displacement of the natural world to the bathroom down to the cult of consumption, the atomic bomb and Marxism."

In his key theoretical essay, "The Power of the Powerless," Havel proposed that resistance to the regime be reflexively grounded in a moral posture of "living in truth" and that truth itself was embodied in nature. Nothing illustrates his theoretical construct of the "pre-political lifeworld" more intimately than the biosphere, and it is no coincidence that some of the first protests against the Communist regime were driven by desperate environmental conditions in northern Bohemia as well as the corresponding endemic social and moral malaise (Vaněk and Kilburn 2004). So, for Havel, the laws of nature and nature's God suggest a kind of ontological baseline by which we can objectively measure the transgressions of the system.

Summary and praxis

Dissident Havel diagnosed the post-totalitarian system as one based on lies and destructive of both the natural order and human dignity and spirit. Demanding and enforcing conformity and complicity with the aims of the system forced people to become alienated from their true selves, from the "order of being," and turned people into unwitting accomplices of the regime, a process he called the "automatism" of the system or "social auto-totality." (Havel 2018:368) His response was to develop and propose a theory of dissent based on the following precepts:

- The system is illegitimate; its truth claims and policies cannot be trusted and should be resisted.
- The aims of the system = living within the lie; living in truth = the aims of life.
- Individual conscience can intuit the truth against the obfuscations and distractions of the system. Shame, guilt, and humiliation are autoimmune responses indicating a lie is being practiced upon the body politic.
- Truth can be philosophically asserted through phenomenological discipline, winnowing a true sense of self from alienating and disorienting "others."
- It can also be avouched, asserting truth on one's honor and staking one's claim with a pound of flesh.
- Truth is absolute, universal, and metaphysical but fragmentarily immanent and graspable in moments of transcendence and breakage, such as those triggered by moments of estrangement, especially in the arts.
- Truth can be dynamically revealed in a dialectical process of contestation.
- Finally, truth is manifest in the state of nature, which organically reflects the aims of life. Disruptions in the ecological balance are a measure of our fallenness.

Havel triangulated these theoretical coordinates to construct a praxis of "living in truth" that would underwrite dissident opposition to the post-totalitarian regime of late socialism[10]. The first test came in 1976 during the trial of the musical underground around the rock group the Plastic People of the Universe, who had been hounded and charged with "disturbing the peace" for their independent, albeit otherwise non-political, cultural activities. Havel saw in the long-haired, marginalized musicians "the unintentional personification of those forces in man that compel him to search for himself, to determine his own place in the world freely, and in his own way, not to make deals with his heart and not to cheat his conscience, to call things by their true names and to penetrate...to the "deeper level of being." (Havel 1991:104)

Havel's recognition of the existential stakes of this minor trial of unknown delinquents and his orchestration of their defense fostered an unlikely coalition of interests among various intellectuals, writers, rockers, workers, philosophers, and even former reform Communists. With Czechoslovakia's recent ratification of the International Covenant on Civil and Political Rights and 1977 being declared the "year of the political prisoner" by Amnesty International (which also won the Nobel Peace in 1977), Havel leveraged the momentum of that historical moment to turn his ad hoc defense coalition into a standing oversight committee on rights, freedom and living in truth under the post-totalitarian regime.

Charter 77 construed itself not as a dissident organization, much less as a political party, but rather a "loose and open association of people" (Charter 77 Declaration 1977/1985) dedicated to the realization of human rights and freedoms in Czechoslovak society. This framing was partly strategic as opposition groups and activities were strictly illegal but also followed Havel's general strategy of "antipolitical politics" to avoid replicating the systems-thinking that had led to the violation of human rights and the "aims of life" in the first place. Traditional political opposition would have entailed recognizing the legitimacy of the regime and engaging in good-faith dialogue that risked reincorporation. Instead, Havel (2010:455) sought to reimagine politics itself "not as the technology of power and manipulation, of cybernetic rule over humans or as the art of the utilitarian, but politics as one of the ways of seeking and achieving meaningful lives, of protecting them and serving them": an "anti-political politics."

---

[10] The essay, "Power of the Powerless" and the example of Charter 77 had a profound influence in the region and beyond, inspiring and providing "theoretical underpinnings" not only to the Solidarity and KOR movements in Poland (see the comments of Zbigniew Bujak in Paul Wilson's translator's preface to the original essay) but also later in Belarus (Charter '97); Iran (see Falk 2013, on Akhbar Ganji as the "Iranian Havel"); and China (Charter '08, for which dissident activist Liu Xiao Bo was awarded the 2010 Nobel Peace Prize).

Despite comprising only a few score, active members, all of whom suffered repression and recrimination for its deceptively modest demands that the regime obey its own constitution and international treaty obligations, Charter 77 persisted as a thorn in the side of the regime, a glitch in the matrix, throughout the waning years of socialism. When widespread protests erupted in 1988 and 1989, however, the Charter was one of the few independent civic associations practiced in self-organization and advocacy. Together with student leaders, they were able to quickly harness, sustain, and direct the energy and momentum of the protests toward the political goal of regime change. Rebranding themselves the Civic Forum, this cadre orchestrated massive demonstrations, developed a political platform of demands and tactics, and began secret negotiations with the regime for a peaceful transfer of power. The anti-politicians had become, willy-nilly, a political opposition.

It was an awkward transition. The entire premise of Havel's dissident praxis had been the debilitating dysfunction of the aims of the system and the necessity to realign one's conscience and everyday conduct with the aims of life, by definition, non-utilitarian and improbable. But society cannot function without systems, and democracy requires compromise and consensus. Politics is the art of the possible, and while Havel may have cheekily titled his 1997 collection of early presidential speeches "The art of the *im*possible," politics as morality in practice would prove difficult to finesse. Tucker describes the dilemma:

> (Havel believed) if people are virtuous and authentic and moral, they will create a great state. ... His conception of democratic policy is not that different from, say, that of Thomas Jefferson, who also believed that we have to build a great republic by having great people. But the opposite concept, the concept that underlies the American constitution, is that people are not virtuous and they're not going to be virtuous. They're corrupt – they always will be corrupt. What you need to do then is construct institutions that will transform the greed, the power-hunger, the narcissism – to refer to contemporary politics – of people into political virtue. You do that by creating a balance of power between those institutions. Havel was completely unprepared to construct institutions, because as a dissident he never had any experience of constructing institutions, of managing anything more than a samizdat publishing house. When the time came he didn't know how to construct institutions that would neutralize the legacies of totalitarianism that still exist in the minds of people, long after the regime is dead. (cited in Willoughby 2016)

When Havel introduced the term post-totalitarianism in the "Power of the Powerless," he was careful to clarify that he did not mean it to indicate the historical closure of the totalitarian project, but rather its refinement and insinuation into a Foucauldian web of power. He probably hoped for the same when, at the end of the essay, he sketched the outlines of a "post-democratic" society, imagining that virtuous democratic practices would be organically inscribed throughout the coming post-Communist system.

> Above all, any existential revolution should provide hope of a moral reconstitution of society, which means a radical renewal of the relationship of human beings to what I have called the "human order," which no political order can replace. A new experience of being, a renewed rootedness in the universe, a newly grasped sense of higher responsibility, a newfound inner relationship to other people and to the human community—these factors clearly indicate the direction in which we must go. (Havel 2018:405)

But attempting to realign an entire society to the aims of life, especially when that society had been complicit in the aims of the former system - and when the public's already unrealistic expectations were focused on the material, rather than the moral, rewards of freedom - was no small order and some of Havel's strained political positions reveal the contradictions of being a philosopher-king. Popescu links this vision to post-modernity as an overcoming of modernity's taint of the ideal of democracy. Havel (cited in Popescu 2011:29) seemed to believe that by privileging truth, morality, and "the care of responsible individuals for one another" above institutions of government, one could create a perfect democracy "infused with personal and civic values." But as post-totalitarianism was just a subtler and more insidious form of totalitarianism, post-democracy might just as well catalyze some of the worst traits of Western liberalism.[11]

Havel's failure to realize the potential of that historical moment and overcome the alienation and technocratic subversion of true democracy has been described as a tragedy (Keane 2001). But it is a tragedy not only in the sense of a lost historical opportunity but also in the original Greek sense of unavoidable

[11] A clever protest sign displayed during the anti-government demonstrations in November 1989 illustrates the naivete of Havel's program. Reading *Jaro 68* (spring 1968, a reference to the Prague Spring uprising), it could be turned upside down to read *Podzim, 89* (Fall, 1989), suggesting a symmetry between the goals of the two uprisings, innocent of the extent to which the post-1989 regime would entail a complete inversion of the original goals of democratic socialism. Seeking capitalism with a human face, little did they suspect what those jangling keys might unlock.

fate. Even as he sought to return to Europe and Czechoslovakia's democratic traditions, Havel believed that socialism and liberal democracy were two sides of the same coin of modernity and that the West, while more stylishly engineered, suffered essentially the same alienation from truth, manipulation by ideology, and distractions of shallow consumerism as the post-totalitarian system. The regimes of East Central Europe were, he wrote, "a convex mirror of all modern civilization…the avant-garde of a global crisis" (Havel 2010:448).

> The post-totalitarian system is only one aspect—a particularly drastic aspect and thus all the more revealing of its real origins—of this general inability of modern humanity to be the master of its own situation. The automatism of the post-totalitarian system is merely an extreme version of the global automatism of technological civilization. The human failure that it mirrors is only one variant of the general failure of modern humanity. (Havel 1978:section XX)

If Western liberal democracy was indeed just as corrupt, just as susceptible to normalization, just as beholden to the aims of the system, then the very nature of Havel's critique of the post-totalitarian system contained the seeds of its own destruction. Attempting to ground a post-modern version of Platonic ideals through "living in truth" ironically opened Pandora's box of post-truth that undermines democracy and bodes a post-liberal future of populism, authoritarianism, and conspiracy theory.

Conclusion

Havel's analysis and critique of the post-totalitarian system was a useful and necessary paradigm for building morale and solidarity in specific and awkward historical circumstances. But the context has shifted in the 30 years since the overthrow of Communism, and his original critique of ideology, technology, and politics has borne strange fruit. Three aspects of his theory bear further scrutiny for the dangers they inadvertently pose to the contemporary liberal democratic order.

Limits of antipolitical politics

First, Havel's rejection of ideology and his notion of anti-politics were effective critical tools in opposition but proved insufficient once in power. The "loose informal association" of the dissident coalition Charter 77, which later formed the core of Civic Forum, was too decentralized for effective governance—and construing the system, even politics itself, as the enemy constructed an antithetical relationship between civil society and the state that hampered efforts at democratic reconstruction. While Popescu (2011) rejects the notion that Havel's version of anti-politics, with its emphasis on "human rights, social justice, and the law," was necessarily deconstructive or negative, she concedes that his political inexperience effectively ceded the terrain to more bullish players.

> In my view, the claim that antipolitical politics did not become the predominant post-communist democratic paradigm in Eastern Europe does not lead to the logical conclusion that it failed. Rather the program was never implemented and there are few conclusions that can be drawn on that basis. In fact, the claim that apolitical politics failed the test of democratic reconstruction has as much support if not less than the reverse claim: that it was precisely the abandonment of the core values of antipolitical politics that led to the rise of the nationalism, partisanship, and corruption that in turn contributed to the negative political climate after 1989. (Popescu 2011:106-107)

The rise of the antiestablishment populist movements across the region in recent years capitalized on the failures, corruption, and incompetence of post-Communist political parties and drew on a deep suspicion of elites, technocrats, and experts, tilling the fertile land cleared by Havel's earlier critique. Perhaps in dreaming of an antipolitical, post-democratic system, he should have been more specific.

Intolerance

Furthermore, Havel's strict, almost Manichean, distinction between the aims of life and the aims of the system, between truth and lies, and his injunction against any sort of consort or compromise with the regime established a kind of moral fundamentalism that worked against diplomacy, accommodation, politics as the art of the possible; that is, against the very premises of liberal democracy itself.

Havel may have been personally motivated by a version of post-democracy as "a rehabilitation of basic human values of trust, openness, responsibility, solidarity, love ...deriving directly from life" (2018: 405, 407). Still, his skepticism toward Liberalism as just a slicker version of post-totalitarianism left the door open for darker visions. Civic (aka liberal) nationalism is based on shared democratic values, tolerance, the rule of law, and other ideological constructs and self-evident truths based on the very Enlightenment values - rationalism, anthropocentrism, positivism - that Havel decries as having triggered the crisis.

Liberal democracy and human rights are as much creations of modernity as technoscience, Marxism, and the atomic bomb. Resorting to organic metaphors and appeals to the "order of Being" rather than rational, empirically grounded common interest opens the door to illiberal *ethnic* nationalism, which has always appealed to romantic notions of primal, organic, pre-political community and grounded itself in a primordial appeal to Blud und Boden. No doubt Havel would have been horrified by the illiberalism of the far-right populist and ethnic nationalist movements that have taken power across the region and around the world. Still, they share core elements of his critique of Communist technocracy.

Truth and consequences

Finally, Havel's rejection of modernity, rationality, and technoscience as handmaidens of alienation and the anthropocalypse seems to cut the empirical roots from any truth claims. Rather than putting cards on the table, he claimed that the scientific method merely shuffled the deck, shifting the scale of inquiry and honing the precision of its instruments while avoiding fundamental questions of meaning.

> Inclination to the so-called scientific interpretation of the world is an escape from truth rather than an inclination towards it… an escape from the obligation to avouch personally for one's truth. (cited in Williams 2023:20)

Grounding authenticity in vague appeals to pre-objective or antipolitical personal experience or intuition, to nature or nature's God, or the vague "Order of Being," is too subjective for any agreed-upon reality or common basis of community or civil society. Under post-totalitarian conditions, the "lived effort of sustaining a separate, unsanctioned reality became the backdrop of Havel's understanding of oppression and the core of his solution for resistance" (Popescu 2011: 4). But unsanctioned reality is not just the refuge of the sovereign citizen and truth seeker, but also hackers, terrorists, and conspiracy theorists who pursue their own version of the truth down rabbit holes, instinctively discrediting any mainstream narrative as a sign of one's own integrity. If each can decide for himself the "real aims of life," then it's open season on the truth. Tucker outlines the perils of relativism,

> Even after we eliminate forms of fanaticism as candidates for authenticity, there may still be competing and inconsistent forms of authenticity. Different people may claim to experience Being differently, claim to listen to different calls for responsibility for Being, and accordingly develop different and conflicting systems of ethics of authenticity. Havel did not address this problem. (Tucker 2000:161)

Each of Havel's truth claims assumed good faith on the part of his fellow citizens and failed to recognize their regressive potential. His proposition that the system was built on lies designed to distract, manipulate, and co-opt the people and should be resisted at every step for the preservation of essential liberty and the very salvation of one's soul leads to a distrust of all authority and a Goldwateresque posture that "Extremism in the defense of liberty is no vice." Furthermore, basing truth on one's personal intuition and willingness to avouch unmoors it from any meaningful referent or common purpose. As Williams laments,

> What distinguishes fanatical attachment from steadfast avouching of a truth? On what grounds can we deny membership in Šafařík's pantheon of slain avouchers to cult leaders such as Jim Jones, David Koresh, or Shoko Asahara? ...Would we have to accord equal validity to Havel's prison letters and Hitler's *Mein Kampf*, written in captivity at Landsberg? (2023:19-21)

Combined with a Heideggerian distrust of science, technology, and expertise, all that is solid melts into the air, and rational policy discussions are laid bare to denialism. Given this, Havel's dialectical approach may just be a recipe for a trial by combat rather than a pursuit of truth through an energetic, public-spirited debate of opposed theses.

Havel was an indispensable figure in the transition out of Communism in Central Europe and an inspiration to anti-authoritarian dissidents everywhere, but his political naivete, philosophical romanticism, and ungrounded optimism in human nature complicate his legacy. As truth appears once again under assault, one can still take courage from his example and continue the struggle. The truth might yet prevail, ale *dá to fušku.*

References

Ash, Timothy Garton. 1985. "Poland: The Uses of Adversity." *New York Review.* June 27. https://www.nybooks.com/articles/1985/06/27/poland-the-uses-of-adversity/

Brennan, Daniel. 2013. "A Reappraisal of the Political Philosophy of Václav Havel".PhD Dissertation, Bond University, Queensland, Australia. https://pure.bond.edu.au/ws/portalfiles/portal/18242703/A_reappraisal_of_the_political_philosophy_of_Václav _Havel.pdf

Charter 77 Declaration. [1977]1985. Pp. 217-21 in *The Power of the Powerless* edited by Václav Havel et al. Armonk, N.Y.: M. E. Sharpe.

Danaher, David. 2015. *Reading Václav Havel.* Toronto: University of Toronto Press.

Dershowitz, Alan. 2004. Rights from Wrongs: The Origins of Human Rights in the Experience of Injustice. Basic Books.

Falk, Barbara. 2013. "Reflections on the Revolutions in Europe: Lessons for the Middle East and the Arab Spring." Pp. 281-315 in *Samizdat, Tamizdat, and Beyond: Transnational Media During and After Socialism,* edited by Friedericke Kind-Kovacs & Jessie Labov. New York: Berghahn Books.

Havel, Václav. 1989. *Living in Truth.* Edited by Jan Vladislav. Translated by Paul Wilson. London: Faber and Faber.

Havel, Vaclav. 1978. The power of the powerless. https://www.nonviolent-conflict.org/wp-content/uploads/1979/01/the-power-of-the-powerless.pdf

Havel, Václav. 1991. *Open Letters: Selected Writings, 1965-1990.* Edited and translated by Paul Wilson. New York: Knopf.

Havel, Václav. 1993. *Summer Meditations.* Translated by Paul Wilson. New York: Vintage Books.

Havel, Václav. 1997. *The Art of the Impossible: Politics as Morality in Practice, Speeches and Writings, 1990-1996.* Translated by Paul Wilson and others. New York, Toronto: Knopf.

Havel, Václav. 2010. "Politics and Conscience." Pp. 440-456 in *The Czech Reader,* edited by Jan Bažant, Nina Bažantová, and Frances Štarn Durham and London: Duke University Press.

Havel, Václav. 2018. "*The Power of the Powerless.*" *Eastern European Politics and Societies: EEPS* 32(2):353–408. Translated from Czech by Paul Wilson doi: 10.1177/0888325418766625.

Keane, John. 2001. *Václav Havel: A Tragedy in Six Acts.* Basic Books.

King, Martin Luther, Jr. 2019. "Letter from Birmingham Jail." In *American Political Thought,* 2nd edition. Edited by Isaac Kramnick and Theodore J. Lowi. New York: Norton.

Kuran, Timur. 1997. *Private Truths, Public Lies: The Social Consequences of Preference Falsification.* Cambridge: Harvard University Press.

Kusy, Miroslav. 1985. Chartism and real socialism. In *The power of the powerless: Citizens against the state in Central and Eastern Europe,* ed. John Keane, 152-177. London: Hutchinson.

McIntyre, Lee. 2018. *Post-Truth.* London, England: MIT Press.

Popescu, Delia. 2011. *Political Action in Václav Havel's Thought: The Responsibility of Resistance.* Lexington Books.

Popescu, Delia. 2018. "The Importance of Bearing Witness." *East European Politics and Societies and Cultures* 32, no. 2 (May): 315-319.

Simecka, Milan. 1984. *The Restoration of Order: Normalization of Czechoslovakia, 1969-76.* Translated by A.G. Brain London, England: Verso Books.

Tucker, Aviezer. 2000. *The Philosophy and Politics of Czech Dissidence from Patočka to Havel.* Pittsburgh, PA: University of Pittsburgh Press.

Vaněk, Miroslav and Michael Kilburn. 2004. "The Ecological Roots of a Democracy Movement" *Human Rights Dialogue* 2(11) http://www.cceia.org/resources/publications/dialogue/2_11/section_1/4443.html

Wikipedia. "Truth prevails." Last modified 12 May, 2022. *https://en.wikipedia.org/wiki/Truth_prevails.*

Williams, Kieran. 2022. "Truth: Pravda." In *Václav Havel's Meanings: His Keywords and Their Legacy* edited by Kieran Williams and David Danaher. Prague: Karolinum Press.

Willoughby, Ian. 2016. "Havel's philosophy was far more important than that of many academic thinkers" Interview with Aviezer Tucker. *Radio Prague International.* August 1. https://english.radio.cz/havels-philosophy-was-far-more-important-many-academic-thinkers-says-aviezer-8218461

Wilson, Paul. 2018. "The Power of the Powerless, Revisited." *East European Politics and Societies and Cultures* 32, no. 2: 232-238. https://doi.org/10.1177/0888325417747971

Chapter 6

Re-conceptualising Scientific Expertise in International Criminal Investigations: An STS Perspective[1]

Karen McGregor Richmond

Abstract: Contemporary societies have become reliant upon the guidance of scientific, and technological, experts whose inputs are utilized - to a hitherto unparalleled degree - by a proliferating array of complex and specialised systems. Paradoxically, the contemporary reliance upon expert knowledge has given rise to a countervailing popular skepticism, which threatens to erode the very foundations of rational discourse, generating developmental obstacles across the panoply of natural, and social scientific domains, and creating tensions within discrete sites of technological application and epistemological uncertainty.

The field of international criminal justice has come to be regarded as a particular site of contestation, the investigation and prosecution of criminal acts - at the international level - being dependent upon the collection, processing, and categorisation of a diverse body of objective evidence, drawn from multiple sources: forensic samples, documentary material, 'open source' data and witness statements, inter alia, are recovered, and evaluated, by a heterogeneous body of institutional actors, drawn from diverse fields and backgrounds, possessed of varying levels of expertise, and increasingly founding upon disruptive new technologies, which themselves emerge across multiple disciplinary boundaries, confounding pre-existing institutional norms and expectations.

Clearly, the articulation of a coherent theoretical foundation for interdisciplinary expertise would serve all disciplines. However, that need is particularly acute within a criminal justice sector facing ethical and epistemological challenges generated by the emergence and confluence of machine learning technologies, biomedical research and the proliferating use

---

[1] STS refers to Science and Technology Studies

of telecommunications data. Meanwhile, citizen participation in open source investigations has grown steadily (at least insofar as public involvement facilitates distributed data collection), offering a direct challenge to scientific and technological experts, trust in whose practices has been further eroded by the politicization of knowledge production and dissemination.

If the international legal system is to maintain a robust and rational approach to the ethical, legal and social challenges engendered by machine learning, biomedical research, and sundry emergent technologies, then its responses must be founded upon a coherent theoretical account of trans-disciplinary scientific and technological expertise: an understanding whose broader application will enable citizens and policy makers alike to answer questions related to the proper function of expertise, its efficient mobilisation, and its limits. This necessary foundational research may thereby serve as a theoretical base which subsequent elaboration may aid institutional agents in negotiating disagreements between experts, serving not merely to justify the decision-making process to the public, but to facilitate their involvement in a dialectic process of policy development. The primary objective of this paper is therefore to develop, articulate, and disseminate, a normatively coherent theoretical account of transdisciplinary expertise, as practiced in the international criminal justice sector. An account which may demonstrate the potential for STS scholarship to address this area of collective concern, resolving the ontological and epistemological tensions which have been generated by the mobilization of trans-disciplinary scientific and technological innovations, deployed across disciplinary boundaries.

Keywords: Science and Technology Studies (STS); Criminal Evidence; International Law; Expertise; Open Source Intelligence (OSINT)

***

Contemporary societies have become reliant upon the guidance of scientific and technologically informed experts, whose inputs are utilized - to a hitherto-unparalleled degree - by a proliferating array of complex and specialized systems. Paradoxically, this technocratic reliance upon expert knowledge, alongside the perceived necessity for epistemic deference, has given rise to a countervailing current of populist skepticism, which threatens to erode the very foundations of rational discourse, propagating developmental obstacles across the panoply of natural and social scientific domains, and generating tensions within discrete sites of technological application and epistemological uncertainty. The field of international criminal justice has become a particularly febrile site of contestation, given that the investigation and prosecution of criminal acts at the international level is dependent upon the collection, processing, and categorization of a heterogeneous array of material evidence drawn from multiple sources, all of which requires interpretation and evaluation by skilled specialists. Forensic samples, documentary material, 'open source' data, and witness statements, *inter alia*, are recovered and evaluated by a diverse body of institutional actors, drawn from multiple fields and backgrounds, possessed of varying levels of expertise, and increasingly founding upon disruptive new technologies, which themselves emerge across multiple disciplinary boundaries, thereby confounding pre-existing institutional norms and expectations.

Clearly, the articulation of a coherent conceptual framework for scientific expertise - a framework that is alive to contemporary currents and developments - would benefit all disciplines. However, that need is particularly acute within the international criminal justice sector, a sector that is facing a cluster of ethical and epistemological challenges, not least those generated by the emergence and confluence of machine-learning technologies, biomedical research, and the proliferating collection of telecommunications and 'open source' digital data. The latter phenomenon raises added concerns. Given that citizen participation in 'open source' investigations has grown steadily (insofar as the collection and analysis of digital evidence now involve non-expert NGOs and allied institutional agents), it is vital that the international criminal justice field is able to offer a systematic and theoretically robust account of what it regards as foundationally valid expertise, in addition to being able to objectively weigh expert outputs. Such an account is rendered exigent by the increasingly distributed nature of data collection and by the proliferation of pseudo-forensic 'analyses' – frequently involving new and untested theories and methods – which together constitute a direct challenge to the epistemic privilege of scientific and technological experts, trust in whose practices is being further eroded by the politicization of knowledge production and dissemination.

Suppose the international legal system is to achieve a more robust and rational approach to the ethical, legal, and social challenges engendered by distributed evidence collection, machine learning, bio-medical research, and sundry emergent technologies. In that case, its responses must be founded upon a coherent theoretical account of scientific and technological expertise. It may thereby benefit the international criminal justice field for actors to acquaint themselves with the perspectives of Science and Technology Studies (STS), whose adherents have proffered comprehensive analyses of the exercise of scientific expertise and experience. Thus, the objective of this chapter is to posit just such a turn towards reflexive social-scientific Studies of Scientific Knowledge (SSK), and to increase lawyer's familiarity with the phenomenon of epistemological contest and closure commonly referred to as 'knowledge controversy': understandings, whose broader application will enable jurists, legal practitioners, citizens, and policy-makers alike to answer questions related to the proper function of expertise, its efficient mobilization, and its limits. Such familiarization may thereby serve as a theoretical and conceptual resource, which subsequent elaboration may aid institutional agents in negotiating and closing controversial disagreements between experts and non-experts, the policy-making establishment, and the *polis*, serving not merely to justify the decision-making process to the public, but to facilitate the latter's involvement in a reflexive and dialectic process of legal policy development. It is hoped that such an account may further demonstrate the potential for STS scholarship to address this area of collective concern, assisting in the resolution of the ontological and epistemological tensions that are increasingly generated by the mobilization of multi-disciplinary scientific and technological innovations across disciplinary boundaries.

In pursuance of this aim, the article proceeds from a social-constructionist ontological perspective, founding upon STS perspectives aligned with an interpretivist epistemological orientation. The author adopts a case study paradigm utilizing a single unit (specifically an instance of 'knowledge controversy' involving expert knowledge in an international criminal case), analyzed in accordance with doctrinal and socio-legal methods. The discussion begins with an account of the development of contemporary theories of scientific knowledge, centrally those associated with Studies of Scientific Knowledge (SSK) and Harry Collins' 'taxonomy of expertises'. This is supplemented by a discussion of professional scientific and legal expertise exercised within (and across) disciplinary boundaries, as typically associated with meso- and macro-sociological neo-Weberian perspectives, and those affiliated with Foucauldian studies of governmentality. The discussion then turns to the STS concept of 'knowledge controversy'. After introducing a basic taxonomy of approaches to knowledge controversy, the study embarks on an STS-infused exegesis on the common origins of law and science, leading to a brief prolegomenon on the use of scientific knowledge and methods in international criminal cases. The foregoing theoretical and conceptual elements are then applied in relation to the case study, which has been chosen due to its paradigmatic, thus generalizable, dimensions.[2] The unit of analysis, or case to be studied (discussed, *supra*) is defined as "a phenomenon of some

---

[2] The case study research paradigm has been described by Robson as "...a strategy for doing research which involves an empirical investigation of a particular contemporary phenomenon within its real-life context using multiple sources of evidence." Therefore, a case study should not be "characterised by the methods used to collect or analyse data but rather its focus on a particular unit of analysis: a case." (Robson 1993: 146)

sort occurring in a bounded context. The case is, in effect, one's unit of analysis" (Miles and Huberman 1994:25). It is therefore capable of maximizing the utility of information from '*small-n*' samples, and single cases. Flyvberg further subdivides information-oriented case selections into four sub-categories: extreme cases, maximum variation (or diverse) cases, critical cases, and paradigmatic cases, the latter composing the unit of analysis in the instant study (Stake 1994). The paradigmatic case discussed is specifically the introduction of emergent trans-generational trauma and immunosuppression theories in the *Germain Katanga*. Having analyzed the case and the knowledge controversy that it generated, in STS perspective, the results are evaluated, with conclusions and implications advanced.

Science and Technology Studies and Studies of Expertise and Experience

In order to pursue socio-legal inquiries related to the development of forensic and biomedical techniques and principles - and their operation within the criminal justice system - it is first necessary to outline a subsisting body of literature broadly classified as Science and Technology Studies (STS). STS is an umbrella term used to describe a corpus of anthropological, sociological, and historical studies of science: one which also contains developed sub-categories associated with Studies of Scientific Knowledge (SSK) and Studies of Expertise and Experience (SEE). The majority of works within this diverse group of studies share a social constructivist perspective and proceed on the understanding that scientific truth claims and technological artifacts are co-produced through a variety of natural, cultural, and social practices. The Science and Technology Studies corpus thus tends to demonstrate - through empirical methods - the extent to which contingent historical factors, social values, and group interests are embedded in the production of scientific practices, artifacts, and truth claims.

Attempts to categorize the field of Science and Technology Studies tend to coalesce around certain common themes, which may be categorized chronologically, doctrinally, or thematically. The following section of this chapter, therefore, describes and analyses the development of that sub-category of STS scholarship devoted to the development of theories of scientific expertise. Beginning with Merton's scientific norms, it charts the rise of social constructivism in the wake of Kuhn's theory of scientific revolutions. It then follows the programmatic shift of sociologies of science from the Weak to the Strong (consisting of a group of interrelated theoretical approaches, most notably the Epistemological Program of Relativism (EPoR, the Edinburgh school, and the Bath school). It critically analyses the Third Wave of STS scholarship - devoted to Studies of Expertise and Experience – and demonstrates the potentials and limitations of the current theory before concluding by suggesting ways in which these limitations may be overcome through a return to the underlying philosophical concepts which undergird contemporary theories of expertise.

In order to offer a more complete analysis of Collins and Evans socio-historically inflected theory of theoretical development, it is first necessary to situate their theory within the Science and Technology Studies corpus. In their 2002 Discussion Paper (Collins and Evans 2002: 240), *The Third Wave of Science Studies: Studies of Expertise and Experience*, the authors suggest that Studies of Expertise and Experience (SEE) constitute a 'Third Wave' of STS scholarship: one which supersedes all prior efforts to analyze the nature of scientific expertise. From whence does this third wave emerge? According to Collins and Evans tripartite account of STS scholarship, Wave One STS scholarship consists of those meager and highly-circumscribed pre-Kuhnian studies of positivist science.[3] Wave Two comprises the post-Kuhnian socially constructivist studies most closely associated with the Edinburgh and Bath schools and associated works.[4] Wave Three constitutes a later attempt to steer the course of STS scholarship toward Studies of Expertise and Experience (SEE). In doing so, Collins and Evans recommend a break with relational approaches, though it may be observed that this approach, somewhat paradoxically, regresses to a normative position which displays some of the distinguishing features of the Wave One studies which they had previously derided. Indeed, given Collins' vociferous criticism of 'First Wave' sociology - particularly its subservience to a positivistic view of the scientific method - it is paradoxical that Wave Three promulgates "...a preference for the norms and culture of evidence-based scientific argument [which] cannot be relinquished in the process of policy-making without giving up much more than [we] would want to surrender" (Collins and Evans 2007: 11).

Nonetheless, Collins and Evans assert that such a move is necessary in order to provide STS with the means to address the problems associated with the epistemological privileging of scientific expertise in wider society. Their stated objective is "to find a rationale which is not inconsistent with the last three decades of work in science studies". This is deemed necessary due to the perception that no STS scholars "have solved [the problem of expertise] in a way that is completely intellectually satisfying" (Collins and Evans 2002: 236). Thus, the requirement for Wave Three "[presented as] one of the ways in which Wave Two can be applied to a set of problems that Wave Two cannot handle in an intellectually coherent way" (Collins and Evans 2002: 240). This is an audacious claim in which elaboration has been the focus of critique from STS

---

[3] These have generally been associated with the Weak Programme.

[4] See, for example, Bloor 1976; 1999.

scholars, most notably Jasanoff, not least due to a number of paradoxical implications.

Thus, Collins proposes that the application of scientific-realist arguments from a socially constructivist perspective would lead to intellectual incoherence: a fundamental incoherence that can only be rectified through a normative regression, which will allow Wave Three to apply normative arguments from the perspective of scientific realism. This paradoxical claim lies at the heart of Collins approach and informs much of Wave Three thinking. "Wave Three involves finding a special rationale for science and technology, even while we accept the findings of Wave Two – that science and technology are much more ordinary than we once thought." (Collins and Evans 2002: 236).

In counterpoint, as stated above, it may be argued that Collins' Wave Three is not a development of Wave Two but simply a regress to Wave One, albeit a Wave One substantially diminished on the shoals of social constructivism. It is, therefore, a regress which, once again, places the sociology of scientific knowledge in thrall to positivist science. While Collins and Evans may have found their theory upon a critique of the shortcomings of Wave Two, their critique may itself exhibit flaws. Indeed, Collins and Evans attribute an extreme epistemological relationalism to social constructivism that it has not generally claimed for itself (save for its stronger proponents and iterations), a misattribution which is freighted with the unnecessary baggage of moral relativism, and which exemplifies their "persistently reductive reading of sources" (Jasanoff 2003: 931). In counterpoint, Jasanoff highlights the rich diversity of 'Wave Two' scholarship, which, "for more than a generation now, has formed part of a project of social theorizing that goes far beyond the relativism which the authors focus on." Jasanoff cautions that "To say that work as diverse as that of Brian Wynne, Michel Callon, Bruno Latour, and for that matter my own represents Wave Two's preoccupation with relativism is to seriously misread this work" (Jasanoff 2003: 931)

A further criticism of Collins and Evans theoretical approach revolves around their exclusive focus on a contemporary body of expertise enacted by a core-set of 'esoteric' research scientists. The emphasis on contemporary expertise flows from their assertion that past scientific disputes are deemed to have been largely settled. Therefore, the knowledge claims that exist behind the wave-front of research science and which compose the greater part of scientific knowledge are accorded a similar status to that which Wave One sociologists and the Weak Programme, accorded to normative, positivistic science. Collins and Evans fail to countenance the argument that the acceptance of such knowledge claims over an extended period of time may not simply entail that such claims represent 'natural' scientific facts. Rather, such acceptance may be the product of continuing social factors associated with the longevity of a dominant social class, and the robustly appealing valencies of a particular paradigm.

In addition to their chronological categorization, Collins and Evans argue for the 'necessary compartmentalization' of a discrete body of esoteric research science. They deem such compartmentalization as necessary on pragmatic grounds. However, it may be argued that the need to compartmentalize is not only pragmatic but crucial to the validity of Wave Three theorizing. In reality, the Wave Three position can only be maintained for as long as it is incubated within the confines of a discrete field of contemporary research science. Collins and Evans strive to demarcate an area for contemporary research science that is epistemologically privileged but, in so doing, limits the application of their theory. As Jasanoff states, activities within the core-set may be only one of a number of modalities that must be accounted for when making decisions relating to the status of scientific expertise: "Just as in sociology of science, 'facts' are depicted as the results (not the causes) of closure of controversy, so 'core-sets' in public domain technical controversies are the results, not the causes, of controversies settling into normalised patterns of debate" (Jasanoff 2003: 395).

Further criticism of Wave Three scholarship extends to Collins and Evans methodological aim, whose objective is to unearth and reify a kernel of epistemologically-privileged contributory expertise. The limitations of Collins methodological approach become apparent when their approach is translated to other domains beyond that of esoteric research science. Expertise, as practiced, is multi-faceted, and may be exercised - not merely through the dissemination and transmission of knowledge - but through; discretion, understatement, deflection, or even silence. These features of expertise are regularly encountered in professional practice within the legal sector. However, methodological techniques, which attempt to unearth propositional knowledge, may lack the ability to capture much that can be classified as practiced expertise. Therefore, more suitable alternatives will be proposed, which refract expertise and measure the flow of knowledge across sub-system boundaries, between modalities, and through time.

If Collins theory of scientific expertise represents the most developed sociological approach. In that case, it is one which, for all of its sophistications, remains flawed in its conception and may be ill-suited to exploring the nature of scientific expertise: limitations which are thrown into sharp relief when attempting to analyze scientific expertise conducted in conformation with accepted natural principles, under the external gaze of regulatory objectivity, in conformation with protocols, as part of a system of interdisciplinary co-production.

The taxonomy of expertises

In order to analyze the nature of scientific expertise, as embodied and enacted by institutional agents in the international legal system, it is necessary to utilize Collins and Evans hierarchical categorization of 'expertises`(Collins & Evans 2007: 14). According to the authors´ schema, the diverse array of 'expertises' can be systematized and tabulated. Indeed, Collins and Evans' taxonomic approach to the sociology of scientific expertise and experience is reducible to a conceptual 'Periodic Table of Expertises'. Progression through the periodic table brings the reader in contact with progressively more specialist and procedural forms of expertise. The table has a somewhat abstruse quality, given the inclusion of two 'dispositions' and a final series of 'meta-criteria', which non-experts may use to judge an expert's credibility and reliability. Therefore, for present purposes, discussion will be limited to the first three categories of the 'periodic' table, as follows.

Ubiquitous expertises

Ubiquitous expertise are those proficiencies which all citizens must possess in order to live in a particular society. It is a category of universal application, which includes the expertise needed to make political judgments or to speak a native language. This ubiquitous expertise emerges from a mass of tacit knowledge (defined by Collins and Evans as "things you just know how to do without being able to explain the rules for how you do them" (2007:13)).

Dispositions

Collins and Evans include personal qualities whilst discounting them as 'not very important'. They include two dispositions: interactive ability (linguistic fluency) and reflective ability (analytic flair). Having placed communication skills and the ability to make political decisions within the category of ubiquitous expertise, Collins and Evans go on to assert that more skillful iterations of these ubiquitous expertise are the product of innate disposition rather than the product of learning and development, which might place them in the category of specialist expertise.

Specialist Expertise

The specialist expertise rests upon a body of ubiquitous expertise. Beginning at the first rung, we encounter what Collins and Evans call 'Beer-mat knowledge' by which they mean those forms of general propositional knowledge which may increase a holder's stock of 'facts' but which are, in and of themselves, insufficient to serve a functional purpose. The next category relates to popular understandings of scientific practices and developments, as derived from the media and books. Knowledge derived from such sources enables the construction of basic inferences and is "transmissible as a set of ideas" (Collins and Evans 2007:20). Popular understandings of science populate the outer ring depicted in Figure One and are marked by relative certainty, faith in a unified community of experts, and a lack of detail.

Collins and Evans then insert a third category of ubiquitous tacit knowledge. Primary Source Knowledge is a somewhat ambiguous category comprising a population of non-institutional actors who outwit the scientific field and nonetheless interact with the primary literature on a particular topic.

At this stage Collins and Evans climb beyond the foothills of ubiquitous tacit knowledge, towards the domain of expertise *in stricto sensu*. According to their thesis, these developed forms of expertise can only be gained through the accumulation of specialist tacit knowledge and "immersion in the specialist culture" (Collins and Evans 2007:24). Thus, they return to the 'enculturation theory' first advanced under EPOR.

Collins and Evans novel contribution to the sociology of expertise and experience derives from their decision to split the traditional category of specialist expertise into two sub-categories: contributory and interactional expertise. Contributory Expertise in a particular field allows the acquirer to contribute to that field through the efficient utilization of an internalized set of skills.

> When expert status is achieved, complete contexts are unselfconsciously recognized and performance is related to them in a fluid way using cues that it is impossible to articulate and that, if articulated, would usually not correspond, or might even contradict, the rules explained to novices. (Collins and Evans 2007: 24)

Contributory experts are seen as needing two additional faculties: 'translation' (the ability to move between different social worlds)[5] and 'discrimination' (the ability to make distinctions between different kinds of claims and sources of credibility). This flexible definition of contributory expertise may serve as a useful tool with which to interrogate the nature of acquired skills and can be applied to a wide range of expertise, from driving a car to gravitational-wave research. However, Collins and Evans categorization fails to account for the conditions under which DNA-profiling experts enact their expertise. Forensic-scientific investigations are the result of co-production with non-experts; they are constrained by regulatory protocols of universal application and are

---

[5] The prospect of interdisciplinary expertise is largely ignored under Collins and Evans' schema.

conducted under the external gaze of auditors, policy-makers, and other institutional actors within the criminal justice system. Therefore, any analysis of forensic-scientific expertise must account for these differences.

The final categorization, introduced by Collins and Evans, relates to 'interactional expertise'. As with contributory expertise, this category is "deeply tacit-knowledge-laden" (Collins and Evans 2007:28). However, interactional experts are merely conversant in the language of a particular specialist field absent of actual experience in its practice. This category would, therefore, apply to professionals within the legal field whose specialist knowledge allows them to interrogate the truth claims of contributory experts from the forensic-scientific domain. Interactional expertise is, therefore, parasitic on contributory expertise and lacks the self-generating capacities and transmissibility of the prior category.

Elaboration of a conceptual framework for understanding the nature of expertise would be incomplete absent a survey of meso- and macro-sociological models. There is a normative and qualitative dimension to Studies of Expertise and Experience (SEE), which finds its expression in avowedly neo-Weberian understandings of technocratic state (and, by extension, legal) governance and administration (Pollitt and Bouckaert 2017). As noted, *supra*, law's interaction with cognate disciplines has been marked by an increasing reliance on the authority of experts to inform (or even to determine) the practical reasoning of the courts in relation to matters of legal fact-finding and proof.[6] The 'rise of the expert' has, from a traditional social science perspective - been viewed as a desirable product of functional specialization within society and the latter's subsequent colonization by a heterogeneous and proliferating array of professions.[7] The necessary conflation of expertise with specialization derives largely from the seminal writings of Weber, who, discussing the role of knowledge in society, uses the terms 'experts' (*Experten*) and 'specialists' (*Spezialisten*) interchangeably. Indeed, Weber's normative account of expertise is central to '*Science as a Vocation*', in which he argues that "only by strict specialization can the scientific worker become fully conscious, for once and perhaps never again in his lifetime, that he has achieved something that will endure. A really definitive and good accomplishment is today always a specialized accomplishment" (Weber 2004).

Weber's view of expertise as being intrinsically bound with specialization finds similar expression in his perspectives on the rationality of Western economics and administration. Resonates with prior discussion regarding the

6 See, for example Woolf 1996: chapter 13.

7 See, for example Porter 1990: 81. See also Luhmann 1982

shared epistemic foundations of law and science, the latter being judged according to the nostrums of legal fact-finding.

> Office management, at least all specialised office management – and such management is distinctly modern – usually presupposes thorough training in a field of specialisation. This, too, holds increasingly for the modern executive and employee of a private enterprise, just as it does for the state officials. (Weber and Tribe 2019)

Thus, Weber relates the increasing colonization of society by experts – as one expression of a macro-sociological process with ramifications that echo beyond the field of scientific endeavor. Weber is not alone. Since the beginnings of social-scientific inquiry, scholars have focused on the development of expertise, its inextricable links to democratic government, and the interactions between expertise and democratic values. Weber's inquiries were preceded by those of Henri de Saint-Simon and Auguste Comte, both of whom argued for the reconfiguration of society under the direct governance of scientific precepts rather than politics. Their technocratic approach to sociology had, in turn, a formative influence on thinkers as diverse as Mill, and Marx (Eriksen 2011).

Returning to the Weberian model, expertise is viewed as providing the government with the required knowledge regarding which means to select in order to reach a political goal; Weber noting that experts make decisions based on verified knowledge, whilst politicians (or citizens) attempt to negotiate goals, based on norms and values. Therefore, the establishment of a tier of knowledge-based administrators, whose competencies were limited by specialty and compartmentalized within a norm hierarchy, would provide political authorities with a powerful instrument with which to attain their goals. It is banal to note that the modern politician has learned to make ample use of such specialist, objective knowledge as the means by which to attain political goals (Turner 2008).

The gradual encroachment of the expert administrator on the democratic sphere by expert administrators was also observed by Jurgen Habermas, who - contrary to Weber - noted the tendency for the language of systems (which he defines in his *weltschauung* as 'steering media') to superimpose themselves over the natural, and spontaneous, communications of ordinary citizens. These judgments and opinions - the discourses of technocrats and experts – were endowed with authority by law to the extent that they came to replace the consensual, ethical communications of the 'lifeworld' (Habermas 2015). In this respect, Habermas' was heavily influenced by Foucauldian post-structuralism, particularly the focus on the instrumental deployment of expertise by liberal government as a means to influence and discipline the minds and bodies of human subjects (Foucault 2008). Foucault noted the ways in which the expert knowledge of the *epistemes* – such as those at play in medicine and science - were disseminated and promoted as incontestable facts. Facts, which form the taken-for-granted boundaries of common discourse. This expert knowledge thus composed the discursive frame, which defined the parameters of power and authority for any given era. Foucault further considered the utility of expertise to the art and practice of liberal governance through a process that he termed 'bio-politics'.[8] Bio-politics thus became a central theme of Foucault's work: an ordering concept that allowed governments to manage and govern populations effectively. Governance was dispersed amongst the *epistemes* of law, medicine and psychiatry, and this expert knowledge - legitimated by their seeming neutrality – would prove highly effective in their ability to penetrate the private family unit. From within, they could then enroll, manage, and reshape individuals as disciplined subjects.

The meso- and macro-theoretical groundwork provided by Weber, Habermas, and Foucault continues to shape discussion of the role of expertise. Neo-Weberian perspectives illuminating the ways in which professional groups further their collective interests, through processes of inclusion and exclusion. These demarcation theories revolve around the central proposition that disciplines and professions themselves define their borders through 'boundary work'. Thus, the practice of inclusion and exclusion comes to determine what is science and what is non-science. This binary demarcation is marked by negotiations between communities who claim scientific expertise and allied institutional actors. Such theories view scientific controversy and consensus-building as activities central to the shaping of group cohesion (Collins and Evans 2007:1; Lawless 2022: 3). However, the macro-sociological perspectives and contemporary demarcation perspectives elide a developed discussion of epistemic hybridity or the cross-fertilization and transposition of expertise across disciplinary boundaries. Further, to the extent that contemporary legal consumption of expert knowledge founded upon monolithic understandings of scientific expertise grounded in Mertonian norms, such understandings will also elide the trier-of-fact. The following section offers a paradigm case study of the difficulties that international courts face when confronted with emerging cross-disciplinary theories and empirical studies.

---

[8] Foucault describes bio-politics as 'a new technology of power... [that] exists at a different level, on a different scale, and [that] has a different bearing area, and makes use of very different instruments.' Bio-politics was predicated on the use of measurements, indices and statistics to create, identify and categorise new populations. Examples included measurements of the 'ratio of births to deaths, the rate of reproduction, the fertility of a population, etc.' *See* Foucault 2004.

Scientific Expertise and Knowledge Controversy

Scientific controversies are an increasingly frequent phenomenon in public discourse (Martin and Richards 1995). Whatmore describes these as 'moments of ontological disturbance', when unexamined nostrums and beliefs are rendered mutable (Whatmore 2009). The appearance of a knowledge controversy is usually defined by the public re-assessment of expert knowledge (Callon 1998; Latour 2004; Stengers 2005:994). And such situations, when boundaries are rendered uncertain are frequently to be found in the field of legal adjudication, where tribunals of fact acting in the public good must adjudicate upon – indeed cannot displace or postpone judgment upon – contentious and emergent scientific claims and precepts. Scientific and technological controversies require to be negotiated and resolved in the instance of the courtroom. The abundant literature on knowledge controversy – so central to STS studies – has been efficiently resolved by Martin and Richards (1995) into four paradigm categories. The first of these distinct approaches is the positivist approach, according to which social scientists – and, by extension, tribunals of fact – should accept orthodox scientific perspectives, analyze particular issues from these perspectives, and resolve scientific debate concerning normative and rational scientific precepts. Thus, the separation of an objectivist and realist science from both social sciences and both legal and public perceptions resonates with Weberian specialization and professionalization, as well as Collins 'wave one' normativity and works best when attempting to distinguish aberrations from standard practice.

The second variant of knowledge controversy is framed around a group politics approach, focusing on the meso-level interactions of courts, NGOs, government departments, corporate interests, expert panels and citizen collectives (Martin and Richards 1995:3; Nelkin 1975; Nelkin 1992; Nelkin 1995: 456; Nisbet, Brossard and Kroepsch 2003). Thus, the knowledge controversy aligns with neo-Weberian perspectives and attributes no special place for knowledge controversy beyond that routinely attributed to group contestation and political maneuvers. Such forms typically involve discussion of comparable resources or, in Bourdeusian terms, competing forms of social and symbolic capital. The group politics approach is particularly prevalent when dealing with issues with a public policy dimension and is therefore predicated upon more overt forms of conflict, losing traction in the face of non-controversial iterations of Kuhnian 'normal science'.

The third knowledge controversy variant is characterized by an avowedly constructivist SSK approach (Pinch and Leuenberger 2006; Mercer 2002a; 2002b). For the social scientist, the plurality of accounts of the natural world and its processes are evidence of the social construction of scientific negotiations and mediations: constructions that are ordinarily occult but made tractable through disputation. Thus, the SSK approach conforms to Collins Wave Three approach insofar as it challenges positivist accounts of science that relegate social scientific analyses to explanations of Lysenko-ist deviations from realist norms. In counterpoint, the SSK approach to knowledge controversy is symmetrical: it refrains from privileging 'normative' science whilst retaining a breadth of focus that encompasses wider social dynamics. Further, SSK empirical studies offer insights into ongoing and unresolved controversies, following these towards closure, the latter imposed not by scientific method but by the adjudication of a wide community of actors representing vested interests and accepted norms. However, such micro-level studies often overlook the importance of professional and structural influences that are better highlighted by competing approaches.

The fourth category - the social structural approach - cleaves to the conceptual imperatives of social structure, foregrounding class, state governance, colonial privilege, and patriarchal dominion (Martin and Richards 1995; Harding 1991). Thus, knowledge controversies come to be understood through, for example, Marxian understandings of the totality of economic relations that structure the relationship between economic base and ideological superstructure, shaping a resultant mode of production. Thus, the actions of a semi-autonomous political group or profession are replaced by a more structured and determinative set of social relations. Thus, knowledge controversy emerges as a product of systemic contradiction of capital or through class struggle. Similarly, feminist analyses fall within the social structural approach, being organized around gender - as opposed to economic - relations. Having laid out the four ideal categories of knowledge controversy, the discussion now turns to the interactions between law and science prior to an analysis of the affordances and dimensions of knowledge controversy at play in the *Katanga* case.

The common origins of law and science

In a seminal STS study devoted to the interplay of legal and forensic-scientific cultures, Lynch contends that law and science are separated by a common language. Their shared vocabularies feature similar terms, including; 'investigation', 'proof', 'law', 'test', 'inquiry', 'fact' and 'evidence' (Lynch 2013). These terms may carry distinct meanings, particular to their respective fields. However, Lynch argues that the use of parallel terms in both the legal and scientific fields indicates a common origin. A number of scholars have attempted to trace this common origin. Their conclusions achieve little by way of consensus, except the fact that such studies may point to the dominant historiographical preoccupations of the time. For example, at the dawn of the Scientific Revolution in the late seventeenth century, religion was widely held to have provided the formative basis of both law and science (Faigman 2000). This view persisted through the Enlightenment into the twentieth century[9], when it was complemented by a similarly determinative theory based on the significance of economic forces (Freudenthal and McLaughlin 2009). In the modern era, scholars have embarked on more radical departures from classical social theory, developing arguments that highlight the significance of disciplinary power and the evolution of governmentality (Foucault 2004). These 'genealogies of fact' have unearthed structurally complex accounts of the development of conceptual frameworks and the power relations that they serve. However, the most trenchant accounts of specialist knowledge and experience have been developed within the branch of sociology referred to as Science and Technology Studies, which serve to illuminate the social construction of the scientific method.

Shapiro, notably, has traced the *fons et origo* of modern scientific inquiry back to legal procedures and shared institutional structures. She posits that the commensurate nature of these corresponding institutional arrangements allowed for the easy transmission of common values between discrete fields, both of which were based on a particular mode of objectivity. According to her thesis it was neither God, nor Mammon - but legal bureaucracy - which provided the modalities for the post-Enlightenment scientific method. Shapin (1995) and Schaffer (2011) take up this perspective, citing the congruence between the courtroom and the scientific experiment, both of which serve as discrete sites of knowledge-making in which disinterested participants apply impersonal rules within the context of a body of transparent and 'rational' procedures, all of which are designed to produce 'morally certain' resolutions on 'matters of fact'(Shapin and Schaffer 2011). By way of example, in 1659, when Robert Boyle completed the construction of his *machina Boyleana*[10], thence embarking on a program of experimentation in order to 'solicit and interrogate the testimony of nature' (Shapiro 2022), his determinations were deemed to lack the epistemological authority of the conclusions derived as a result of inferential legal reasoning. Remarkably, this conjoining of legal and scientific epistemology subsisted until the Victorian era. Indeed, Shapiro (2022) notes that it was not until the mid-eighteenth century that a distinct rupture occurred between the moral certitude of the courtroom (expressed as 'a satisfied conscience', or proof 'beyond reasonable doubt') on the one hand and

---

[9] Thomas Merton's 1936 doctoral dissertation – 'The Merton Thesis' - outlines the links between a particular mode of religious pietism and early experimental science. The thesis continues to form the subject of academic debate. *See* Shapin 1988

[10] The *Machina Boyleana* was a pneumatic engine designed to facilitate experimentation.

on the other, the 'mathematical certainty' to which contemporary science aspires (Shapiro 1986: 153; Shapiro 2022).

This bifurcation was to become more pronounced throughout the course of the twentieth century as science became increasingly characterized by a distinct set of norms. The 'norms of science', as described by Merton (1979), may be particularized as follows: communalism, universalism, disinterestedness, and organized skepticism. Paradoxically, it should be noted that Merton's normative categorization - which views science as striving towards an ideal of impartiality and rationality - was heavily influenced by Boyle, who, as demonstrated above, had been heavily influenced by the 'rational bureaucracy' of the courtroom. Thus, it may be posited that the normative roots of contemporary science may be traced to the fountainhead of legal inference.

Turning to Mertonian normativity, it may be argued that, in common with legal positivism, the scientific method - as Merton describes it – generates an agglomeration of normative outputs, which – in respect of their social valences- combine to produce a privileged scientific *leitkultur*. The scientist thus comes to embody the authoritative representative of an epistemically privileged set of truth claims, which are both esoteric (being within the province of expert knowledge) and positivistic in inflection. Describing this widely-held Mertonian view of the scientific method, Collins goes as far as to assert that, up until the 1960s, "it was inconceivable that decision-making in matters that involved science and technology could travel in any other direction than from the top down" (Collins and Evans 2002:239). Further, the role of social science was thereby restricted to reinforcing the nostrums of positivist science rather than questioning their foundations.

Merton thus established the normative characteristics through which to recognize a socially-legitimated form of scientific expertise. However, studies of normative science were also faced with the requirement of accounting for the changing - and often contradictory - nature of scientific truth claims, a task taken up by Thomas Kuhn, whose work provided the first comprehensive account of the nature of scientific revolutions (Kuhn 2012). Kuhn argued that paradigm shifts in the natural sciences are best viewed as tectonic conceptual changes, which are at first resisted, later achieving the momentum necessary to overthrow the prevailing dogma, thus gaining widespread acceptance. The classic example is provided by the Scientific Revolution, which swept away the earth-centered Aristotelian universe and replaced it with a Heliocentric model,[11] thus inculcating a mode of inquiry through which all natural

[11] The poet John Donne, for example, wrote of the social anomie precipitated by the Copernican Revolution.

phenomena could be tested, in conformity with the broader modalities of skepticism, empiricism, and mathematization. In addition to generating epistemological changes, Kuhn demonstrated that scientific paradigm shifts may inculcate profound, frequently anomic, social changes.[12] Kuhn attributed such anomic effects to the incommensurability of competing paradigms: the 'old' and 'new' paradigms would stand in opposition until - following a brief contretemps - one view would prevail, to the exclusion of the other. The dominant view, once embedded, would thereafter generate its own corpus of 'normal science'.

Ironically, Collins claims that Kuhn's work would itself be responsible for a paradigm shift in Science and Technology Studies, breaking the wave of 'positivism' and ushering in the era of 'unbridled social constructivism' (Jasanoff 2003: 391). However, Collins 'First Wave of Social Studies of Science' must be approached with caution. It is a neat categorization which, Collins himself concedes, may do violence to the subject through the compression of the 'work of many authors and thinkers into a few simple formulae' (Collins and Evans 2002: 237). Indeed, it may be argued that any attempt to periodize this diverse body of studies - particularly by way of socio-historical treatment of the problematic development of 'scientific legitimacy' - may be misleading. However, it should be noted that Collins follows none other than Kuhn in approaching the topic in broadly Hegelian terms (thesis, antithesis, and synthesis). Further, Collins and Evans socio-historical categorizations serve a useful purpose, not least since their recognition of the first wave of normative formed the backdrop against which they were able to instantiate their own programmatic paradigm shift in Science and Technology Studies. Collins and Evans 'Wave' theory, will form the subject of further discussion below.

Scientific Expertise in International Criminal Proceedings: Case Study and Analysis

Psychological evidence is frequently submitted in the courtroom, particularly in the form of testimony relating to the accused's mental state and as a mitigating factor relevant to culpability (Lingaas 2021). However, such is the lack of admissibility criteria at the International Criminal Court (ICC) that novel forms of 'scientific' evidence may be brought before the court absent of any demonstration of foundational validity. One such example occurred in the context of the *Prosecutor v Germain Katanga*, trial, in which a clinical psychologist was asked to report on the possible psychological effect of

---

[12] *See* John Keats' *Lamia* (1884), in which the poet laments the fact that Sir Isaac Newton had 'unweaved the rainbow' by using a prism to split light into its constituent parts.

traumatization on behalf of the accused may have had upon victims. The expert gave testimony on two causal theories, both of which posit that trauma may be passed from trauma survivors to subsequent generations. The first mechanism is (behavioral attachment theory) and well-established. The psychologist gave further testimony on the relevance of recent biomedical research, which purports to show that trans-generational trauma negatively affects the epigenetic development of the children and grandchildren of trauma survivors (ICC: Prosecutor v Germain Katanga: 2018*)*.

> (T)rauma doesn't change your genetic make-up, but it alters your epigenetic development. That means if you have been traumatised you might give the impact of this traumatisation not just to your children but to your children's children. So there seems to be a change in the stress-regulating system of offspring over two generations which prepares these children to live in hostile environments. So it's almost as if I'm born with the capacity to survive better. However, if you change the stress-regulating system in a human, you also expose them to a great degree of vulnerability for other…other illnesses. (ICC: Prosecutor v Germain Katanga: 2018)

Genetic factors associated with early-life stress contribute to the pathogenesis of mental disorders in adulthood. Stress, immunity, and disease are interrelated. The immunological consequences of chronic stress. The suggestion is that exposure to stress in an individual (A) may affect the epigenetic development of any children born subsequent to the stressful event. These children (B), and grandchildren (C), will develop a more robust response to environmental stress based upon a rapid release of cortisol.

> What we have also found is a variation in gene [mutation]. That means the way these nerve cells are built, and we find that heightened cortisol level in babies of traumatised mothers, and more suicide and depression in those babies when they have become mothers, without adding extra traumatic stressors in their life. (ICC: Prosecutor v Germain Katanga: 2018)

Indeed, high cortisol levels may also promote immunosuppression in addition to better stress regulation, thus resulting in exposure to disease. However, as presented, the testimony appears to conflate three distinct physiological and genetic mechanisms in a way that renders the testimony open to misinterpretation. For disambiguation purposes, it should be noted that cortisol is a hormone that acts as an inhibitory regulator of cytokines. The latter class of biochemical compounds stimulates the human immune response, thus helping to combat disease. However, these same compounds also create an inflammatory response which, whilst necessary to combat disease, is damaging to the circulatory system. Thus, cortisol regulates cytokine expression in order to prevent damaging inflammatory events (hence the use of corticosteroids to treat joint inflammation). Cortisol is also released in response to stress, and the release of cortisol in response to stress is natural and beneficial but may be harmful when faced with prolonged and excessive exposure.

It is notable that in dealing with this testimony, the court dismissed requests for reparations made by the five applicants, finding that the applicants had not established the requisite standard of proof. Specifically, the applicants failed to establish a causal nexus between the psychological harm they claimed to have personally suffered and the crimes for which Mr. Katanga was convicted. However, in resolving this issue, the court appeared unable to engage with expert scientific evidence in any systematic manner or to proffer a comprehensive analysis or elaborated reasoning on the issue. In order for international courts to engage more fruitfully with the increasing need for expert opinion to resolve knowledge controversies - particularly those that emerge from cross-disciplinary research (such as that led in the *Katanga* claim, which embraces genetics, biochemistry, and clinical psychology) - a more developed understanding of knowledge controversies, and their resolution, will be necessary. Thus, the following section offers an analysis of the various elements of the *Katanga* controversy in relation to the four idealized forms of knowledge controversy previously isolated.

Analysis of Knowledge Controversy

It is established from the foregoing discussion that the tribunal of fact - acting as a bounded and professionally embedded species of social analyst - was charged with the task of closing an emergent knowledge controversy over the reliance on studies purporting to demonstrate a biochemical mechanism for the epigenetic generational transmission of trauma. In this section, the analysis isolates elements of the legal discussion and resolution of the question, evaluating these in relation to the four main categories of knowledge controversy (Martin and Richards, 1995). This necessarily begins with a discussion of ontological and epistemological perspectives. In relation to such issues, the group's political and social structural approaches are largely silent on account of their lack of conceptual frameworks to approach such questions. It follows that the positivist approach holds to an objective and realist position regarding onto-epistemological matters, whilst the SSK approach is aligned with interpretivist and social-constructionist precepts. In relation to the knowledge controversy around epigenetic trauma transmission, the testimony and jurisprudential analysis bear the marks of scientific positivism (hence onto-epistemological realism). However, despite the legal and scientific fields sharing the same foundational methods of analysis, hypothesis testing, and logic, the tribunal of fact chose not to engage with the evidence in a systematic manner, instead retreating into legal - as opposed to scientific - framings of causality. A more scientifically literate tribunal of fact may have been able to resolve the controversy in conformity with positivist precepts. Specifically, it may have been noted that, although cortisol is released in response to stress, and that release of cortisol in response to stress is natural and beneficial., the physiological response to chronic stress may result in immune system resistance, which may potentially lead to the onset of inflammatory disease through an inability to regulate cytokine expression. Thus, through engaging with the cited literature, it could be argued that the report – as presented - misrepresents the epigenetic mechanism by conflating the findings of heightened and comparatively rapid cortisol expression in children of PTSD sufferers, on the one hand, with immunosuppression in response to chronic stress on the other. As the expert rightly states, 'babies born to traumatized mothers may have higher and faster arousal peaks'. However, it would be hazardous to draw from the foregoing causal mechanism an inference to the effect that "trauma is passed on at a biological level which *perpetuates traumatic experiences* in a trans-generational manner" (emphasis added). Thus, it is possible to resolve the issue in concordance with positivist understandings of science.

Turning to the SSK approach, what does the foregoing case entail in terms of expertise? Whilst no criticism of the expert is implied in the foregoing discussion, it may be argued that, in circumstances wherein the court defers to an expert to provide specialized opinion evidence, the court itself bears a responsibility to ensure that expert evidence can be delivered with sufficient focus to avoid the specialist from being drawn out of their proper area of expertise. Further, that such evidence can be delivered with a sufficient level of granularity that allows for the communication of nuance, detail, and necessary contextual elaboration. Crucially, for the international courts to engage systematically with expert testimony requires a procedurally-grounded theory of expertise, alive to the disciplinary embedding of particular *epistemes*, and able to distinguish between expertise in its contributory and interactional forms, thus possessed of the ability to recognize the instances when expert testimony crosses disciplinary boundaries.

It is posited that one way of resolving the issue of the merits of the alternative claims, based upon a developed body of literature on behavioral attachment theory and a novel epigenetic theory, might be to approach the matter from a procedural perspective, weighing the testimony itself in terms of valid expertise. In so doing, triers-of-fact might lean on the distinction between contributory and interactional experts, as discussed above. Collins' approach is informed by the works of both Polanyi and Heideggerian phenomenology. Hence the central claim that genuine contributory expertise is grounded in practice and is founded upon a body of 'tacit knowledge,' and that the comparatively developed forms of contributory expertise can only be attained through an accumulation of specialist tacit knowledge based upon an 'immersion in the specialist culture.'

In circumstances such as those encountered in the Katanga case, an appreciation of the distinction between contributory and interactional expertise (as between the clinical psychologist's tacit understanding of behavioral attachment theory compared with the theoretical knowledge of novel epigenetic theories) would not only allow the court to define the borders of the expert's field but also to assess when, and to what degree, if any, those borders can safely be transgressed. It is posited that such theoretical insights, derived from the sub-field of Science and Technology Studies devoted to Studies of Expertise and Experience, may provide the necessary foundational insights that might allow the courts to account for the increasingly cross-disciplinary nature of expertise and thereby deliver rational and coherent judgments on matters of expert opinion within the international criminal justice arena.

The closure of the controversy will be achieved in differing ways. Most straightforwardly, the positivist approach necessitates that problems emerging from scientific disputes are resolved through scientific means. Logical analysis based upon the evidence and facts will direct inquiry to the correct answer, which will resonate with the wider scientific community. However, social pressures from beyond the boundaries of the scientific realm may serve to close down the controversy whilst the issues remain unresolved, as was the case in Katanga. Thus, the legal system may be viewed as an interposing social force that may interrupt of distort the correct closure of scientific knowledge controversies. Likewise, the SSK approach looks to closure within the scientific community but in relation to the ability of scientific factions to persuasively dominate the marketplace of competing ideas. Such approaches are echoed by the group politics approach and the social structural approach, albeit the latter two approaches admit a wider range of actors, all of whom are engaged in hegemonic struggles. Such divergences in approach also shape understandings of who is the best place to make decisions regarding the resolution of scientific disputes. Whilst the positivist approach places decision-making firmly within the ambit of the scientific community, a view supported by the legal community in its positivist orientation towards scientific knowledge is at odds with the strategies for resolution employed by the international courts. Given the differing affordances at play in relation to the resolution of knowledge controversy, it is therefore vital that the international criminal justice field develops the reflexivity necessary to propound a more systematic and theoretically robust account of what it regards as foundationally valid expertise, in addition to being able to objectively weigh expert outputs using logic and evidence, in concordance with positivist science. Failure to do so may result in the deterioration of epistemic value placed on scientific and technological experts, trust in whose practices is being further eroded by the politicization of knowledge production and dissemination.

References

Bloor, David. 1991. *Knowledge and Social Imagery.* 2nd ed. Chicago, IL: University of Chicago Press.

Bloor, David 1999. "Anti-Latour." *Studies in History and Philosophy of Science* 30(1):81–112. doi: 10.1016/s0039-3681(98)00038-7.

Callon, Michel. 1998. "An Essay on Framing and Overflowing: Economic Externalities Revisited by Sociology." *The Sociological Review* 46(1_suppl):244–69. doi: 10.1111/j.1467-954x.1998.tb03477.x.

Collins, Harry, and Robert Evans. 2002. "The Third Wave of Science Studies: Studies of Expertise and Experience." *Social Studies of Science* 32(2):235–96. doi: 10.1177/0306312702032002003.

Collins, Harry, and Robert Evans. 2007. *Rethinking Expertise.* Chicago, IL: University of Chicago Press.

Faigman, David. 2000. "LECTURE: Legal Alchemy: The Use and Misuse of Science in the Law." http://hdl.handle.net/20.500.13051/7821

Foucault, Michel. 2008. *The Birth of Biopolitics: Lectures at the College De France 1978-79.* Basingstoke: Palgrave Macmillan.

Foucault, Michel. 2004. *"Society Must Be Defended": Lectures at the College De France 1975-76* Reprinted ed. London: Penguin Books.

Freudenthal, Gideon, and Peter McLaughlin, eds. 2014. *The Social and Economic Roots of the Scientific Revolution: Texts by Boris Hessen and Henryk Grossmann.* 2009th ed. Dordrecht, Netherlands: Springer.

Habermas, Jürgen. 2015. *The Theory of Communicative Action: 2. Lifeworld and Systems a Critique of Functionalist Reason.* Cambridge: Polity Press. http://rbdigital.oneclickdigital.com.

Harding, Sandra. 1991. *Whose Science? Whose Knowledge?: Thinking from Women's Lives.* Buckingham, England: Open University Press.

ICC: *Prosecutor v Germain Katanga.* 2018. Case No.: ICC-01/04-01/07. Decision on the Matter of the Transgenerational Harm Alleged by Some Applicants for Reparations Remanded by the Appeals Chamber in its Judgment of 8 March 2018: Date: 19 July 2018

Jasanoff, Sheila. 2003. "Breaking the Waves in Science Studies: Comment on H.M. Collins and Robert Evans, 'the Third Wave of Science Studies.'" *Social Studies of Science* 33(3):389–400.

Kuhn, Thomas S. 2012. *The Structure of Scientific Revolutions*. Chicago, IL: University of Chicago Press.

Latour, Bruno. 2004. "Why Has Critique Run out of Steam? From Matters of Fact to Matters of Concern." *Critical Inquiry* 30(2):225–48. doi: 10.1086/421123.

Lawless, Christopher. 2022. *Forensic Science: A Sociological Introduction*. 2nd ed. London, England: Routledge. https://doi.org/10.4324/9781003126379

Lingaas, Carola. 2021. "Dehumanising Ideology, Metaphors, and Psychological Othering as Evidence of Genocidal Intent." *International Criminal Law Review* 22(5–6):1044–67. doi: 10.1163/15718123-bja10098.

Lynch, Michael. 2013. "Science, Truth, and Forensic Cultures: The Exceptional Legal Status of DNA Evidence." *Studies in History and Philosophy of Biological and Biomedical Sciences* 44(1):60–70. doi: 10.1016/j.shpsc.2012.09.008.

Luhmann, Niklas. 1982. *Luhmann: The Differentiation of Society*. Translated by S. Holmes and C. Larmore. New York, NY: Columbia University Press.

Martin, Brian, and Evelleen Richards.1995. "Scientific Knowledge, Controversy, and Public Decision-Making." Pp. 506-526 in *Handbook of science and technology studies edited by* Sheila Jasanoff, Gerald E. Markle, James C. Petersen, and Trevor Pinch. Newbury Park, CA: Sage

Mercer, David. 2002a. "SSK and law/science encounters involving controversial science and technology: a brief critical overview" Pp.254-272 in *2002 Yearbook of the Institute of Advanced Studies on Science and Technology* edited by A. Beal. Graz: Profil.

Mercer, David. 2002b. "The Intersection of Sociology of Scientific Knowledge (SSK) and Law: Some Themes and Policy Reflections." *Law Text Culture* 6.

Miles, Matthew B., and A. Michael Huberman. 1994. *Qualitative Data Analysis: An Expanded Sourcebook*. 2nd ed. Thousand Oaks, CA: SAGE Publications.

Merton, Robert K. 1979. "The normative structure of science." Pp. 267-278 in *The Sociology of Science: Theoretical and Empirical Investigations* edited by N. W. Storer. Chicago, IL: University of Chicago Press.

Nelkin, Dorothy. 1975. "The Political Impact of Technical Expertise." *Social Studies of Science* 5(1):35–54. doi: 10.1177/030631277500500103.

Nelkin, Dorothy, ed. 1992. *Controversy: Politics of Technical Decisions*. 3rd ed. Thousand Oaks, CA: SAGE Publications.

Nelkin, Dorothy. 1995. "Science Controversies the Dynamics of Public Disputes in the United States." Pp. 444–56 in *Handbook of Science and Technology Studies*. California: SAGE Publications.

Nisbet, Matthew C., Dominique Brossard, and Adrianne Kroepsch. 2003. "Framing Science: The Stem Cell Controversy in an Age of Press/Politics." *Harvard International Journal of Press/Politics* 8(2):36–70. doi: 10.1177/1081180x02251047.

Eriksen, Erik Oddvar. 2011. "Governance between Expertise and Democracy: The Case of European Security." *Journal of European Public Policy* 18(8):1169–89. doi: 10.1080/13501763.2011.615206.

Pinch Trevor, and Christine Leuenberger. 2006. "Studying Scientific Controversy from the STS Perspective". *Paper presented at the EASTS Conference "Science Controversy and Democracy"*. National Taiwan University, Taiwan. http://sts.nthu.edu.tw/easts/conference.htm

Pollitt, Christopher, and Geert Bouckaert. 2017. Public Management Reform: A Comparative Analysis-into the Age of Austerity. Oxford: Oxford University Press.

Porter, Roy. 1998. *England in the Eighteenth Century*. First Edition. The Folio Society.

Shapin, Steven. 1988. "Understanding the Merton Thesis." *Isis; an International Review Devoted to the History of Science and Its Cultural Influences* 79(4):594–605. doi: 10.1086/354847.

Shapin, Steven. 1995. *A Social History of Truth: Civility and Science in Seventeenth-Century England.* 2nd ed. Chicago, IL: University of Chicago Press.

Shapin, Steven, and Simon Schaffer. 2017. *Leviathan and the Air-Pump: Hobbes, Boyle, and the Experimental Life*. Princeton, NJ: Princeton University Press.

Shapiro, Barbara J. 1986. "To A Moral Certainty: Theories of Knowledge and Anglo-American Juries 1600-1850." *Hastings Law Journal* 38(1):153.

Shapiro, Barbara J. 2022. *Beyond Reasonable Doubt and Probable Cause: Historical Perspectives on the Anglo-American Law of Evidence*. Berkeley, CA: University of California Press.

Stake, Robert E. 1995. *The Art of Case Study Research*. Thousand Oaks, CA: SAGE Publications.

Stengers, Isabelle. 2005. "The cosmopolitical proposal" in *Making Things Public: Atmospheres of Democracy* edited by Bruno Latour, and Peter Weibel. London, England: MIT Press.

Turner, Stephen. 2008. "Blind spot? Weber's concept of expertise and the perplexing case of China." Pp. 121-134 in *Max Weber matters: Interweaving past and present* edited by D. J. Chalcraft, F. Howell, M. Lopez Menendez, and H. Vera. London, England: Ashgate Publishing.

Weber, Max. 2004. *The Vocation Lectures: "Science as a Vocation"; "Politics as a Vocation."* edited by D. S. Owen and T. B. Strong. Cambridge, MA: Hackett Publishing.

Weber, Max. 2019. *Economy and Society: A New Translation*. edited by K. Tribe. London, England: Harvard University Press.

Whatmore, Sarah J. 2009. "Mapping Knowledge Controversies: Science, Democracy and the Redistribution of Expertise." *Progress in Human Geography* 33(5):587–98. doi: 10.1177/0309132509339841.

Lord Woolf. 1996. *Access to Justice: Final Report*. London: Her Majesty's Stationery Office.

PART 3. Participatory research as a panacea

Chapter 7

Alternative science communication in the post-truth moment: case of grassroots innovations and Shodhyatra in India

Hemant Kumar and Rachan Daimary

Abstract: In this post-truth moment, when the fields of Science, Technology, and Society (STS) and Science Communication Studies (SCS) are facing a number of challenges, it is difficult to comprehend how different knowledge systems may coexist. Knowledge from the informal sector, particularly grassroots innovations (GIs) in India, has captured the interest of policymakers, practitioners, and academics alike since the turn of the twenty-first century. Honey Bee Network, an informal network comprised of several organizations from the formal sector, aims to foster grassroots innovations and the propagation and exchange of knowledge about GIs. This study explores the various avenues of communication that are present in GIs. The participation of the common people is essential to the development of GIs at the local level. This is because multiple actors from both the informal and formal sectors are engaged in the production of knowledge. In this study, ethnographic strategies are employed in order to gain an understanding of the communication procedures that led to the participation of the actors. Individuals and community members engage in the early phases of the GI process and share their knowledge. Later on, participants from the formal sector, such as scientists and entrepreneurs, are involved in the diffusion of these ideas to a larger audience. In addition, this highlights the difficulties that are linked with communication and public participation in GIs. Also, the study investigates the plausibility of several knowledge systems coexisting together.

Keywords: Grassroots innovations, public engagement, science communication, Shodhyatra, India

***

Introduction

The post-truth world has been a point of concern as a result of global events such as the rise of right-wing authoritarianism (Hoffman 2018; Söderberg 2022) and the COVID-19 pandemic (Parmet and Paul 2020; Nally 2022), among others. This discussion has highlighted doubts regarding scientific institutions and the media (Mcintyre 2018; Iyengar and Massey 2018), and it has also been debated how Science, Technology, and Society (STS) studies are responsible for this predicament (Hoffman 2018; Söderberg 2022). Hoffman (2018), however, contends that STS is not to blame because it is grounded in 'knowledge democracy' and 'citizen scientists' and instead studies the process of 'truth' formation rather than the 'truth' itself. In a similar vein, Jasanoff and Simmet (2017) propose that STS must not only explain the current predicament but also offer alternatives. Hoffman (2018) consequently believes that "post-truth politics" is nothing more than 'demagoguery,' which demonstrates how 'ordinary people' are mobilized to undermine established standards of behavior, institutions, and the law (Hoffman 2018: 449).

Although the STS tradition acknowledges that 'ordinary people' actively contribute to the process of knowledge production, such as indigenous and local knowledge, and may offer an alternative to the knowledge the scientific institutions possess (Dusek 2006). This understanding will also aid in the exploration of the heterogeneous knowledge systems that comprise science and alternate forms of knowledge, which is one of the goals of STS. Therefore, one might argue that these 'ordinary people' are not simply demagogues; instead, we ought to focus our attention on their knowledge. This may resolve some of the issues that were left unanswered about how STS can continue to function in the era of post-truth and also what we can learn from this situation. In addition, STS and science communications appear to be distinct, yet Davies (2022) argues that STS approaches aid science communication studies (SCS) in investigating the field of science.

On the other hand, Chilvers and Kearnes (2020) claim that the existing models of public representation and participation in science failed to address the concerns of diverse and emerging publics in this post-truth era. Davies (2022) maintains that the development of new methodologies and mobilizing numerous paradigms and disciplines will aid in enhancing the field of SCS. In other words, in this era of post-truth, STS and SCS disciplines may explore alternative knowledge forms that belong to 'ordinary people' as well as the methods by which individuals and groups may reflect their subjective realities. Although this makes the process complex, it is the researcher's obligation to establish a strong empirical foundation for efficient scientific communication so that the public can make informed decisions.

Consequently, it is crucial to comprehend the context and nature of the alternate forms of knowledge, which are predominantly produced in the informal sector. It is also vital to highlight that the informal sector employs more than two billion people worldwide, representing more than sixty percent of the global workforce primarily residing in low and middle-income countries (LMICs) (ILO 2018). Thus, such knowledge forms represent the workers in the informal sector, which is local and contextual. It is known by various names, including *recycling, do-it-yourself* (D-I-Y) in the United States, *Systém D* in France, *Gambiara* in Latin America, *Jua Kali* in Africa, and *grassroots innovations* (GIs) in India and the United Kingdom (Kumar and Bhaduri 2014; Oldenziel and Trischler 2016). In this chapter, we will concentrate on GIs in the Indian context, which includes individual or community attempts to solve local-level problems without the backing of the formal sector, particularly in their early stages (Gupta 2016; Bhaduri and Kumar 2011). GIs are novel technologies or procedures developed at the base of the pyramid as a direct result of exigencies, difficulties, or difficulties (Hilmi 2012). Gupta (2016) asserts that GIs are "by, for, and at'' the grassroots. Scholars have investigated GIs from a variety of perspectives, including the motivation behind their innovations (Bhaduri and Kumar 2011; Kumar and Namrata 2022), networks and their impacts (Kumar 2014), diffusion (Abrol and Gupta 2014), and GIs movement (Smith et al. 2016), amongst other themes. However, at this point, no one has investigated the communication dynamics in GIs. Therefore, we strive to explore the various facets of communication in GIs and how popular models of science communication might answer the issues of the post-truth era.

As a result, this chapter is divided into five sections. The second section examines a brief history and the theoretical concepts of science communication and GI literature. This section also looks at some of the features of GIs and how the formal sector supports forming various communication channels. In the third section, the technique and case chosen for the study are presented. The fourth section examines the case of *Shodhyatra*. The conclusion is the final piece, and it explores some of the study's policy implications.

Science communication in the post-truth world

Science communication is the study of systematic communication for the general public in a society to inform them about science in a non-scientific manner. More specifically, Horst et al. (2017 cited in Davies 2022:883) define science communication as “actions that aim to communicate scientific knowledge, methodology, processes, or practices in settings where non-scientists are a recognized part of the audiences”. Therefore, it is a multidisciplinary field that includes science outreach, science popularization, science public relations, and even scientific marketing. This field began in the early nineteenth century and was most likely more inclusive in its early years, encouraging lay people to participate, utilize, and produce scientific knowledge (Sheets-Pyenson 1985; Broks 2014). In the context of science communication, Broks (2014) gives several examples and argues that even editors and publishers of scientific literature supported the engagement of "lay people" in scientific endeavors. The nature of scientific communication, on the other hand, shifted in the latter half of the 19th century as the discipline of science became increasingly dominated by specialists and developed into a more skilled field of knowledge (Broks 2014).

Therefore, there was a clear distinction between 'experts' and 'lay people,' Science communication started to simplify information for persons who do not belong to the 'expert' class (Fischhoff and Davis 2014). As a consequence of this, the general public is being kept out of scientific endeavors. Or, to put it another way, science was presented to the people rather than being done by them (Broks 2014). This has further grown into models of science communication and is commonly referred to as the Public Understanding of Science (PUS). It was the duty of the scientists, in their capacity as experts, to communicate science to the general public so that individuals may gain an understanding of what science is and how it operates.

However, PUS has come under fire and been labeled as a "deficit" model due to the fact that it places lay people at the receiving end and gives them minimal opportunity to contribute to scientific knowledge (Schäfer 2011; Davies 2013; Broks 2014; Bensaude Vincent 2014). This led to the development of several different models, including the Public Engagement with Science and Technology (PEST) model, the web model, and the mediatization models, among others (Schäfer 2011). These models give scope for the 'publics' to engage with science and technology, making it more democratic and accountable, especially when public funding is involved (Selin et al. 2016; Davies 2013; Wilkinson 2013). These PEST models are widely used in science communication and policy formulation worldwide. However, Davies (2013) says that defining PEST is challenging and that there may be numerous nuanced understandings among researchers.

Broks (2014) contends that PEST models are simply more complex versions of deficit models. As it stands, policymakers still determine the agenda for dialogue, and the voices of lay people are given less consideration than they deserve. He suggests that when we try to conceptualize science communication as a process of creating meanings rather than transmitting information, we bring up a new challenge for the field, especially when user-driven science and technology are getting popular (Broks 2014). These issues are referred to as the "residual realist" approach by Chilvers and Kearnes (2020).

They propose four paths based on the relational coproductionist approach used in STS, namely: "reflexive participatory practices," "ecologizing participation," "responsible democratic innovations," and "reconstituting participation" for science and democracy.

Furthermore, they suggest that there is no need to replace existing models of engagement in science; instead, these suggested paths can be employed in addition to the existing ones. Possibly Davies (2022) also argues that SCS would benefit from incorporating insights from STS and other interdisciplinary methodologies. These issues are even more pressing in the post-truth era when the public has less faith in scientific knowledge, fake news is more prevalent, and uncertainty plays a role in decision-making. Given this context, GIs working in the informal sector become an interesting case study to investigate the process of giving meaning to their knowledge in communication. We will do this in the following section of this chapter.

GIs: knowledge, characteristics and context

A variety of distinct kinds of knowledge can be found in a society. The activities of the modern or scientific era are the foundation of the majority of the knowledge that is generated by the formal sector. On the other hand, Cunningham and Williams (1993) contend that such scientific knowledge is limited by the values of space, time, and culture. Therefore, in order to comprehend the various forms of knowledge that exist in today's society, one needs to investigate several other types of knowledge. Alternative knowledge forms include traditional knowledge, indigenous knowledge, pseudo-science, and GIs, among others. According to Cunningham and Williams (1993), historians of science 'invented' modern science in the first half of the twentieth century when they presented the 'scientific revolution' of the seventeenth and eighteenth centuries as a noteworthy event. Later, in the early nineteenth century, the discipline of science communication emerged (Broks 2014). However, science experienced tremendous growth, and its name, nature, and understanding all began to change around the end of the nineteenth century. By the second half of the twentieth century, science had undergone a complete transformation (Cunningham and Williams 1993; Aggasi 1988; Broks 2014). For instance, science is frequently described as 'small science,' 'little science' (Aggasi 1988), 'low cultures of science' (Sheets-Pyenson 1985), or 'thrifty science' (Werrett 2019) in the early 19th century. Some features of this type of science included the fact that anybody could participate, raw materials were accessible, and research tools were inexpensive (Broks 2014; Werret 2019). Because this 'low' or 'small' science was open to anyone and not limited to elite scientists with specialized knowledge at the time, it was also known as the *Republic of Science* (Broks 2014). This facilitated maximum engagement from a vast pool of interested individuals, each of whom had the potential to contribute to and benefit from the development of knowledge (Secord 1994).

This was also based on the principle of 'utilitarianism,' i.e., the greatest happiness of the greatest number (Broks 2014). Perhaps this was why Sheets-Pyenson (1985) claims that publishing houses simplified science communication to engage the public. Despite this, at the close of the 19th century, things had progressed to the point where science had become more sophisticated, formal, and exclusive to specialists (Broks 2014). Science, which is formal and universal nowadays (Dusek 2006), transformed in the twentieth century and is now referred to as "big science" due to its nature and scale (de Solla Price 1963). This shift can also be seen in the field of science communication (Broks 2014). Today's science, or big science, is distinguished by specialist knowledge, distinguishing it from lay people. This was not the case in small science, where practitioners of science and laypeople were closely associated (Broks 2014). Both experts and lay people were users and producers of knowledge, unlike today's sciences, in which lay people only receive knowledge and do not contribute to its production. However, society still includes numerous individualistic and community initiatives that reflect 'small' or 'low' cultures of science. This is especially visible in low- and middle-income countries (LMIC), where science is not a dominating paradigm or has not reached the masses. There are various other forms of knowledge, and it is necessary to acknowledge this fact to construct heterogeneous knowledge systems. Considering this discussion in a post-truth context, such exploration gives scope for expressing individuals' multiple subjective realities and reminds us of the early nineteenth-century 'small cultures of science.'

As mentioned, these alternative knowledge forms are generally found in the informal sector, and the environment in which they operate is distinct from that of knowledge in the formal sector (Dusek 2006). Kumar and Bhaduri (2014) attempted to draw a distinction between knowledge produced in the informal sector and the formal sector regarding the economic and socio-psychological values they created. These values diverged from the standards and ideals of knowledge produced in the formal sector. Basole (2014) cites Sahasrabudhey and Sahasrabudhey (2001) to refer to this type of knowledge as "Lokavidya," which roughly translates to "common peoples' knowledge." Based on informal learning, Kumar and Bhaduri (2014) categorized it as tacit or experiential knowledge. In this context, GIs are individual or community initiatives to solve either long-standing or newly-emerging problems in the informal sector (Bhaduri and Kumar 2011). There is a vast list of such innovations[1], including the mitticool, mini tractor, onion transplanter, windmill made of bamboo, tree

[1] For a detailed list of GIs, please visit www.nif.org.in

climber, and machines for making sanitary napkin pads. The formal sector either ignores these issues or provides unaffordable solutions. Contrasting the high costs and complicated nature of "planned obsolescence" innovations in the formal sector, GIs are developed from routine tasks, emergent issues, and a refusal to accept the status quo (Kumar and Bhaduri 2014). Kumar and Bhaduri (2014) and Gupta (2016) contend that GIs are movements in India founded on Gandhian ideology and projecting the knowledge of innovators for a better society within their familiar environment. Gupta (2016) further suggests that it might be possible that people do not even realize that their efforts are innovative or that they have struck across something that others can learn from. However, some innovators put in a lot of effort to systematically find or create an innovative approach to tackling an issue they face. Even the United Nations (2020) suggests that these innovations may appear in a variety of ways, such as through co-creation, encouragement, or spontaneity.

It is noteworthy that these innovators do not have formal education or drop out of school early for various reasons (Bhaduri and Kumar 2011). Their knowledge is not 'scientific' but comes from field experiences and through their networks (Basole 2014; Kumar and Bhaduri 2014; Wangare 2015). Thus, GIs represent the knowledge generated in the informal sector, which has frequently been disregarded by the formal sector for various reasons (Dusek 2006; Kumar and Bhaduri 2014). Nevertheless, with the efforts of a few individuals and institutions, GIs have been recognized, and there is evidence to suggest that they can tackle a variety of issues prevalent in the informal sector (Ustyuzhantseva 2015). Furthermore, expanding literature on studying knowledge and its processes in the informal sector leads to a deeper understanding of informal sector innovations (Hossain 2016; Gupta 2019; Kumar 2020).

In India, the efforts of the Honey Bee Network (HBN) movement, an informal network, brought the attention of policymakers and researchers to innovations at the grassroots level (Kumar and Bhaduri 2014; Smith et al. 2016). HBN helps GIs through identification, value addition, commercialization, patenting, and information dissemination (Gupta 2012). The primary agencies of this network are government entities such as the National Innovation Foundation (NIF) and non-government organizations such as the Society for Research and Initiative for Sustainable Technologies and Institutions (SIRSTI) and the Grassroots Innovation Augmentation Network (GIAN) (Kumar 2014). These key agencies collaborate with a wide variety of other organizations, including educational institutions, research institutes, media groups, non-governmental organizations (NGOs), and so on. Thus, this network is an innovation in and of itself, serving to foster knowledge and innovations in the informal sector. HBN has also embraced several innovative techniques in its support of GIs. For instance, to identify GIs coming from various regions of the country, they use national campaigns, the *Shodhyatra*, biennial awards, the participation of the media, kite, and food festivals, and many others.

Similarly, the NIF possesses fully developed procedures for value addition, patenting, commercialization, and dissemination. It has been supporting GIs under various programs with the assistance of other formal sector organizations (Ustyuzhantseva 2015). The following section will detail the methodology that was employed for the objective of gaining an understanding of the communication channels that are present within GIs.

Exploring GIs in India

This exploratory study follows a descriptive and analytical approach using qualitative methods. One of the authors has been working on GIs for almost a decade now, and over that time, we have compiled data from a wide variety of original sources. This entails conducting interviews with innovators as well as other players engaging with GIs, such as officials from the NIF, entrepreneurs, the common public, and students, amongst others. The collection of data from GIs is difficult because innovators are dispersed throughout the country, their addresses are inaccessible, and the majority converse in vernacular languages, which poses a linguistic barrier (Bhaduri and Kumar 2011; Joshi, Chelliah, and Ramanathan 2015). We have interacted with these innovators and other actors at several events made possible by NIF, such as award functions, *Shodhyatra,* and its follow-up visits, conferences, and food festivals. HBN and its affiliated agencies have established several channels to facilitate communication in GIs. For example, print and electronic media, platforms for social media, award functions, exhibitions at the President's House, workshops and conferences, local festivals, *Shodhyatra*, and others. HBN also publishes newsletters in Kannada, Malayalam, and Oriya in addition to *Honey Bee Magazine* in English and Chinese, *Loksravani* in Gujarati, *Sujhbujh* in Hindi, *Palle Surjana* in Telugu, *Num Vali Velanmai* in Tamil, and others.

The objective of publishing such a newsletter is to recognize and disseminate the common people's knowledge globally and in their own language. Thus, these secondary sources also aided in our ability to comprehend GI communication in greater detail. The majority of these channels are also utilized in the formal sector regarding science communication. However, our primary focus is on *Shodhyatra*, a unique platform that opens up multiple communication channels in GIs. Now, we will delve into the philosophical underpinnings and key constituents of *Shodhyatra*.

Shodhyatra: Journey in search of knowledge

The HBN hosts *Shodhyatra* twice a year, and translates to "journey in quest of knowledge" and is based on Gandhian philosophy.[2] This event is exclusive to the GIs, and it brings together a wide variety of people to walk together in remote locations to share and exchange the experiences and knowledge of local communities. Until now, a total of forty-nine *Shodhyatra* have been organized and have traversed around 6000 km in various locations across India (Figure 7.1).

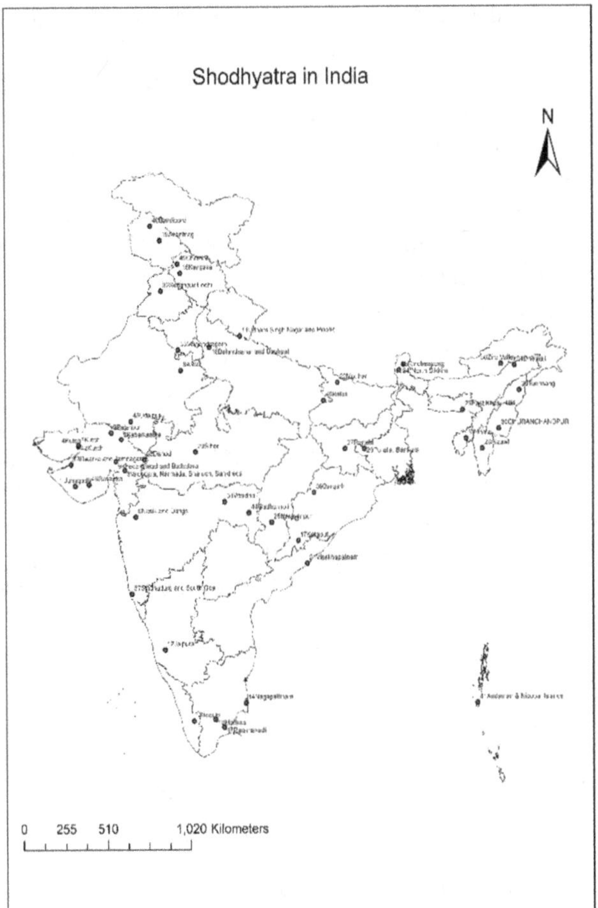

Figure 7.1. *Shodhyatra in India: Yatra serial number and beginning point* (Please note that this map is for illustrative purposes only and is not drawn to scale).

---

2 For details see https://www.sristi.org/shodhyatra/

Shodhyatra began in the late 1980s in Gujarat with the efforts of a few farmers to recognize and share their experiences, particularly in organic farming. Later, a workshop was held in Ahmedabad, Gujarat, under the "Management of Natural Resources" project on "how to make contact between experimenting farmers and scientists". Farmers from HBN attended the workshop. Following extensive consultation with farmers, it was decided to begin the *Shodhyatra* on foot in quest of knowledge, creativity, and innovations at the grassroots level.

The decision to start on foot took a lot of work. However, given that Gandhi had experimented with it several times during the freedom struggle movement, it was one of the obvious choices. Amato (2004) maintains that a journey on foot, often known as walking, has historical and cultural relevance because it has been the primary form of human mobility ever since the dawn of humanity. It is a metaphor for everything involved in living a life, from the mundane activities of daily existence to wars, religion, and knowledge. According to Wallace (1993), walking has been associated with an affinity to nature and an understanding of the environment, which are believed to have eroded in the modern day. Thus, Lee and Ingold (2006) view it as an action that can be used to challenge or reconstruct modernity.

Amato (2004) contends that walking is a form of communication in and of itself because it has its language, replete with vernacular dialects and idioms. In addition, he asserts that it communicates a wide range of information about the identity, significance, circumstances, and purpose of the individual doing the walking. The walk can also be interpreted by onlookers, who can receive a message from the walker. Lee and Ingold (2006) argue that walking *with* the community as a researcher helps to involve and connect with people in shared circumstances. Furthermore, they assert that walking as a researcher helps explore three types of interaction with the environment: opportunities to look around, time for thinking, and sensory experiences. So, the *Shodhyatra* is more than a journey undertaken in quest of knowledge. As a result, we also participated in eight such *Shodhyatra* to better grasp the background and philosophy of it, as well as to collect data.

Shodhyatra's constituents include HBN-affiliated organizations, its officials, grassroots innovators, farmers, students, academicians, entrepreneurs, the media, foreigners, and members of the local community, among others. On average, Shodhyatri walks 12 to 20 kilometers daily over a week, depending on the terrain and weather. Another significant characteristic of it is that it takes place during times of severe weather, such as when the summer *Shodhyatra* takes place in extremely hot regions and when the winter Shodhyatra takes place in extremely cold parts. This allows Shodhyatris to better understand how local communities adapt in such harsh situations. As a result, we may contend that *Shodhyatra*, an original and creative method of searching for GIs in India, presents a chance to investigate communication-related advances in the informal sector. Using theories of scientific communication, we can examine the dynamics of communication in GIs as a comparison of "small scientific cultures."

Communication in GIs

Therefore, based on the earlier discussion, GIs are parallel to 'small cultures of science,' and *Shodhyatra* provides an interesting platform to investigate communication in this context. As previously stated, the HBN uses other communication channels to communicate information on GIs in India and elsewhere. Nonetheless, our focus is *Shodhyatra* as a distinguishing phenomenon of the communication platform in GIs. *Shodhyatra* is an amalgamation of knowledge from various individuals, particularly farmers, grassroots innovators, and researchers, where multiple forms of knowledge and experiences are exchanged. This information is derived primarily from local experiences and one's own environment. In addition, when shodhyatris travel to other locations, they frequently share their knowledge and learn from the local populace.

Such a communication structure displays multiple communication layers in which entities from the formal sector, such as HBN associates SRISTI and NIF, define the communication agenda. Local non-governmental organizations (NGOs) are also involved since they have better access to the community and may assist with community engagement. There are different phases of the *Shodhyatra*, namely pre-*Shodhyatra*, *Shodhyatra*, and post-*Shodhyatra*, and each phase employs a different strategy. This helps to establish a rapport with the community and a lasting relationship with them. The walk is organized so that Shodhyatris has ample time to explore the surroundings, converse with locals, explain to bystanders the purpose of their visit, schedule a competition to learn about local food and its values, a competition for students to demonstrate their environmental knowledge, etc. In addition, meetings are organized to address the community's difficulties and obstacles and their solutions with the community. Usually, these solutions are formed on the basis of the resources that are readily available in the area, and they reflect the locals' understanding of sustainable practices. However, because these people are so humble in their attitude, they do not value their effort; hence, *Shodhyatra* empowers them to realize what they have done that they would not have done otherwise. Apart from looking into innovations, *Shodhyatra* recognizes seniors for their survival instincts, local herbal medicine practitioners for their services, farmers for their creative techniques, and women and children for their contributions to the community. These communication exercises underline biodiversity and the locals' knowledge and sustainable use of the environment's abundant resources. For example, Shodhyatris visits schools in those places to encourage competition among students about their environment. They also ask teachers to maintain a biodiversity register with the school about their environment. Students must fill out the register based on their home and community learning. HBN may later publish that knowledge in its newsletter. Furthermore, food competitions among women for local recipes indicate the biodiversity and nutritional foods accessible in communities. Later on, HBN may provide help for the commercialization of these recipes on a larger scale, which may require the addition of value or the patenting of the recipes in the name of the community.

During the walk, Shodhyatris also explored each other and discussed their sustainable habits, such as the usage of plastics, water, and plants, that they have embraced in their context. Before the walk begins, they resolve to seek knowledge in the community, and before retiring for the night, each Shodhyatri reflects on the day's learning. In addition, the secondary literature, such as the HBN newsletter, is disseminated across the community so that people who read it can acquire knowledge from multiple sources. Also, the *Shodhyatra* receives extensive coverage from the regional media, which helps to convey knowledge about the event and its purpose to the larger community in those areas.

However, some difficulties to be faced in *Shodhyatra* are connected to logistics. Organizing and running a large gathering for an entire week is challenging. There is a possibility that a Shodhyatri may fall sick or arrive late because of the difficult terrain or adverse weather conditions. Occasionally, local residents may need to wait due to these factors. In addition, the people have a limited understanding of the events that occur after the *Shodhyatra*, despite the fact that HBN has developed a follow-up mechanism for after the *Shodhyatra*. Nevertheless, in terms of communication, it is continually evolving, and every time a new method is designed to improve the communication process among and with the community, it is further strengthened.

Therefore, this event is filled with communication exercises focused on learning from the community and nature. *Shodhyatra* symbolizes Wallace's (1993) belief that walking brings us closer to nature and enhances our investigation of the surroundings. Shodhyatris visit a particular village and observe its environs. If they encounter a new innovation or technology, they discuss its characteristics, utility, experiences, and development with the community. Even if a possibility exists, they also offer ways to enhance it and how it can be implemented in other contexts. Thus, what Lee and Ingold (2006) suggest walking as a methodology helps to explore the environment, time to reflect on it, and sensory experiences is true in *Shodhyatra*. Also, Amato's (2004) argument of walking as communication is reflected in *Shodhyatra*, and multiple communication methods happen during the journey.

Furthermore, these activities satisfy the PEST criterion in an informal setting. It also answers the communication problems Broks (2014) raised in the context of PEST, since all types of knowledge are given an equal place in informal contexts, where science, technology, and other subjects can be discussed without hierarchy. In light of Davies's (2022) argument it provides an opportunity to explore various choices, insights, and impacts that one may observe in society, which has become necessary in this post-truth moment.

Conclusion

Despite worries about trusting science, its methods, and institutions, it is the preeminent mode of knowledge in the present era. Although multiple forms of knowledge exist in society, and GIs are a combination of scientific, traditional, and innovative ideas that reflect the 'small cultures of science' of the early nineteenth century. It offers an opportunity for grassroots communities with their understanding of sustainability and biodiversity to seek solutions to their problems. This displays the ability to encourage community empowerment rather than treating common people as demagogues in the classic sense, in which they are used as a way of subverting established standards of behavior, institutions, and the law. Instead, by expressing multiple subjective realities, lay people's knowledge may provide a different perspective on post-truth challenges of climate change, gender issues, and others.

Multiple channels of communication are available for observation in the GI system. A variety of formal sector organizations and institutions are involved, necessitating the use of several formal communication mechanisms. The *Shodhyatra* provides a unique perspective from which to examine the communication dynamics of GIs. The walk, as a unique method, appears to be an approach that is both realistic and effective in its goal of communicating information regarding GIs. In *Shodhyatra*, communication does not adhere to a hierarchical structure; the line dividing scientific and lay people's knowledge is hazy, and the idea of experts and expert knowledge no longer exists.

In its place, there is a greater emphasis placed on the democratization of knowledge that is "by, for, and at" the grassroots. Every person has the opportunity to interact with others and gain knowledge while they are outside in nature. In its true sense, public engagement with science and technology may be contextualized during the *Shodhyatra*. Scholars and practitioners working in STS and SCS may benefit from a learning opportunity presented by GIs to democratize the rapidly evolving institutions that make up "big science." In addition, it is crucial to preserve and bring innovative communication strategies to enrich and democratize alternative knowledge spaces.

Additionally, this demonstrates the viability of a democratic worldview because diverse knowledge systems are able to coexist rather than compete with one another. This study opens up more avenues to investigate communication in alternative knowledge forms by investigating other communication channels in different contexts, especially at the intersection of STS and SCS. It would also be fascinating to explore communication dynamics in the process of knowledge production in the informal sector and how it would shape, particularly in light of the rapidly evolving nature of both technology and the communication media. In this age of post-truth, when people's faith in science is eroding for a variety of reasons, it is essential to broaden one's perspective beyond the scientific discourses that are now being discussed. The various alternative forms of knowledge could shed some light on how to approach and overcome these obstacles.

Acknowledgments

We are grateful to the HBN for providing us with the opportunity to engage and communicate with grassroots innovators and other actors. We would also like to thank Mr. Hiranmoy Chetia for his assistance in creating Figure 7.1.

References

Abrol, Dinesh, and Ankush Gupta. 2014. "Understanding the Diffusion Modes of Grassroots Innovations in India: A Study of Honey Bee Network Supported Innovators." *African Journal of Science Technology Innovation and Development* 6(6):541–52. doi: 10.1080/20421338.2014.976974.

Agassi, Joseph. 1988. "The Future of Big Science." *Journal of Applied Philosophy* 5(1):17–26. doi: 10.1111/j.1468-5930.1988.tb00225.x.

Amato, Joseph A. 2004. *On foot: A history of walking*. New York and London: New York University Press.

Basole, Amit. 2014. "The Informal Sector from a Knowledge Perspective." *Yojana*. 8-13. https://cse.azimpremjiuniversity.edu.in/wp-content/uploads/2019/08/Basole_Yojana_Article-205ka80.pdf [Accessed on 01-08-2023].

Bensaude Vincent, Bernadette. 2014. "The Politics of Buzzwords at the Interface of Technoscience, Market and Society: The Case of 'Public Engagement in Science.'" *Public Understanding of Science* 23(3):238–53. doi: 10.1177/0963662513515371.

Bhaduri, Saradindu, and Hemant Kumar. 2011. "Extrinsic and Intrinsic Motivations to Innovate: Tracing the Motivation of 'Grassroot' Innovators in India." *Mind & Society* 10(1):27–55. doi: 10.1007/s11299-010-0081-2.

Broks, Peter. 2014. "Science Communication: A History and Review." Pp. 307-329 in *All about Science: Philosophy, History, Sociology & Communication* edited by M. Burguete and L. Lam. Singapore: World Scientific Publishing.

Chilvers, Jason, and Matthew Kearnes. 2020. "Remaking Participation in Science and Democracy." *Science, Technology & Human Values* 45(3):347–80. doi: 10.1177/0162243919850885.

Cunningham, Andrew, and Perry Williams. 1993. "De-Centring the 'Big Picture': *The Origins of Modern Science* and the Modern Origins of Science." *British Journal for the History of Science* 26(4):407–32. doi: 10.1017/s0007087400031447.

Davies, Sarah R. 2013. "Constituting Public Engagement: Meanings and Genealogies of PEST in Two U.K. Studies." *Science Communication* (35) 6: 687–707. https://doi.org/10.1177/1075547013478203

Davies, Sarah R. 2022. "STS and Science Communication: Reflecting on a Relationship." *Public Understanding of Science* 31(3):305–13. doi: 10.1177/09636625221075953.

De Solla Price, Derek J. 1963. *Little Science, Big Science.* New York: Columbia University Press.

Dusek, Vaul. 2006. *Philosophy of Technology: An Introduction.* Malden, Oxford, Victoria: Blackwell Publishing.

Fischhoff, Baruch and Alex L. Davis. 2014. "Communicating scientific uncertainty." *Proceedings of the National Academy of Sciences,* 111(S4): 13664–71. https://doi.org/10.1073/pnas.1317504111

Gupta, Anil K. 2012. "Innovations for the poor by the poor." *International Journal of Technological Learning, Innovation and Development* 5(1/2): 28-39.

Gupta, Anil K. 2016. *Grassroots Innovation: Minds on the Margin Are Not Marginal Minds.* Noida, India: Random Business.

Gupta, Shaphali. 2019. "Understanding the Feasibility and Value of Grassroots Innovation." *Journal of the Academy of Marketing Science* 48(5):941–65. doi: 10.1007/s11747-019-00639-9.

Hilmi, Mohd Faiz. 2012. "Grassroots Innovation from the Bottom of the Pyramid." *Current Opinion in Creativity, Innovation and Entrepreneurship* 1(2). doi: 10.11565/cuocient.v1i2.5.

Hoffman, Steve G. 2018. "The Responsibilities and Obligations of STS in a Moment of Post-Truth Demagoguery." *Engaging Science, Technology, and Society* 4: 444-452. DOI:10.17351/ests2018.259

Hossain, Mokter. 2016. "Grassroots innovation: A systematic review of two decades of research." *Journal of Cleaner Production* 137: 973-981. https://doi.org/10.1016/j.jclepro.2016.07.140

ILO. 2018. *Women and men in the informal economy: a statistical picture 3rd edition.* International Labour Oraganization, Geneva.

Iyengar, Shanto and Massey, Douglas S. 2018. "Scientific communication in a post-truth society." *Proceedings of the National Academy of Sciences* 116, no. 16: 7656-7661. https://doi.org/10.1073/pnas.1805868115

Jasanoff, Sheila and Hilton R Simmet. 2017. "No funeral bells: Public reason in a 'post-truth' age." *Social Studies of Science* 47, no. 5: 751-770. DOI: 10.1177/0306312717731936

Joshi, Rajul G., John Chelliah, and Veera Ramanathan. 2015. "Exploring Grassroots Innovation Phenomenon through the Lived Experience of an Indian Grassroots Innovator." *South Asian Journal of Global Business Research* 4(1):27–44. doi: 10.1108/sajgbr-01-2014-0003.

Kumar, Hemant, and Saradindu Bhaduri. 2014. "Jugaad to Grassroot Innovations: Understanding the Landscape of the Informal Sector Innovations in India." *African Journal of Science Technology Innovation and Development* 6(1):13–22. doi: 10.1080/20421338.2014.895481.

Kumar, Hemant, and Namrata. 2022. "Socio-Economic Environment and Motivation to Innovate: Exploring Grassroots Innovations Process in India." *Technology Analysis and Strategic Management* 1–14. doi: 10.1080/09537325.2022.2088345.

Kumar, Hemant. 2014. "Dynamic Networks of Grassroots Innovators in India." *African Journal of Science Technology Innovation and Development* 6(3):193–201. doi: 10.1080/20421338.2014.940170.

Kumar, Hemant. 2020. "Publication Trends in the Informal Sector Innovation Research." *Journal of Scientometric Research* 9(2s):s5–13. doi: 10.5530/jscires.9.2s.32.

Lee, Jo, and Tim Ingold. 2006. "Fieldwork on Foot: Perceiving, Routing, Socializing." Pp. 67–85 in *Locating the Field: Space, Place and Context in Anthropology.* edited by Simon Coleman and Peter Collins. Oxford and New York: BERG

Mcintyre, Lee. 2018. *Post-truth.* Cambridge, Massachusetts: The MIT Press.

Nally, David. 2022. "Theorising post-truth in the COVID era." *Journal of Educational Change* 23: 277–289. https://doi.org/10.1007/s10833-022-09457-3

Oldenziel, Ruth and Helmuth Trischler. 2016. "How old technologies became sustainable: An introduction." Pp. 1-12 in *Cycling and recycling: histories of sustainable practices* edited by R. Oldenziel and H. Trischler. New York, Oxford: Berghahn.

Parmet, Wendy E. and Paul, Jeremy. 2020. "COVID-19: The First Posttruth Pandemic." *American Journal of Public Health* 110, no. 7: 945–6. doi: 10.2105/AJPH.2020.305721.

Sahasrabudhey, Sunil and Chitra Sahasrabudhey. 2001. The Lokavidya Standpoint [Hindi]. Varanasi; Lokavidya Pratishtha Abhiyan.

Schäfer, Mike S. 2011. "Sources, Characteristics and Effects of Mass Media Communication on Science: A Review of the Literature, Current Trends and Areas for Future Research: Mass Media Communication on Science." *Sociology Compass* 5(6):399–412. doi: 10.1111/j.1751-9020.2011.00373.x.

Secord, Ann. 1994. "Science in the pub: Artisan botanists in early nineteenth century London." *History of Science* 32(3): 269-315. https://doi.org/10.1177/007327539403200302

Selin, Cynthia, Kelly Campbell Rawlings, Kathryn de Ridder-Vignone, Jathan Sadowski, Carlo Altamirano Allende, Gretchen Gano, Sarah R. Davies, and David H. Guston. 2017. "Experiments in Engagement: Designing Public Engagement with Science and Technology for Capacity Building." *Public Understanding of Science (Bristol, England)* 26(6):634–49. doi: 10.1177/0963662515620970.

Sheets-Pyenson, S. 1985. "Popular Science Periodicals in Paris and London: The Emergence of a Low Scientific Culture, 1820-1875." *Annals of Science* 42(6):549–72. doi: 10.1080/00033798500200361.

Smith, Adrian, Mariano Fressoli, Dinesh Abrol, Elisa Arond, and Adrian Ely. 2016. *Grassroots Innovation Movements.* London, England: Routledge.

Söderberg, Johan. 2022. "The moment of post-truth for Science and Technology Studies." In *Post-Truth Imaginations: New Starting Points for Critique of Politics and Technoscience*, edited by Kjetil Rommetveit, 86-110. London and New York: Routledge.

United Nations. 2020. *Policies and Strategies to Promote Grassroots Innovation: Workbook.* United Nations Economic and Social Commission for Asia and the Pacific (UN ESCAP). https://www.unescap.org/sites/default/files/Workbook%20Policy%20Support%20Grassroot%20ESCAP%20GIAN%20HBN_FINAL.pdf.

Ustyuzhantseva, Olga V. 2015. "Institutionalization of grassroots innovation in India." *Current Science* 108(8): 1476-1482.

Wallace, Anne D. 1993. *Walking, Literature and English Culture: The Origins and Uses of Peripatetic in the Nineteenth Century.* Oxford: Clarendon Press.

Wangare, Boniface J. 2015. "Knowledge management in the informal sector: A review of Kenya's Jua Kali industry." *International Journal of Economics, Commerce and Management* 3(10): 257-278.

Werrett, Simon. 2019. *Thrifty Science: Making the Most of Materials in the History of Experiment.* Chicago, IL: University of Chicago Press.

Wilkinson, Clare. 2014. "Engaging with Strangers and Brief Encounters: Social Scientists and Emergent Public Engagement with Science and Technology." *Bulletin of Science, Technology & Society* 34(3–4):63–76. doi: 10.1177/0270467614552841.

Chapter 8

Reflexive philosophical dialogue about science, technology and society: tackling the post-truth condition

Lynda Dunlop and Joshua Stubbs

Abstract: Global social networks have facilitated the proliferation - especially in relation to climate change - of views based on what people believe rather than the best available disciplinary knowledge. This post-truth condition is accentuated by inaccurate and misleading but popular perceptions of scientific evidence as the source of objective truth on the one hand and the triumph of extreme skepticism, which dismisses it altogether, on the other. This polarization presents a risk to societies, especially as climate change becomes more extreme and potential responses more reliant on new science and technologies. In this chapter, we argue that public philosophical dialogue is necessary to achieve science and technology for society in a post-truth age. Through reflection on research with European youth on large-scale climate interventions, we illustrate the affordances of reflexive philosophical dialogue about science, technology and society. We argue that philosophical dialogue - an exchange of ideas focused on a search for meaning and truth - develops scientific and political criticality and engagement which is necessary to construct and deconstruct arguments and communicate sensitively through challenge, disagreement and disjuncture, which counteracts the processes leading to polarization. Decisions about climate science and technology are made in a social and political context. Science can indicate potential solutions but democratic decision-making requires reflexive, critically-informed public dialogue and recognition of the strengths, limits and limitations of scientific knowledge and methods.

Keywords: polarization, post-truth, climate change, dialogue, critical thinking

***

Introduction

Public dialogue about science, technology, and society is increasingly polarized, particularly in the context of complex issues, such as the climate crisis, which by their very nature do not lend themselves to simple solutions (Lucas 2018). Science communication has been criticized for an overemphasis on knowledge dissemination over critical dialogue with different publics (i.e., those with different stakes in the issue and those with different beliefs and knowledge systems). In mass media, reporting is increasingly polarized, and where there is an aim for 'balance' (leaving aside the extent to which this is possible or desirable), which is often interpreted and enacted by giving a platform to representatives of opposing positions. Increasing political polarization, which has coincided with increasing social inequalities and declining trust in science, among other global societal trends, has been connected to the rise of the post-truth condition (Lewandowski, Ecker, and Cook 2017); that is, the "circumstances in which objective facts are less influential in shaping public opinion than appeals to emotion and personal belief," as the Oxford English Dictionary (OED) put it in denoting post-truth word of the year in 2016 (OED 2016). This polarization, which tends to oversimplify complex issues, provides fertile ground for the political disenchantment, disengagement, and distrust that incubates and animates the post-truth condition. Indeed, it is this that needs to be redressed in order to enable more meaningful democratic deliberation about inextricably political, scientific, and technological debates to take place.

Features of the post-truth condition can be seen in relation to climate change in the media, among lobbyists, and even in governments. Indeed, post-truth has been described as "a political operation aimed at generating uncertainty in the public's mind in order to preserve the status quo and to delay climate-change mitigation" (Lewandowski et al. 2017:361). This has real-world impacts, which can be seen in the US, such as filing of its intention to withdraw from the Paris Climate Accord in 2019 when Donald Trump was President and the government of Jair Bolsonaro cutting the budgets and programs of Brazilian environmental ministries and agencies. The post-truth condition is exacerbated by popular but inaccurate perceptions of the strengths, limits, and limitations of science and by the nature of public discourse around complex issues that involve disagreement along different axes (e.g., economic, ethical, environmental). In conceptualizing the post-truth condition, it is important to distinguish between the constructive, healthy skepticism that should (ideally) direct, inform and guide scientific and philosophical inquiry, on the one hand, from unrestrained cynicism that is predicated on rejecting evidence or arguments that are contrary to one's beliefs and commitments, on the other.

A shift in how different publics interact with science and politics is needed to achieve science and technology *for* society in a post-truth age. While science and technology may always be for society in intent, 'society' is diverse, and the risks and benefits of innovation are distributed unevenly. The extent to which science serves the needs of a diverse society is a political issue involving discussion about power, democracy, and values (Owen et al. 2012). Owen et al. call for the democratization of the governance of intent whereby the 'right' impacts are underpinned by deliberative democracy, research trajectories are determined with society, and there is reflection on the roles and responsibilities of those enabling, leading, and implementing research. The shift in emphasis towards science *for* society requires an approach to science, technology, and society that opens up rather than closes down discussion and creates spaces for people from different cultural and political groups to share, exchange, and critique ideas and, significantly, disagree. Interaction is important in itself, not just because it produces a quantifiable opinion (Eliasoph 1996).

Discussion about complex and, at times, emotive issues need to be held in an environment that is not predicated upon winning debates but on understanding perspectives that are not one's own and developing a more analytically reflexive and critically informed understanding of the strengths and limitations to particular lines of action, and to seeing disagreement as a valuable state for constructing shared futures. As "the current state of public discourse can no longer be examined through the lens of misinformation that can be debunked but as an alternative reality that is shared by millions" (Lewandowski et al. 2017:360), what is needed is not an approach that imposes science on the public, but an approach which works *with* different publics to both determine and answer questions that are "in principle open to informed, rational, and honest disagreement" (Floridi 2013:195). Public philosophical dialogue - by which we mean an exchange of ideas focused on a search for meaning and truth - can play a role in building bridges between people inhabiting alternative or opposed cultural and political realities by identifying common ground in spite of important differences. In this chapter, we examine how this might work, beginning by discussing the concept of cultural cognition linked to polarization and the post-truth condition before outlining the role that public philosophical dialogue can play in discussing controversial emerging science and technology relating to geoengineering (sometimes referred to as climate engineering).

The problem of polarization

Polarization, whether in politics or beliefs, erodes democratic capacities (Talisse 2021). At one level, polarization causes stalemates, negative campaigning, and hostilities in political contexts, but it is also troubling at a deeper level. Talisse argues that polarization in belief pushes those with similar beliefs together, resulting in a stronger commitment to and entrenchment in their ideas and reducing their capacity for nuance. Polarization in belief also reduces trust in others who have different beliefs, leading to the 'other side' being seen as rivals to be defeated rather than political equals with whom to create a shared future. In the context of climate change, this can be seen in climate change denial, which closes down and misdirects public dialogue. Closing down the discussion moves society further away from being able to respond to climate change, respond to legitimate concerns about the implications of taking the climate crisis seriously in policy and society, and to being able to determine the role (or otherwise) of technologies associated with large scale climate intervention in contributing to this, and maintains the (unsustainable) status quo.

Polarization on the issue of climate change has become more pronounced and visible over the past decade, often along party political or ideological lines, but also depending on a sense of personal identity (Lucas 2018). On the one hand, it could be considered surprising that political polarization has coincided with increased access to the internet via digital technologies and, therefore, a wide(r) range of sources of information. On the other hand, such technologies may have made it easier to only be exposed to ideas and information that support one's own beliefs, and a variety of social media platforms have made connections between people with shared beliefs easier. Polarization in public opinion on a range of issues, from sexuality to income inequality, is associated with political protest behaviors (Kleiner 2018) and makes it difficult for those trying to create policies to attend to problems such as climate change (Brewer 2012).

Research on polarization can entrench oppositional positions. The problem with many correlational studies is that positions on climate change and political beliefs are reduced to over-simplified categories, with little light shed on why these beliefs are held or on how people understand different positions. In contrast, Lucas (2018) correlates concern for climate change with different human values rather than using measures such as affiliation by political party or newspaper. This approach opens up the potential for discussion about what it means to hold certain values, how different values come into contact with each other, and what these values mean for climate action. It also allows for areas of mutual concern to be identified, which can bring people holding different values or different knowledge bases together to solve real problems.

Increasing political polarization may simultaneously exacerbate and be intensified by processes of 'cultural cognition.' Cultural cognition refers to instances whereby individuals' group commitments, shared values, and sense of belonging unconsciously influence their perceptions of facts and thereby facilitate politically biased information processing (Kahan et al. 2012). Forming an unvirtuous cycle, it is plausible that increasing political polarization and cultural cognition are mutually reinforcing. Therefore, a focus on imparting scientific knowledge about important but politically contested issues (such as climate change) is not enough to cultivate a critically informed public understanding of them, even when a scientific consensus has been reached (Kahan, Jenkins-Smith and Brahman 2011; Kahan et al. 2012). In a study based on a nationally representative survey of 1540 adults in the USA, for example, respondents' concerns about climate change did not increase as their scientific literacy increased. Furthermore, those who belonged to cultural groups with opposing political predispositions to climate change grew further apart on this issue as their scientific literacy increased (Kahan et al. 2012). These findings suggest that as worthwhile and important as improving the clarity of scientific communication may be, it is insufficient in isolation to 'dispel public conflict' on scientific issues that 'feature cultural meanings' that divide those of traditionally opposed political groups (Kahan et al. 2012: 734). It is, therefore, vital that spaces and techniques aimed at effectively disrupting the unvirtuous cycle of political polarization, cultural cognition, and public misunderstanding of issues such as climate change - which threaten everyone and require consensus-based collective action to be addressed - are identified, honed, and disseminated. This is where public philosophical dialogue has an important role to play.

More important than knowledge in countering polarization seems to be disposition. It is thought that 'science curiosity' may counteract politically biased information processing; that is, a general disposition "that reflects the motivation to seek out and consume scientific information for personal pleasure" (Kahan et al. 2017:180). It has been found that science-curious individuals who express a particular preference for watching scientific documentaries and are drawn towards newspaper headlines that contradict their beliefs "form a more balanced, and *across the political spectrum* a more uniform, understanding of the significance of [scientific] information on contested societal risks" (Kahan et al. 2017: 192, emphasis added). Indeed, their curiosity and willingness to be surprised and have their preconceptions defied are not tuned down or turned off even when engaging with scientific information that is inextricably linked to culturally divisive political topics. Such evidence points to a need to discover how best to promote science curiosity, particularly among educators and science communicators. Research into how to promote scientific curiosity remains at a formative stage (Kahan et al. 2017), but open questions (Lindholm 2018) and identifying inconsistencies in observations (Luce and Hsi 2015) are considered important in stimulating curiosity about science in educational settings. These are both activities that can be facilitated through political dialogue.

Dialogue and scientific decision-making

In our discussion about the causes and consequences of polarization of public dialogue on climate change, we have identified ways in which polarization can be addressed: through working with values and promoting curiosity. Groves (2019) has argued that the linear relationship between evidence and decision making perceived to have been eroded by the post-truth condition never really existed and that the focus ought to instead be on reframing the discussion around the outcomes of uncertainties in the climate system - including unacceptable losses - which brings values into the discussion and creates better conditions for collaboration and consensus for action. It is important for democratic societies to cultivate populations with capacities to reflect, think independently, be critical of tradition, empathize with others, and communicate their ideas effectively (Nussbaum 2010).

However, there is still a tendency for both the social sciences and the public interest to play a subordinate role to the sciences in shaping scientific discourse - i.e., defining problems and future research trajectories - including in relation to geoengineering - with the social sciences and humanities filling the gaps, measuring or describing societal changes or public attitudes, with science deciding the problems that are to be prioritized (Szerszynski and Galarraga 2013). In the context of geoengineering, there has been substantial attention in the literature to 'upstream' engagement (i.e., when public engagement occurs prior to significant research into new technologies) and to avoid a deficit view of the public (Corner and Pidgeon 2014), but there remains a need to resist subordination and integration of disciplines in favor of epistemological responsibility (Szerszynski and Galarraga 2013). Scientists do not, and should not, exercise absolute authority over how and when scientific methods should be used, even if they are responsible for producing scientific evidence to inform such decision-making. Whilst science communication has arguably moved to a more dialogical approach, it is not clear that scientific or technological decision-making can be considered participatory or dialogic. This would require greater involvement of decision makers (funders, learned societies) in public dialogues about science prior to research being commissioned and more value given to multi-disciplinary teams. As Eliasoph (1996) argues, the very creation of space for open-ended discussion can implicitly call into question power relations and thus make public life a potential source of power.

Philosophical dialogue

When we look at methods or approaches to working with values, promoting curiosity, cultivating critical and independent thinking, and empathy with others, the community of inquiry associated with philosophical dialogue offers potential. Philosophy is based on critically questioning whether our beliefs are justified and creating shared meaning. The aim of the philosophical dialogue is not to encourage debate purely for the sake of attempting to win arguments, regardless of whether one agrees with them, but to encourage people to consider the implicit and explicit assumptions that underpin their beliefs and whether they make logical sense and have a strong foundation in scientific (or other) evidence. In so doing, philosophical dialogue can improve people's capacity to reason well and thereby help people to construct less politically biased, fragile, and sensitive belief systems. Importantly, philosophical dialogue can help participants to identify better versus worse reasons for endorsing particular arguments.

In this sense, philosophical dialogue addresses some characteristics of the post-truth condition. This is because while the post-truth condition does "not seek to establish a coherent model of reality [... but to] erode trust in facts and reality, to the point where facts no longer matter or are not even acknowledged to exist" (Lewandowsky et al. 2017:361), practicing philosophy is centered on attempting to establish clearer, scientifically informed and better justified arguments with attention to underpinning values and assumptions, through thinking in community with others. Philosophical dialogue may be a particularly potent technique because it is predicated upon and actively encourages those participating in it to question their taken-for-granted values and inquire into their assumptions and beliefs, as well as those of others. Moreover, philosophical dialogue involves sharing questions, ideas, and arguments publicly, meaning that a multiplicity of perspectives are heard, reflected upon, and interrogated. While philosophical inquiry is predominantly concerned with fundamental concepts such as truth, justice, and knowledge, it usually begins with fun or playful questions, which can ease people into feeling more comfortable about asking questions and sharing thoughts about contested but important issues. Therefore, we argue that promoting public reflexive philosophical inquiry could prove to be a key ingredient in the treatment of the post-truth condition, particularly in relation to climate change and geoengineering.

Why geoengineering?

Geoengineering refers to a range of methods of large-scale intervention in Earth's climate system in order to either remove carbon dioxide from the atmosphere or reflect solar radiation away from the planet. These range from nature-based solutions like afforestation at scale to highly technological responses, including stratospheric aerosol injection. Risks associated with geoengineering vary with the specific method and how it is implemented, and there are differential impacts with geography, which means that economic, ethical, and environmental questions about geoengineering need to be considered extremely carefully. Societies need to think about if or how, to research, deploy, and regulate proposed approaches to geoengineering, what the impacts are on people in different parts of the world, and about how decisions can be made in situations with high levels of complexity and uncertainty. Philosophical inquiry is one means by which these discussions can be had.

Geoengineering is an appropriate context to illustrate the value of philosophical dialogue for a number of reasons: it is an issue on which people can, given the same information, hold different positions; it impinges on concepts such as certainty, responsibility, and risk; and the consequences of decisions has an impact on people and the planet. The climate crisis presents a social and political challenge to societies. Currently, the options open to humanity are adaptation to climate change (e.g., changing how and where we live), mitigation, i.e., reducing emissions of greenhouse gases, including carbon dioxide, in order to prevent the climate from heating further, and geoengineering.

Philosophical inquiry into geoengineering

We have argued above that philosophical dialogue offers the potential to counter polarization by providing a place to identify shared values and work on questions of common interest, and, importantly in the context of the post-truth condition, to reflect on our own and others' perspectives and reasons for holding different positions. Talisse (2021) argues that there is a need for public spaces for public cooperation not organized around political allegiances. Creating opportunities for public philosophical dialogue in which there is facilitated open-minded engagement with, and critically-informed deliberation about, contested topics such as climate change and geoengineering may help to cultivate the individually and collectively enriching disposition - or virtue - of science-curiosity, in addition to enhanced reasoning. Our work with colleagues and young people across Europe (Dunlop et al. 2022) incorporated philosophical dialogue into a series of workshops on climate engineering which supported participants to investigate scientific, technological, ethical, social, and political dimensions of different responses to the climate emergency.

Philosophical dialogue, usually facilitated by a philosopher trained in managing discussion, is perhaps most common in educational settings but is also used in counseling, philosophy cafes, workplaces, prisons, festivals, hospitals, and refugee centers (Goucha 2007; The Philosophy Foundation 2022) and on television and radio e.g., Michael Sandel's The Public Philosopher broadcasts for the BBC (BBC, 2023). Anyone can participate in philosophical inquiry. In contrast to debate and other methods of exchange of ideas which favor 'winning' arguments, philosophy in these non-academic settings attempts "to make sense of observed phenomena and to encourage people to express, compare and analyze ideas, while accepting that these ideas are relative, imperfect or subjective...they question the reality of what is known or thought, they investigate causes thoroughly, they test the possibility of entertaining opposite views, and in they constantly reconsider the criteria for legitimacy " (Goucha 2007:174).

Philosophical dialogue can be uncomfortable, and participants may find it frustrating to deal with uncertainty and lack of resolution of philosophical questions, and they may find it personally challenging when they change their position in response to a question (Dunlop et al., 2020). Therefore, essential to the quality of philosophical dialogue is the skill of the facilitator (Kennedy 2004). The facilitator is needed to support the group to create principles or ground rules for discussion (and to hold the group to account for these), identify and ask open questions and probe for truth without leading participants; to manage dominant voices and perspectives, and enable participants to feel comfortable and confident to disagree with each other on sensitive issues (Kennedy 2004). The facilitator also ensures the dialogue remains relevant to the topic or questions at hand and centers on an appropriate level of abstraction (Floridi 2013); a skilled facilitator will also identify the most appropriate questions, times, and ways to intervene without being too leading or overbearing (Kennedy 2004), and address bad faith actions (e.g., by encouraging member of the group to identify and challenge inconsistencies in argument). In this way, it allows participants to explain why they think as they do, to hear alternative perspectives, and reflect on what new insights are meaningful. This allows the reasons for skepticism (e.g., on climate change) or lack of trust (e.g., in scientists or governments) to be identified and for the root causes of the post-truth condition to be identified, creating possibilities for a response. Further, it builds relational connections between people, allowing them to experience empathy and compassion through the process.

Before considering the nature of philosophical dialogue, it is worth reflecting on Midgley's analogy between philosophy and plumbing. She describes these activities as similar in the sense that they support complex structures and are often unseen - until a problem (here, the post-truth condition) is evident:

> Plumbing and philosophy are both activities that arise because elaborate cultures like ours have, beneath their surface, a fairly complex system which is usually unnoticed, but which sometimes goes wrong. In both cases, this can have serious consequences. Each system supplies vital needs to those who live above it. (Midgley 1992: 139)

Midgley points out that philosophy is just as necessary as plumbing. It is important to recognize that, although most people may not think explicitly about philosophy regularly or read philosophical texts, our individual and collective decision-making processes and actions are constantly and unavoidably informed by philosophical standpoints. It is important that people recognize this and learn to think philosophically, or else sleepwalk through life before waking up to a flooded or crumbling house. Engagement with philosophy is therefore needed in order to ensure that the public, politicians, and scientists remain cognizant and questioning of the philosophical principles and values that inform decision-making processes in relation to important topics such as climate change and geoengineering.

Philosophical dialogue on geoengineering

Philosophical dialogue tends to start with a question, such as 'should geoengineering be deployed?' This type of question cannot be answered by scientists or using scientific evidence in isolation, even though both must inform the potential answer. This is because they are *philosophical* rather than scientific questions in the sense that they are "open to informed, rational and honest disagreement" (Floridi 2014: 195) - as such, they are portals to considering the ethical, social, and political dimensions of geoengineering and climate change. While it is important that scientists and scientific evidence inform potential answers to this question, it is also important to recognize that the desirability of such answers will depend on the different values and perspectives that people have.

Furthermore, a hallmark of a philosophical question is that any meaningful or worthwhile attempt to answer it automatically produces a cascade of underlying, often non-scientific questions that need to be answered first. In this instance, for example, the question 'Should geoengineering be deployed?' leads to other, unavoidable questions such as: 'What types of intervention are geoengineering?' 'How much evidence is required on the potential effectiveness or harm of geoengineering before it can be considered worthwhile?'; 'Whose consent is needed before geoengineering can be researched or used?'; "Who should pay for geoengineering, countries who have the biggest past or contemporary carbon footprint?', 'How, if at all, should those who are detrimentally affected by geoengineering be compensated?'. Answers to these and other questions that are inseparably bound to questions about geoengineering cannot be provided by scientists or scientific evidence alone because they are inherently philosophical and political questions; they require critical consideration of more than just scientific evidence: they also require consideration of culturally contested conceptions of power, justice, values, and nature, among other concepts. Table 8.1 below suggests some philosophical questions that can open dialogue on different methods of geoengineering.

Geoengineering classifications	**Examples of approaches**	**Philosophical questions to consider**
Carbon dioxide removal addresses the root cause of climate change.	Land-based carbon dioxide removal includes afforestation, air capture and bio-energy with carbon sequestration.	Is it preferable to deal with the causes or consequences of climate change?
	Ocean ecosystem methods of carbon dioxide removal include ocean afforestation and ocean fertilization.	What risks are acceptable when trying to prevent further climate change?
Solar radiation management attempts to offset the effects of increased greenhouse gas concentrations.	Surface albedo enhancement could make the Earth's surface brighter and, therefore, reflect more solar radiation. These approaches include painting white roofs in urban areas, planting more reflective crop varieties, and covering deserts with a reflective surface.	How can individual freedoms be reconciled with the collective action needed to respond to the climate crisis?
	Reflectors located in space could reduce the proportion of sunlight reaching the Earth.	How can technologies that have an impact on global temperatures be regulated fairly?
	Injection of sulfate (or other) aerosols into the lower stratosphere to reflect sunlight before it reaches the Earth's surface.	If a process mimics nature, is it inherently desirable?

Table 8.1. Philosophical questions about geoengineering proposals

Having identified philosophical questions, the community of inquiry is the method used to answer them. This is an approach to doing philosophy based on scientific inquiry (Peirce 1960; Dewey 2007) but applied to philosophy education by Lipman (2003) in a Philosophy for Children (P4C) program. The community of inquiry involves all participants in the use of mental resources to solve problems or answer questions in common. Rather than learn the final products of science, participants examine their own ideas and explore new ideas. The freedom and control it offers enables people to connect science with their prior knowledge and interests and pursue their own curiosities.

One of the key distinctions between science education and communication, on the one hand, and humanities education and communication, on the other, is the way in which knowledge is presented (Lipman 2003; Nussbaum 2010). According to the philosopher Matthew Lipman, for example, science textbooks often present (scientific) knowledge as though it is settled and unproblematic (Lipman 2003). In the humanities, in contrast, knowledge is treated as though it is intrinsically problematic: people are encouraged to actively seek and consider new problems of interpretation or conceptualization rather than apply standard methods of problem-solving to new problems and establish different ways of looking at the issues (Donnelly 2004). Teaching about the nature of science helps to clarify and contextualize the processes used to obtain reliable knowledge. It tackles themes such as the tentativeness of scientific findings, the subjective role of the researcher, or the existence of different scientific methods. In short, it tackles the special way in which scientific knowledge is formed, it is concerned with science as a human enterprise and discusses the tools and products of science (McComas 2015). It also highlights that the implications of scientific findings or methods need to be thought about and discussed by democracies. In this vein, it has been argued that the aim of education, be it related to science, social science, or the humanities, should be to "present situations where problems are *relevant to the problems of living together*, and where observation and information are calculated to *develop social insight and interest*" (Dewey 1916: 226, emphasis added).

In other words, it is important that participants are encouraged to think about how to live peacefully and cooperatively with one another, even - or, indeed, *especially* - when they are presented with shared problems (such as climate change) and a host of potential solutions, each with different implications (such as adaptation versus mitigation versus geoengineering). Because science can give us answers to specific scientific questions but rarely *the* answers to philosophical or political questions, it is important that scientific knowledge is presented as such. Through philosophical inquiry, the principles around which scientifically-informed political decision-making is organized can be identified and critiqued and ultimately adhered to or adjusted, but for this to happen, it is important for people to learn how to communicate their ideas with: "[o]penmindedness, singlemindedness, sincerity, breadth of outlook, thoroughness, [and an] assumption of responsibility for developing the consequences of ideas which are accepted" (Dewey 1916: 414). In other words, people should be encouraged to be open to exploring unfamiliar ideas; remain focused during discussions about particular topics; be honest with themselves and those they are engaging with; be critical of information that they are presented; and be willing to act on their beliefs. These are all attributes that practicing philosophy cultivates - and which people do not necessarily encounter in day-to-day life.

In addition to prompting people to reflect critically on fundamental questions about justice, values, and what actions align with these, practicing philosophy also assists in the development of better reasoning. Indeed, as practicing philosophy involves attempting to construct logical arguments, questioning whether the evidence supports claims, and identifying logical fallacies, whereby rhetorical devices are used to make illogical arguments seem convincing, it can help people to identify misinformation, superficially appealing but poorly supported arguments; and think more clearly about information and arguments related to climate change and geoengineering. Practicing philosophy allows us to engage with fundamental questions about knowledge, truth, reality, experience, and justice, and therefore reflect on how to act. Questions about these topics underpin all human knowledge and activity, including scientific practices.

The limits and limitations of philosophical dialogue

Philosophy is often perceived to be antagonistic to science, even by high-profile scientists of the past and present. Richard Feynman is quoted as saying, "philosophy of science is about as useful to scientists as ornithology is to birds." We have outlined above the potential value of philosophical dialogue in informing science and technology *for* society, but what are the limits and limitations of the approach?

There is a complex relationship between knowledge, arguments, and behavior, as outlined above. Philosophical questions cannot be answered by scientific methods, but a knowledge of how science works and relevant scientific and technological knowledge are important in constraining the questions and possible answers (Floridi 2013). Philosophical dialogue, therefore, needs to be scientifically informed, with awareness of where uncertainties are, and be alert to misunderstandings and misinformation in order to identify desirable outcomes.

Also important is the composition and disposition of participants in philosophical dialogue. There needs to be a diversity of views present in order to identify where likely convergences and conflicts in decision-making might arise and to attempt to understand how constructive compromises might be made. Importantly, where a diversity of views is concerned, there is a need for sincerity and genuine interest in working together in order for dialogue to create capacity for change (Dewey 1916). In this sense, it is important for those who participate in philosophical dialogue to recognize and embrace the need to take the behavioral implications of their beliefs and values seriously and engage with philosophical dialogue in a spirit of actionable intent.

There is a need to be cautious when thinking about how scientists participate in philosophical dialogue: researchers have found deference to scientific authority, even on non-scientific issues, to influence opinion-forming (Ho et al. 2011). It is therefore important to ask:

- Whose perspectives are important to hear (i.e., who are the participants?)
- What conditions are needed to ensure that all participants are able to make a contribution (i.e., what are the principles for discussion, and how are they to be enacted)? These can be created with participants.
- Who should hear (and act) on public(s) perspectives on the issue (i.e., who should listen or receive the outputs of a philosophical dialogue?)
- How will scientific misconceptions be dealt with? Here, there is an important role for the facilitator in correcting through hypothetical reasoning, for example, by asking, "would your ideas change if X was true...?"
- How can scientists contribute without leading the dialogue or influencing the focus?

In our case, we were interested in hearing the perspectives of young people. Policy-makers were identified as among those who needed to listen, so they were sent a guide and policy brief resulting from the dialogue, and young people participating used it in various national and international contexts. Scientists were involved through the research they had presented and disseminated online to avoid deference to authority on non-scientific matters. Philosophical dialogue challenges the post-truth condition in two ways: (i) amongst participants by identifying shared values, points of disagreement, and tentative conclusions that respect differences, as well as identifying further questions for inquiry, and (ii) where incorporated into education and public life more generally, by creating a better quality of argument focused on truth and meaning rather than headline-grabbing polarization.

Conclusion

We have demonstrated how philosophical dialogue can create conditions for criticality through reflexive philosophical inquiry into science, technology, and society. The non-oppositional community of inquiry can be used to examine issues and enables those taking part to identify common ground based on shared values, which can contribute to a response to the post-truth condition. This has relevance for issues beyond climate engineering and environmental issues, including but not limited to health, artificial intelligence, and space tourism.

The philosophical dialogue begins with an open question that, as highlighted, cannot be answered (but can be constrained) by scientific knowledge. It, therefore, lends itself to helping people to engage in authentic and reflexive - rather than politically biased – exploration of topics and their thoughts in relation to them through reasoning. As one question leads to another and participants are obliged to question the justification for their beliefs, entrenchment in particular beliefs can be undone as nuances are revealed, and those of opposing views are no longer seen as an 'other' who simply needs to be defeated. Philosophical dialogue also encourages people to share their thoughts and arguments and, therefore, subject them to public scrutiny. Subjecting particular arguments to public scrutiny forces them to be questioned, defended, or changed in light of new information or perspectives. It also means that the connections that people make between science and their arguments about how it should be used are considered more critically so that poorly constructed arguments can be identified, abandoned, or revised. This can build the communicative competencies to hold science to account as climate change - and the responses to it - becomes more extreme.

References

BBC. "The Public Philosopher." BBC Radio 4. August 2023 https://www.bbc.co.uk/programmes/b01nmlh2/episodes/player

Brewer, Paul R. 2012. "Polarisation in the USA: Climate Change, Party Politics, and Public Opinion in the Obama Era." *European Political Science* 11(1):7–17. doi: 10.1057/eps.2011.10.

Corner, Adam, and Nick Pidgeon. 2014. "Geoengineering, Climate Change Scepticism and the 'Moral Hazard' Argument: An Experimental Study of UK Public Perceptions." *Philosophical Transactions. Series A, Mathematical, Physical, and Engineering Sciences* 372(2031):20140063. doi: 10.1098/rsta.2014.0063.

Dewey, John. 1916. *Democracy and Education.* New York: The McMillan Company.

Donnelly, James F. "Humanizing science education." Science Education 88, no. 5 (2004): 762-784.

Dunlop, Lynda, Elizabeth Rushton, Lucy Atkinson, Eef Cornelissen, Jelle De Schrijver, Tetiana Stadnyk, Joshua Stubbs, Chrissy Su, Maria Turkenburg-van Diepen, Fernanda Veneu, Celena Blake, Saul Calvert, Clémentine Dècle, Kirndeep Dhassi, Rosalind Edwards, Greta Malaj, Jovana Mirjanić, William Saunders, Yara Sinkovec, Suzan Vellekoop, and Xinyue Yuan. 2022. "Youth Co-Authorship as Public Engagement with Geoengineering." *International Journal of Science Education Part B* 12(1):60–74. doi: 10.1080/21548455.2022.2027043.

Dunlop, Lynda, Annie Hodgson, and Joshua Edward Stubbs. "Building capabilities in chemistry education: happiness and discomfort through philosophical dialogue in chemistry." Chemistry Education Research and Practice 21, no. 1. 2020: 438-451.

Eliasoph, N. 1996. Making a fragile public: A talk-centered study of citizenship and power. *Sociological Theory, 14*(3), 262. https://doi.org/10.2307/3045389

Floridi, Luciano. 2013. "What is A Philosophical Question?" *Metaphilosophy,* 44(3): 195-221

Goucha, Moufida. 2007. *Philosophy, a School of Freedom: Teaching philosophy and learning to philosophize: Status and prospects.* Paris: UNESCO. https://unesdoc.unesco.org/ark:/48223/pf0000154173

Groves, Christopher. 2019. "Post-truth and Anthropogenic Climate Change: Asking the Right Questions." *Wiley Interdisciplinary Reviews. Climate Change* 10(6). doi: 10.1002/wcc.620.

Ho, Shirley S., Dietram A. Scheufele, and Elizabeth A. Corley. 2011. "Value Predispositions, Mass Media, and Attitudes toward Nanotechnology: The Interplay of Public and Experts." *Science Communication* 33(2):167–200. doi: 10.1177/1075547010380386.

Kahan, Dan M., Hank Jenkins-Smith, and Donald Braman. 2011. "Cultural Cognition of Scientific Consensus." *Journal of Risk Research* 14(2):147–74. doi: 10.1080/13669877.2010.511246.

Kahan, Dan M., Ellen Peters, Maggie Wittlin, Paul Slovic, Lisa Larrimore Ouellette, Donald Braman, and Gregory Mandel. 2012. "The Polarizing Impact of Science Literacy and Numeracy on Perceived Climate Change Risks." *Nature Climate Change* 2(10):732–35. doi: 10.1038/nclimate1547.

Kahan, Dan M., David A. Hoffman, Donald Braman, Danieli Evans, and Jeffrey J. Rachlinski. 2012. "They Saw a Protest': Cognitive Illiberalism and the Speechconduct Distinction." *Stanford Law Review* 64(4):851–906.

Kahan, Dan M., Asheley Landrum, Katie Carpenter, Laura Helft, and Kathleen Hall Jamieson. 2017. "Science Curiosity and Political Information Processing" *Political Psychology* 38:179–99. doi: 10.1111/pops.12396.

Kennedy, David. 2004. "The Role of a Facilitator in a Community of Philosophical Inquiry." *Metaphilosophy* 35(5):744–65. doi: 10.1111/j.1467-9973.2004.00348.x.

Kleiner, Tuuli-Marja. 2018. "Public Opinion Polarisation and Protest Behaviour." *European Journal of Political Research* 57(4):941–62. doi: 10.1111/1475-6765.12260.

Lewandowsky, Stephan, Ullrich K. H. Ecker, and John Cook. 2017. "Beyond Misinformation: Understanding and Coping with the 'Post-Truth' Era." *Journal of Applied Research in Memory and Cognition* 6(4):353–69. doi: 10.1016/j.jarmac.2017.07.008.

Lindholm, Markus. 2018. "Promoting Curiosity?: Possibilities and Pitfalls in Science Education." *Science & Education* 27(9–10):987–1002. doi: 10.1007/s11191-018-0015-7.

Lipman, Matthew. Thinking in Education. Cambridge University Press, 2003.

Lucas, Chloe H. 2018. "Concerning Values: What Underlies Public Polarisation about Climate Change?: Concerning Values for Climate Change." *Geographical Research* 56(3):298–310. doi: 10.1111/1745-5871.12284.

Luce, Megan R., and Sherry Hsi. 2015. "Science-Relevant Curiosity Expression and Interest in Science: An Exploratory Study." *Science Education* 99(1):70–97. doi: 10.1002/sce.21144.

McComas, William F. 2015. "The nature of science & the next generation of biology education." The American Biology Teacher 77, no 7. : 485-491.

Midgley, Mary. 1992. "Philosophical Plumbing." *Royal Institute of Philosophy Supplement* 33:139–51. doi: 10.1017/s1358246100002319.

Nussbaum, Martha C. 2010. *Not for Profit: Why Democracy Needs the Humanities.* Princeton, NJ: Princeton University Press.

OED. 2019. "Oxford Word of the Year 2016." *Oup.Com.* (https://languages.oup.com/word-of-the-year/2016/).

Owen, R., P. Macnaghten, and J. Stilgoe. 2012. "Responsible Research and Innovation: From Science in Society to Science for Society, with Society." *Science & Public Policy* 39(6):751–60. doi: 10.1093/scipol/scs093.

Peirce, C. S. 1960. "Lessons from the History of Science The Scientific Attitude." Collected Papers of Charles Sanders Peirce, Volumes I and II: Principles of Philosophy and Elements of Logic Charles Sanders Peirce. Harvard University Press, Cambridge MA.

Szerszynski, Bronislaw, and Maialen Galarraga. 2013. "Geoengineering Knowledge: Interdisciplinarity and the Shaping of Climate Engineering Research." *Environment & Planning A* 45(12):2817–24. doi: 10.1068/a45647.

Talisse, Robert B. 2021. "*Problems of Polarization*" in *Political Epistemology* edited by E. Edenberg and M. Hannon. Oxford: Oxford University Press.

The Philosophy Foundation. 2022. "Philosophy for Adults." https://www.philosophy-foundation.org/

Chapter 9

Towards the PERIpatetic approach: epistemology of close encounters within contemporary STS

Matjaz Vidmar

Abstract: This chapter outlines key features of an emerging shift in research philosophy within science and technology studies (STS), involving greater proximity to participants through longer-term interaction and embeddedness in various contexts of Participatory Action Research (Darrouzet, Wild and Wilkinson 2009). Hence, in this chapter, I outline the changing nature of social scientific inquiries and the related re-calibration of researchers' philosophical framework.

Outlining my experience of a multi-level study of innovation processes and practices within the (New) Space Sector in Scotland, I am combining the understanding of abductive epistemology (Blaikie 2004), best practices in researching professional elites (Mikecz 2012), and multi-method research design and data collection, it develops the "participatory strategic ethnography of innovation" inspired by the Biographies of Artefacts and Practices (Hyysalo, Pollock and Williams 2016) and frames the "uninformed insider" positionality as critical advances within STS research.

This is a reflection on half a decade of engagement with the studied community and resolving critical practical and theoretical challenges. In particular, I have been working on a problem-solving-driven study and conducting it in close cooperation with collaborators in the field, having been embedded within an innovation team at a leading research organization.

Put together, I argue that the set-up of this work reflects a new methodological framework of the "PERIpatetic Approach" - a Practical Epistemology for Researching Innovation (PERI). The acronym PERI also (sub-)refers to the ambition that the philosophy behind the PERIpatetic Approach is based on re-interpretation of the core tenants of the strong program in

Sociology of Scientific Knowledge (Bloor 1991) to base the future STS research on Perspectival, Embedded, Responsive and Introspective principles.

Keywords: Participatory Action Research, Abductive Epistemology, Participatory Strategic Ethnography, Practical Epistemology for Researching Innovation, PERIpatetic Approach

Introduction: new approaches for new modes of research

The "common wisdom" on the epistemological, methodological, and practical set-up of social-scientific research is fast departing from the "established norms" of twentieth century sociology. Of particular note is the engagement with professional elites in the scientific, policy, and business communities and the changing role of the researcher in this environment (Mikecz 2012). The key notable trends within it are the emergence of abductive/problem-solving epistemologies, in-depth long-term/longitudinal presence, deployment of mixed methods, and extensive reflexivity. These changes crucially bring about a different set of requirements for the researchers in terms of their positionality when accessing and investigating such environments, which are critically understudied (Berger 2015). The PERIpatetic Approach outlined in this chapter has been carefully developed prior to an extensive field study in the Scottish Space Sector; however, its analysis presented in this chapter is also based on post-research reflection. This is especially relevant for entering into a dialogue with earlier STS theories, such as the Strong Programme, encouraging the application of the same methodology to the study of science as well as social science. Though primarily based on my personal experiences, I believe my reflections and conclusions here apply much more broadly in a variety of social scientific inquiries in professional environments, particularly in contentious or high-stakes areas and in work identified as participatory action research (McIntyre 2007).

In particular, many authors have noted the (increasing) importance of "insider" status for research within professional elite settings. In fact, even the proponents of the "informed outsider" approach directly acknowledge: "researcher has the insight of an insider but the neutrality of an outsider" (Welch et al. 2002:625). However, my experience and observation tend to contradict this conclusion – the researcher can never really have the exact same insight of an "insider," nor can they be as neutral as an "outsider," having committed to a close-quarters longitudinal research project. I argue that with a new research paradigm(s) emerging, focusing on more detailed and prolonged multi-sited ethnographic studies, the researcher's positionality is realigning away from the "informed outsider" approach outlined by Welch et al. (2002) and towards an "uninformed insider" approach, based on meaningful engagement and embeddedness in the studied environment, yet retaining the critical analytical distance. Such a closer relationship acknowledges the need for a significant amount of trust required for access to, and maintenance of, prolonged engagement with the professional elite while recognizing the inherent knowledge asymmetry in this relationship and accommodates more inductive/abductive information gathering and knowledge co-construction (Morrisey and Morrissey 1998). My solution to this challenge is to "embrace" the position of an "uninformed insider," acknowledging a degree of limitation in our insight and a degree of partiality. This researcher position is also much more closely aligned with abductivist epistemology, i.e., the bottom-up problem-solving approach to building theories, where researchers deliberately ensure that they enter the field with as little prior knowledge as possible in order to truly recognize all points of contention and hence limit any bias (Blaikie 2004; Shank 2008). I further argue that this is a recognition of existing practices to both further establish the trust between the various stakeholders and increase confidence in research processes and findings.

Specifically, this chapter outlines an open-ended, interdisciplinary, participatory, and problem-driven approach to empirical work. It begins by presenting the foundations of my ontological and epistemological position – the PERIpatetic Approach – and outlines the empirical context in which my research was conducted. Then, it examines, in turn, my reflections on the four key aspects of the PERIpatetic Approach, in particular, the methodological principles behind participatory strategic ethnography of innovation, the "uninformed insider" positionality, abductive analytical work, and ethical introspection. Finally, I discuss the three stand-out elements of my experience of deploying the PERIpatetic Approach as an "uninformed insider": a) the importance of becoming an insider by building trust and gaining access to the community, b) the importance of "uninformedness" for deploying the strategic ethnography and using abductive research epistemology, and c) the dynamic and fluid nature of this philosophy and researcher's position.

My philosophical position – The PERIpatetic approach

Peripatetic philosophy has a strong historical resonance, as it is often used to denote the Aristotelian approach to conceptual development and pedagogy, its direct meaning of "walking about." However, it is also a very suitable symbolic summary of the break with the "traditional," more dispassionate, and "static" research philosophies deployed in many social science fields in the past, including in Science and Technology Studies (STS). Specifically, often borrowing insights from more close-quarter anthropological studies, a bigger array of ethnographic methods and a move towards situated empirical sites are changing the way in which research is designed and conducted. Through my work analyzing innovation processes using STS principles, I have been particularly drawn to examining multiple perspectives and deploying a mix of different methods, all grounded in being in a constant and close relationship with the key participants in the field – in my case, a gatekeeper organization.

This approach seemed to resonate with some well-established epistemological positions in sociology, particularly participatory action research (McIntyre, 2007) and abductivism (Blaikie, 2004). However, the predominantly single-sited methodological setup within past inquiries often lacked the multi-perspectival flexibility I required. Here, I was inspired by the Biographies of Artefacts and Practices (BOAP) Approach (Hyysalo, Pollock and Williams 2016), putting forward a more strategic view on ethnography by focusing on following subject matter across a multitude of sites (design, production, use of innovation, etc.), extended over time (from years to decades) and deploying a variety of data collection methods (secondary data/document analysis, participant observation, interviewing, etc.). Combining these two starting points, I argue that a new comprehensive framing is required, formalizing the peripatetic nature of contemporary STS innovation research.

Strong Programme Tenet	**Brief Description**	**Dimensions of the PERIpatetic Approach**	**Research Theory and Practice**
Causality	it examines the conditions (psychological, social, and cultural) that bring about claims to a certain kind of knowledge	Perspectival	*participatory strategic ethnography of innovation*
Impartiality	it examines successful as well as unsuccessful knowledge claims	Embedded	*action research as a uniformed insider*
Symmetry	the same types of explanations are used for successful and unsuccessful knowledge claims alike	Responsive	*abductive epistemology with critical realism*
Reflexivity	it must be applicable to sociology itself	Introspective	*ethics-reflection-acknowledgement.*

Table 9.1. Breakdown of PERIpatetic Approach, built on Strong Programme tenets.

Hence, I propose the PERIpatetic Approach as a way forward, combining four elements to bring together a Practical Epistemology for Researching Innovation (PERI), fit to match the advances in the mode(s) of social science research as well as the evolution of the subject-matter fields (see Figure 9.1). It combines an interest in multi-perspectival knowledge-making, acknowledges the embedded researcher's positionality, applies a responsive analytical framework, and deploys introspective mechanisms to legitimize its findings (as presented in Table 9.1). These four elements are in dialogue with the original tenets of the Strong Programme of Sociology of Scientific Knowledge (Bloor 1991), particularly aiming for causal explanations, impartial position, symmetrical analysis, and applying reflexivity to our work.

Firstly, the PERIpatetic research is moving away from producing causal generalizations and towards describing perspectives by recording and analyzing narratives, developing concepts, and integrating individual accounts into community-wide trends. It is particularly concerned with nested multi-layer contexts and is grounded in researchers' experiences rather than a priori intellectual projects. This is a significant departure from the past normative interest in causality, which could be attacked either as inherently positivist or when interpreted through relativism/constructivism, it may be seen as lacking in auto-reflexivity (i.e., if scientific knowledge is socially conditioned, then social scientific knowledge is also socially conditioned. Hence, correlations are only pertinent to the specific context, rather than truly "causal").

Secondly, by being closely embedded in the field, a PERIpatetic researcher is an active co-creator of the studied environment and develops a level of presence and interaction akin to a field's "insider." The applied and empowering nature of action research is in stark contrast with past notions of researchers' impartiality. However, it is more in line with both the de-facto influence we have in the field as well as the current (political) agenda within social science, including innovation studies, to create meaningful and impactful research. However, in order to obtain legitimate and credible findings, strategies for data collection must reflect the intellectual independence of analytical work, perhaps through enacting "uninforemedness."

Thirdly, the epistemological position thus developed combines field insights, in particular, input about the challenges and interests of the various participants, with an iteration of a multitude of viewpoints. Hence, the PERIpatetic approach adopts an abductive epistemological position, i.e., deriving theoretical models from interrogating empirical data through a real-life problem-solving process. Thus, it is less concerned about the traditional measures of theoretical coherence, for instance, the symmetry of explanations, and more focused on being responsive to the effect of the proposed models. This is consistent with the critical realism approach to accept knowledge claims validity through their application rather than derivation.

Fourthly, the PERIpatetic Approach fully adopts and extends the importance of the researcher's reflexivity through a clear and comprehensive introspective process. Specifically, when working embedded in the field of study as an active participant, applying abductive epistemology and collecting data via participatory methodology, understanding positional challenges and intellectual biases is vital. A three-stage solution is suggested here: having a rigorous ethics framework, engaging in constant reflection, and (publicly) acknowledging the participatory action nature of such work.

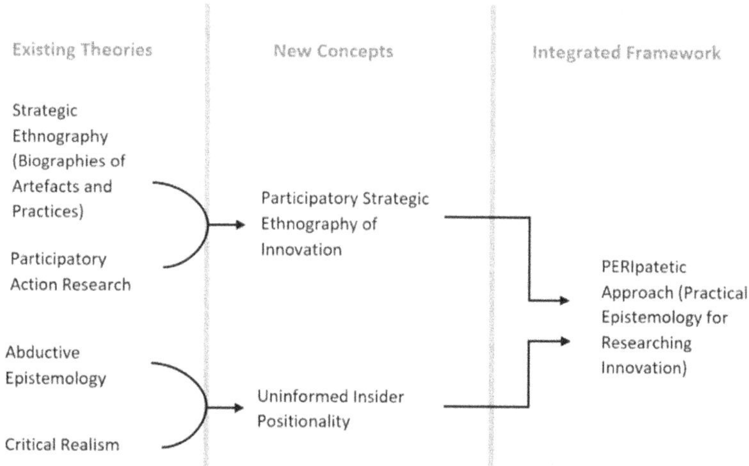

Figure 9.1. The conceptual framework behind the formulation of the PERIpatetic Approach

The framework draws on multi-sited longitudinal ethnography (Biographies of Artefacts and Practices), embeddedness through participatory action research, epistemological abductivism through iterative problem/issue identification, and the critical realism of incremental participatory system understanding and change.

Later, each of these elements will be examined in greater detail, but in order to provide some context for the analysis of their deployment, the next section first outlines the empirical environment I encountered through my research of the evolving networks, practices, and institutions in the emerging Scottish (New) Space Sector.

My Time with the Scottish (New) space sector

The makeup of the Scottish (New) Space Sector is appropriate for intensive qualitative and quantitative research of innovation processes and associated "culture," i.e., the social phenomena related to techno-scientific and economic development, which was the subject of my research. This is due to the following:

1. A good mix of the different types of actors (upstream/downstream, software/hardware, spin-off/start-up)
2. Manageable size for complete study: currently <20 SMEs with <200 employees
3. Strong research base (in particular Dundee, Strathclyde, and Edinburgh Universities) and highly educated workforce
4. Strong institutional presence (university tradition, many institutes, and research stations
5. An existing rich historical context in science, technology development, and innovation (in particular through Scottish engineering legacy)
6. Significant current political and economic interest, noted in particular in roadmaps adopted by the UK government - 8 Great Technologies (Willetts, 2013), Space Innovation and Growth Strategy (Space IGS, 2011), Health and Size of the UK Space Industry (London Economics, 2019), etc.
7. Predicted and part-evidenced rapid growth at a major industry transition

My research (Vidmar 2019b) has benefited from a specific time context, linking the past and present (and looking into the future). While studies of innovation are by their nature historical, the significant (political) emphasis on technology/knowledge transfer and commercialization of scientific research is a very contemporary concern (Nutley, Walter, and Davies 2007). This was helpful both in terms of there being a significant appetite for collaboration from different research institutions, an increased openness (due to perceived benefits) on the part of the Space "community of practice," and a generation of a significant amount of documents, events, discourse, and literature.

Crucially, the Space Sector is currently undergoing a major transition (Adlen 2011), which is bringing the issues of innovation and technology transfer, as well as changes to all levels of its socio-political and economic structure, into focus. For an abductivist and participatory action research rationale, this proved very fruitful, as problems and questions of understanding are emerging in the field, and various stakeholders became increasingly open and interested in working with me to develop both new knowledge and operational solutions. As a way in, I benefited from having studied natural sciences (Physics and Astronomy, with a specific interest in technology development for scientific research) in my first degree. In addition, I have developed close ties with the UK Astronomy Technology Centre's (UK ATC) Innovation team (ATC Innovations), a Science and Technology Facilities Council (STFC) establishment, all of which is now part of the UK Research and Innovation. UK ATC is the national laboratory for the development of Astronomy instrumentation, located at the Royal Observatory in Edinburgh (ROE).

Having had a prior interest in science and technology policy and philosophical, socio-political, and economic aspects of techno-scientific development, I professionally transitioned into Science, Technology Studies (STS) research. However, I retained these previous associations and contacts, as I was particularly interested in working within the fields I already knew and understood. In particular, through both public engagement as well as research activities, I was well-embedded in the Astronomy and Space Science work in Edinburgh and Scotland. Approaching ATC Innovations with an initial interest to work on a research project in this arena led to a research partnership agreement, which enabled a longitudinal embedding in the field by becoming an informal member of the ATC Innovation's team and colleagues there became my gatekeepers for further access to stakeholders.

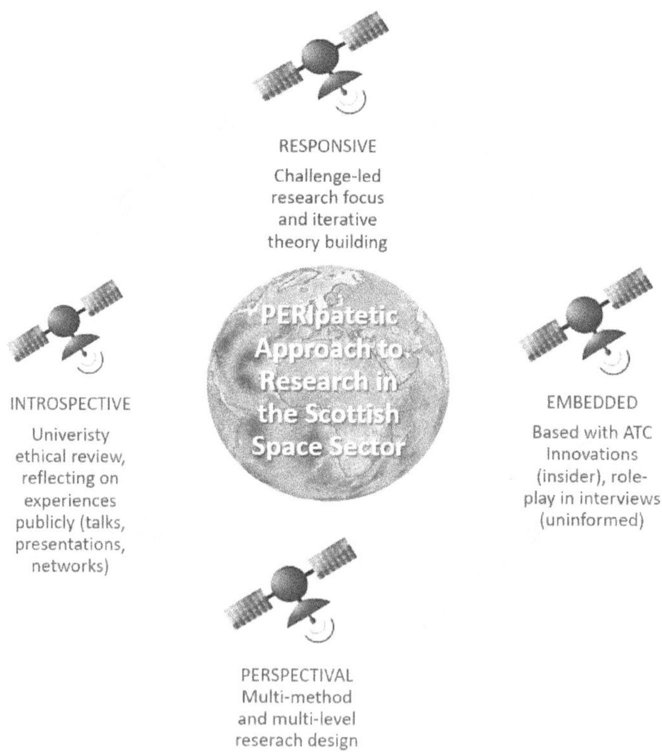

Figure 9.2. Relational positioning of the different frameworks within the empirical work

Additionally, my abductivist problem-solving agenda was well met by my research partners through the development of the Higgs Centre for Innovation. This is a new building and engagement program at UK ATC at ROE, as the UK government's direct investment into the Scottish New Space sector and its development coincided perfectly with my project's timeline. As part of this significant undertaking, outside of the innovations team "normal operation" several questions and challenges emerged, with an acute need arising to understand the multi-layered and complex landscape in which this new Centre needs to be successfully introduced. Having so easily found a significant source of interesting starting points for what shaped up to become my active participatory research into innovation culture within the Scottish New Space Sector, I designed a research project to match my overall interests, inspired by problem-solving demand (broad alignment with the PERIpatetic approach elements is summarized in Figure 9.2).

The first stage of my research project concerned the embedding in the research partner's team, framing of the inquiry by "abducting" a set of research questions and developing methodological and theoretical framing for my study within a pilot setting. These "prompt questions" spanned three dimensions: "How is the Scottish Space Sector formed?"; and "What can innovation intermediaries do to support (Scottish Space) sector growth?"; as well as a reflective question on "How can we integrate the best insights and practices in research and its application?" (Vidmar 2019b:17). Crucially at this point, only the micro-level of the analytical approach was being developed, as in abductivist setup, design flexibility must be ensured to be able to respond to early findings and their reception by the study's participants. In my case, it turned out that the first phase validated the methodological standpoint and theoretical framework for this part of the inquiry and pointed towards a second stage roll-out of micro-level data collection, as well as exposed additional key questions (moving from micro- to meso- and macro-levels). Consistent with other modes of research design, working through the first stage also involved an extensive review of the existing literature and an in-depth analysis of the subject matter (the Space Sector)'s background framing[1].

In the critical second stage, a detailed qualitative inquiry was undertaken to assess not only the descriptive narratives emerging from the data collected in the first stage but also to develop a set of new theoretical models and framings. Specifically, these concerned the understanding of micro-level changes to innovation processes and practices, meso-level innovation networks

[1] See my publications on development policy (Vidmar, 2020b) and opening up of innovation process (Vidmar, 2019a).

emergence and structures, and macro-level geo-sectoral policy and its translation through innovation intermediation[2].

In the third stage, these theories were iterated with the help of research partners and were deployed as key pieces of new understanding in order to solve their real-life problems of establishing a new innovation center and integrating it into the existing ecosystem[3]. From a participatory action research standpoint, the timing of this study is very conducive to having a really meaningful impact on the researched community as findings are available right from the start of the Higgs Centres of Innovation program (in Spring 2018), meeting less entrenchment of practice often present in more mature projects (Zeitz, Mittal and McAulay 1999).

The final fourth stage is about formalizing the insights through (academic) publications and dissemination at events and through smaller meetings and briefings. Importantly, where possible, this should include the development of toolkits and translational activities whereby theories, insights, and frameworks are (re-)deployed in follow-up studies and/or in operational and strategic documents[4].

Having thus set the scene, in the next four sections, I analyze in greater detail my experience with the four research principles of the PERIpatetic Approach - being Perspectival, Embedded, Responsive, and Introspective researcher - and highlight the features of the emerging participatory strategic ethnography methodology and my "uninformed insider" positionality.

Perspectival: a participatory strategic ethnography of innovation

Learning from the biographies of artifacts and practices (BoAP)

The Biographies of Artefacts and Practices (BoAP) perspective (Pollock and Williams 2010) emerged amongst an informal community of scholars seeking to move away from the atomistic perspectives and 'snapshot' studies of particular moments of technology design or use (Hyysalo 2009; Pollock and Williams 2010; Williams and Pollock 2012). The key premise is to go beyond the single-site case study or actor-centered accounts that prevailed within recent STS research and develop more effective methodological templates based

---

[2] See my publications on open innovation and organisational learning (Vidmar *et al.*, 2020) and innovation intermediation (Vidmar 2020a).

[3] See my publication on development of comparable projects in innovation intermediation in Scotland and Slovenia (Vidmar 2019d).

[4] I have written a series of briefing notes, strategic documents and delivered several high-profile presentations, such as work on Edinburgh Living Lab (Vidmar 2019c), Citizen Data Exchange, Innovation Caucus, HEinnovate, Data-Driven Innovation, etc.

upon a longitudinal and multi-site study in the shape of a "strategic ethnography" (Williams and Pollock 2012). BoAP's aim is to evolve a practice more appropriate for effective research of systemic cultural nature (of innovation processes), which can produce findings refuting the field's recent tendency to accept 'flat' ontologies (i.e., lacking temporal and spatial depth and/or analytical clarity of any generalizable phenomenological structure, beyond the apparent descriptive accounts).

Ground-breaking longitudinal studies within the BoAP perspective detailed the complexity of innovation processes involving diverse arrays of players (engineers, users, managers) interacting over protracted periods and across many locales. Work conducted within this framework has highlighted the various ways in which users contribute to the development of products and the role of various forms of innovation intermediation, which may bridge contexts of technology supply and use (Williams and Pollock 2012). As BoAP authors recently pointed out:

> Every research design involves choices about where to address research effort. New sites and relations become visible in the course of fieldwork. [...] This calls for flexibility in research design coupled with the willingness to keep on pursuing the line of investigation beyond the single setting and project funding.

The biographical approach to the study of systemic features of innovation

Empirical work of this type can only be done in a more "engaged" manner; hence, researchers such as myself have moved away from the "one-way" econometrics and survey-based approach traditionally used to map innovation and business-led organizational behavior (Markman, Siegel, and Wright 2008). It is important to note that this is not a break from the key methods used in these fields; rather, it adds depth to (meso- and macro-scale) quantitative data by interpreting it through (micro-scale) qualitative analysis, leading to a more holistic mixed methods approach. For instance, my methodological set-up is built around three methods: quantitative and qualitative (secondary) data/document analysis; qualitative data from interviews, observations, and surveys; and (auto-)ethnographic reflection (Sedlack and Stanley 1992; Richards 1996; Odendahl and Shaw 2002; Delamont 2013; Bryman 2016). Hence, I have labeled this collectively as a "participatory strategic ethnography of innovation," a systematic longitudinal iterative study of the culture of innovation building on a multi-perspective and multi-level inquiry.

As such, this research project is also moving the BoAP methodology into a new direction – towards understanding systemic practices and artifacts instead of specific ones. In particular, the aim of this research was to map out a multi-level perspective on the emergence of innovation systems through the prism of open innovation transition. Methodologies for dealing with the study of systemic elements in the macro-level context are well-developed (and contested) within the STS literature on innovation. In particular, a very successful paradigm, namely Innovation Systems (IS), has been noted for its limited reach in terms of addressing the social phenomena in innovation (Williams and Velasco, 2016).

For instance, linking those top-level approaches to specific instances of innovation in actual product development is methodologically underdeveloped. To bridge this gap, I proposed that the main unit of empirical data collection (in my study's second phase) was an "innovation project," i.e., the development of a (single) product within a studied SME, and then analytically move to a meso-level network of actors well as macro-level policy (Green et al.1999). This was framed using a sensitizing concept of an "innovation moment" (Vidmar et al., 2020), which was developed to combine core theoretical concepts in new product development (Bhuiyan 2011), knowledge management, and organizational learning (Sun and Anderson 2010), as well as frame the evolutionary nature of the innovation process (Edwards, Delbridge and Munday 2005).

Epistemologically, this strategy also follows the CEN/TS standard for innovation management (CEN, 2014), which states:

> The organization should define two main responsibilities in the context of IMS (innovation management systems):
> — responsibilities for the specific innovation projects;
> — responsibilities for the general innovation management.(CEN 2014, 11)

Thus distinguishing between "specific innovation projects" and "general innovation management." It is also consistent with the Oslo Manual (OECD 2007) separating "object" and "subject" approaches: the "object" approach is based on a single business innovation project, e.g., the development of a new product, and the "subject" approach, which looks at a firm in its entirety. Specifically, this enabled me to link SMEs' behavior in innovation process management, rooted in specific examples of new product development and the network interaction based on acquired practices. To turn an "innovation project" into an understanding of the innovation processes and their dynamics requires a strategic ethnographic methodology, which enables a reasoned selection of several ethnographic research components and mixed-method analysis to perform a longitudinal study of innovation as a socio-technical phenomenon.

Of particular interest is a multi-level understanding of the changes in the development of new products due to network-mediated inputs from external partners and ways in which organizations performing new product development responded to the engagement of external sources of knowledge, be those the lead customers, academic researchers, or the wider market. More specifically, recording and analyzing the changing structure of new product development processes and the co-creation of collaborations within the innovation networks were the main targets of the second part of the study, though their interpretation hinged on policy-level dynamics, inter-organizational relationships and learning and the operationalization of innovation intermediation (the latter was examined in the third part of my study).

Beyond the single object of study

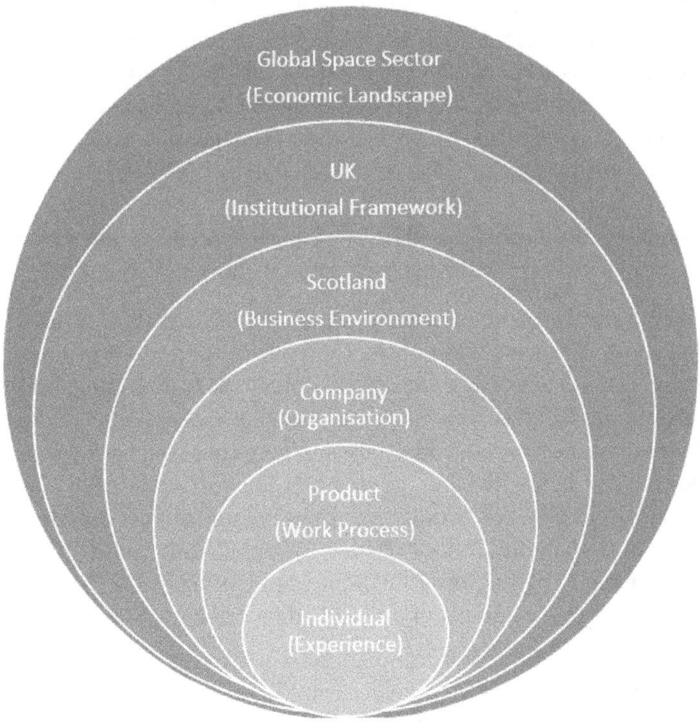

Figure 9.3. The ontological underpinning of a multi-level epistemology as applied to my research of the (New) Space Sector in Scotland.

This means that in practice, the variety of data collected was organized in a series of case studies, joined together into one broader multi-sited and temporally extended case study (Yin 2009) - addressing the emergence and evolution of innovation practices and routines. Such research clearly features a multilevel design structure (see the breakdown in Figure 9.3 above) whereby individual accounts (actor's experiences) are integrated via organizational narratives (companies) further into regional sectoral dynamics, which are, of course, part of an even wider national and international/global landscape. In these higher-level frames, micro-level social science within professional communities of practice becomes linked with econometrics, policy and politics, and legal frameworks, hence leading to a much more complete picture of the studied environment.

However, studying innovation in SMEs is challenging due to several factors, such as short timeframe, unclear boundaries, the informality of operations, and access difficulties. Hence, as noted above, this type of research can only be done in much closer proximity to the studied community. Embeddedness in the field allows the researcher to gain access to all relevant data and informants and develop a longitudinal and multi-sited overview of transformation and change by being present as it (gradually) occurs. However, this impacts the researcher's position within the studied field, the challenges we might be facing, and how to avoid certain major pitfalls. Hence, in the next section, I briefly outline the "uninformed insider" positionality as a PERIpatetic researcher.

Embedded: the "uninformed insider" approach and action research within a professional elite

When researching a niche or small professional environment, it is often the case that most, if not all, of the actors - research subjects - are found to be (inter)connected in some way (Odendahl and Shaw 2002). From a methodological point of view, this has several benefits. For example, it is easier to find research participants and the needed data. However, there are several downsides, too; a breakdown of trust with one of the participants can have significant repercussions across the whole community, and first impressions are important to gain "admittance." Hence, when planning and carrying out research, one has to be very mindful of the power structures and "positioning" of the researcher in the context of the studied community (Mikecz 2012).

On the one hand, in my case, some of these issues were mitigated from the start by opting for a collaborative research project and working closely with a prominent organization in this sector, namely ATC Innovations. Though this addressed the issue of access, it created new problems in terms of my identity in the field. On the other hand, my past Physics / Astronomy background epitomized the tension traditionally experienced by STS scholars - many of whom are social scientists with a natural science background - studying the very professional elite they have once been (or aspired to become) a member of.

Consequently, this means that the access into the field and the design of the data collection must address two significant challenges within research design: the interviewees are the author's peers (in terms of "shared knowledge and/or understanding") or even social superiors (hierarchical status) (Platt 1981), and they represent different types of (professional) elites (Mikecz 2012). In gaining entry into such an environment, it is useful to mention one's natural science credentials and association with reputable organizations in the field. This establishes the researcher as a member of the same professional elite – "one of us" or a "peer." However, my professional and social standing in this elite is precarious at best, as I was a junior researcher and approached the subject from a social scientific point of view, including using "alien" social science methodology. Consequently, there was a clear power hierarchy, and I was at the bottom of it!

I propose that this issue is best addressed by a split approach to the interview, whereby I played up my credentials to organize the interview and establish trust with the informant (Welch et al. 2002; Mikecz 2012). However, I later suggested to the interviewee to enact a "role-play" of a lay interviewer with the peer interviewee (Platt 1981), ensuring a degree of serendipity and spontaneity afforded by a wholesome "alien" data collection. This also enabled me to give myself an alibi for exposing my true ignorance of some of the particulars while at the same time subtly demonstrating mastery of the conversation and key topics. Exhibiting the latter quality is particularly important as the interviewee will likely be assessing the "intellectual legitimacy" of the interviewer (Platt 1981) and, to an extent, the researcher's collaborating organization, which could present reputational risk. Background information about the research (type, format, institutional connections, topic, etc.) has to be provided to the interviewees (to establish an intellectual connection) prior to the interview (Platt 1981), but not the (theoretical) assumptions, predicted outcomes or any provisional results, which will ensure that the respondents do not attempt to "assist" the interviewer in providing the "right" answer (Haidt 2012). This can be addressed by the interview format, for instance, using more open-ended questions, hence preventing the codification of any intentions (Harvey 2014).

Before entering the field, I considered that this positioning was best summed up as the "informed outsider" technique (Welch et al. 2002), whereby the researcher positions themselves as being a well-informed yet non-expert observer of the studied subject (or group). This technique is widely used in journalism; however, it is also an (increasingly) prominent feature in parts of social science research, particularly in STS (Stephens 2007). However, the more I reflectively examined the "informed outsider" position, the more I became convinced that, in truth, I was more of an "uninformed insider" than the other way around.

Though the design and the actual experience of the interviews were close to Mikecz's (2012) own experience, which he describes as:

> my positionality has been somewhere in the middle on the "insider–outsider" continuum. Perhaps the term "concerned foreign friend" would describe my position in this research best. I am not perceived as a local but I am not viewed as an outsider either. (Mikecz 2012: 490)

My overall experience of positionality would be best described as an intern or a new member of staff who is accepted and trusted as a (new) insider in the professional community. This is backed by my having some relevant background knowledge, though I was assumed largely ignorant of the nuances and details of the goings-on. This experience is also very consistent with my interview strategy of "role-playing" lay uninformedness to ensure as complete and unbiased (as possible) data collection after establishing access and trust based on institutional and personal credentials by (insider) association.

This position, however, comes with specific ethical considerations and risks, which have so far not been consistently documented or well understood. In particular, the key considerations are to do with, on the one hand, reputational risks to the researcher and their academic impartiality through bias and conflict of interests, and on the other hand, reputational and other risks to host and partner organizations by dealing with sensitive information and association with the research project due to the embedded insider context. I turn to strategies to address these issues later, after first exploring abductive critical realism ontology and highlighting additional ethical challenges of potential bias.

Responsive: abductive epistemology with critical realism

Positioning oneself as a critical realist doing abductive research (Blaikie 2004; Ong 2012) implies the following research design. Starting from researchers' immersion in the field and interaction with a variety of actors in their environment, an existing "problem" (of understanding) is identified. Subsequently, a theoretical framework is proposed to address such a gap, supported by collecting and interrogating empirical data. This is distinct from both (traditional) deductive approach, which is characterized by defining the theoretical framework upfront and proving it by empirical research, as well as inductive (grounded theory) approach of focusing entirely on (empirical) theory development from collected data (Thornberg and Charmaz 2012). The issues found within these two distinct approaches are two-fold. On the one hand, deductive modes of research are often out of sync with the needs and realities of the studied field and produce less applicable and impactful findings. On the other hand, the inductive approaches often ascribe too strong a value to specific situated and contested data, leading to either narrowly localist findings or largely descriptive accounts.

Abductive reasoning has been recently explicitly shown to be compatible with the critical realist position as the optimal mode of research (Schurz 2018), concluding that

> "[...] we can argue that by performing abductive inferences we always take the advantage of explaining and representing our system of experiences by the best available theoretical model, i.e., by the most simple and most unified theory. [...] Should some part of our theoretical model be false, one of two cases may obtain. Either we observe this in the form of an incorrect prediction; as soon as this happens we will take steps to correct our theory. In other words, abductive inferences are self-corrective and have an inbuilt learning ability. Or we never observe it (because our experiences are limited); then nothing happens and we continue to operate with an instrumentally optimal theory, although it is false, but in a way that cannot be empirically detected by us and, thus, will not practically harm us. Thus by performing abductive inferences to unifying theoretical models, we can only gain but not lose something." (Schurz 2018: 3894–3895)

Hence, adopting the abductivist conceptual framing, one is concurrently collecting and analyzing data and reviewing the literature, with in-situ validation of the proposed theoretical framing (Ong 2012). This leads to generalizable, applicable, and meaningful answers to real pains relevant to the studied field's stakeholders. Many STS scholars have argued for similar epistemologies; for instance, Pickering describes what he terms as "pragmatic realism" (Pickering 1995:183) as "machinic performances and representational chains, and how they are aligned with one another in time." In Pickering's analysis, such an approach is the best option for exploring "the mangle" of "resistance" and "accommodation" within the practice of scientific research and technological development, with the aim to expose the "reconfiguration and extension of scientific culture" (Pickering 1995:21).

This was also a very intuitive mode of research for me, as I previously studied in natural sciences (Physics and Astronomy), and I transitioned into STS primarily to work on deepening the understanding of social, economic, and political factors affecting the processes of innovation, knowledge and technology exchange and public engagement. These topics of interest sprung from identifying past and current issues and challenges with the effectiveness of these processes and their valorization, both within science as well as in wider society. Such motives and questions already set me on a path towards abductive research, with further research design considerations only reinforcing its suitability, in particular with respect to longitudinal study and access to the field.

Furthermore, in contrast to "fully-inductive" approaches, such as grounded theory (Thornberg and Charmaz 2012), abductive epistemology does not presume that there is an unbiased "naive" researcher, which makes it very suitable for theorizing participatory action research with obvious vested interests. However, it is also not fully normative. A deductive approach would, in a similar programmatic research context, likely work to reinforce the researcher's bias by a priori limiting the scope of the study, reducing the richness of the collected data, and narrowing the theoretical toolbox available. By advocating an immersive bottom-up analysis of the field to identify the research questions and then moving to a discursive theorization and data collection, abductive epistemology is also much more flexible and iterative.

It is important to note, however, that such research philosophy leads to very close co-dependence and symbiosis between the researcher and the studied community. This way of conducting social science is often referred to as Participatory Action Research since it is based on extensive and active participation within the studied community (McIntyre 2007). Such study design often contains a prolonged relationship with key partners, building trust and great depth of insight and (in line with abductivism) adopting a detailed understanding of the present issues/challenges. Though Sociology (and in particular Anthropology) scholars often consider immersive, active, and participatory research a "mainstream" research position, STS's relationship with this type of research is uncomfortable due to our core experience with science being socially and politically biased, identified, for example, in the seminal work of Gieryn (1983).

The key problem for the researcher in this situation is the (degree of) compromising of objectivity in the face of going "native" in the field (i.e., adopting an inherent bias shared by the researched community). I propose to tackle this possibility in three ways:

1. *Appeal to professionalism within the community.* The researched community is professional; hence, it is imperative for their professional integrity to be objective in their interaction with the researcher.
2. *My Reflexivity.* Whilst some forms of bias and even "boundary work" (Gieryn 1983) might be present in this community, I, as a researcher, have to recognize it and address it openly in my writing.
3. *Embracing the bias.* Any subconsciously remaining bias (Platt 1981) should be, in my opinion, embraced. As this is an instance of active participatory research, I do not want to hide the fact that my research

has an agenda. However, this agenda is not incompatible with my ethical and scientific values, as though the research is there to describe the world in order to influence it, the potential political decisions about the "influencing" are not taken as part of the research itself.

I believe that such an approach, to immerse oneself in the field, engage with stakeholders for a prolonged period, and abductively develop an understanding of their practices - by actively participating in dealing with their challenges - combines the best of scientific rigor and pragmatism with producing applicable, relevant and meaningful insight from a social scientific point of view. Reflexive and analytical accounts of the experiences in the field based on these premises will be analyzed in the discussion section as I now address ethical considerations of the PERIpatetic Approach.

Introspection: ethics of being on the inside[5]

The position as a participant researcher and the elite nature of the studied community bring about ethical considerations due to the potentially problematic positioning of the researcher as a participant in a (politicized) business development operations and collaboration with a key partner(s) in the field and as a peer/insider researcher. This carries significant reputational risks due to the following:

1. possible perceived conflicts of interest, which could be damaging to the researcher's academic integrity through bias and conflict of interest, and
2. potential damage to the collaborating organization by the researcher's conduct to third parties during the affiliation.

My (partial) "embeddedness" in a key player in the sector meant access to (high-profile) actors was relatively easy; however, engaging them and maintaining their trust was an even more precarious task, as any potential misstep on my part may damage the reputation of my collaborating partners. This includes maintaining good personal relationships in the face of asking challenging questions and scrutinizing individuals' expertise and work practices, which can prove difficult (Ostrander 1993). In my opinion, these considerations can be resolved with a constant careful situating of me, as the researcher, and my project with respect to the collaborating partner and (home) academic institution. The uninformed insider approach allows for this by enabling the role-play of gaining access to the field via my research

[5] This approach was guided by the University of Edinburgh Research Ethics framework, including the Policy on Conflict of Interest (The University of Edinburgh, 2002) and the School of Social and Political Science ethics self-audit procedures. The research project obtained Level 1 Ethical Review Approval from the University of Edinburgh.

partners/gatekeepers (becoming an insider) but conducting it as an (uninformed) academic. Importantly, this approach requires careful interaction with the other actors and constant revisions and improvements to the researcher's strategic positioning.

However, my collaboration with ATC Innovations could be perceived as potentially creating a conflict of interest and introducing bias in my work. For instance, given my close association with the collaborating partner, it might not be unreasonable to assume a conflict of interest on my part. However, closer examination renders such concern unfounded. The key reasons for the dismissal of these concerns are, foremost, the nature of the collaboration is significantly different from a commercial investment into a research program, as this project is part of a mutually beneficial research partnership without any stipulated outcomes nor any restrictions outside general confidentiality of sensitive data and respective policies regarding ethics and intellectual property.

Furthermore, the research partner and gatekeeper I was embedded with, ATC Innovations, is part of a public body, UK ATC, a subsidiary of the Science and Technology Facilities Council (STFC), themselves part of the UK Research and Innovation (UKRI). This means they are bound by the same high ethical standards and accountability as myself through my funding from the Economic and Social Research Council (ESRC) – another UKRI council. As public bodies are mandated to pursue benefits to society without any profit arising to any individual (save from intellectual property), no relative advantage should be gained by any party in this process, eliminating the possibility of an interest-based bias.

In introducing bias, a further challenge in all interviewing is to contextually recognize how competing interests influence the informants' points of view, both during the interview and in data analysis. This is to ensure, with as much rigor as possible, soundness and validity of the obtained information. As such, an "insider" interviewer has to be even more considerate of the smallest nuances in the tone of voice, hesitation, evasion of the question, using a specific turn of phrase, sarcasm, or statements with double meaning when answers are given (Knapp, Hart and Dennis 1974).

However, such issues are not unique to "insider" positionality; it is a feature in all social-scientific research and is always a complex and challenging balancing act of examining, extracting, and assessing obtained data. As outlined earlier and explored further in the discussion, using the positionality of "uninformedness" was a great asset here. By engaging in a role-play, I tried to abstract my "insider" status from the data collection exercise and hence encourage as much as possible the expression of an un-altered point of view by the participants. I have then used iterative abductive epistemology approaches to analyze the data through narrative examination, in particular by contextualized trend analysis. This, however, does not remove all bias, in particular as related to the overarching framing of the inquiry, due to embeddedness in the field. Hence, proponents of the participatory action research approach sometimes suggest the acceptance and acknowledgment of such partiality as a means to effect change (Lundy and McGovern 2006).

In parallel to addressing the concerns regarding academic (im)partiality, due to the nature of the data collection environment (i.e., a very competitive business sector), special care was taken to ensure this project protects commercially sensitive data collected from informants as well as from collaborating partners. I considered this my ethical obligation towards the interviewees, as my research should create any undue advantage or disadvantage to the informant. Though no questions were asked about any specific intellectual property or product feature, some information about particular products and their development and the companies' business partners was collected. This issue was discussed at length with the informants prior to the interview to ensure the informants were comfortable with the steps taken to protect the sensitive information. The two key steps were:

1. After the interview, data was carefully extracted from the data matrix with the removal of any identifiable features of products, companies, or people (which could trace the information supplied back to the company or person who supplied it).
2. Additionally, all other parties mentioned in the conversation were also anonymized, with the exception of some key players due to their unique position. (There is, after all, only one national (UK) and one European Space Agency.)

Even for a very immersive research mode, like the one outlined in this paper, I believe informed participant consent is vital. Per the standard University of Edinburgh procedure, an information and consent form has been issued to all participants, read and signed by them, counter-signed by the researcher, and deposited in my archive. The form also offered an option of choice of the level of anonymity - general anonymity of participants was ensured as standard; however, in case a quote of significant importance would arise from the interview, the researcher reserved the right to use it when presenting research findings. Participants were offered to opt out of this arrangement (by ticking a box) and be granted complete anonymity. Very few informants selected this option, indicating they were happy to "be on the record" with their answers. This further confirmed my belief about their participation being professional in its nature, agreeing with my previous assumptions about their answers being full and honest and all relevant substantive data being provided verbally.

Discussion: "Not all who wander are lost!"

In the above auto-reflexive and analytical account of my initial conceptualization and in-the-field experience of abductivist participatory action research of professional elites with the Biographies of Artefacts and Practices-inspired participatory strategic ethnography, I have been guided by and/or have noticed three key elements which underpin my application of the PERIpatetic Approach as an uninformed insider. These three key elements are:

1. The crucial role of building trust and a meaningful relationship with the studied community is to gain access to data and a level of a comprehensive and holistic understanding of the environment to enable abductivist epistemology - i.e., the need to be an "insider."
2. In order to deploy participatory action research practices in a neutral and inclusively helpful way, one needs to be as intellectually flexible, open-minded, and unbiased as possible; hence the need to be "uninformed."
3. Being an "uninformed insider" is a dynamic relationship with the field, requiring role-play, careful interaction with stakeholders, and constant re-evaluation and repositioning in order to maintain the inside-outside and informed-uninformed balance.

Starting from the bottom of the list, it is self-evident that a researcher's positionality is tied in with very dynamic positions and processes, and only constant reflexive (re)evaluation and (re)configuration can enable a researcher to keep within the boundaries they set and reap their benefits. As noted by Mikecz:

> positionality is not solely determined externally in the context of an insider/outsider dichotomy but is on a continuum that can be proactively influenced by the researcher. (Mikecz 2012: 482)

Of particular importance is the everyday work that a researcher needs to do to maintain this balance. It is in this context that the aforementioned ethical and practical challenges of the uninformed insider approach come to the fore. Rhetoric and communication is the sole most important factor here, as well as setting up a clear framework of rules and behaviors, which is being constantly updated. Though the outsider-insider and informed-uninformed might at first glance look like dichotomous relationships, they are, of course, very spectral in nature. This makes the role-play and interaction more broadly very fluid, as in certain cases, the uninformedness and insiderness need to be played up (or down) more than in others. The experience with the project outlined here was such that rhetorically crossing these two divisions was not too difficult or uncommon, but the overall balance of my positioning in interactions was firmly on the side of being an insider and uninformed.

Moving to trust-building and access, I found that being embedded in the field meant contacts with professional elite informants were not difficult to establish, but they had to be approached carefully. In particular, this relates to a personal introduction to (new) contacts and specifically noting affiliation status (with the University, research partners, project funders, etc.) and research type (i.e., academic). In my view, it is important to be honest and not omit or deny the entirety of the context/situation but to frame its presentation to stress those credentials, associations, and features of the research that are assumed to be most trustworthy to a particular informant. For example, in studying a business community, early and proactive highlighting of care and concern for protecting sensitive information is critical. This often spills over in the explanation of required information into a justification of research as a whole and, more specifically, the themes and questions in the interview. Like the personal introduction, I argue that an honest but carefully framed approach works best and is the only ethically acceptable and practically available solution. In my case, this meant stressing the applied and developmental nature of my project while being very clear; there are also theoretical (academic) interests involved.

I found that it was a deeper relationship and investment I had in the community that enabled me to develop a holistic understanding of the field needed for abductivist knowledge-making process. It led to the identification of real-world issues and pains and made possible continuous iterative dialogue with a multitude of actors (through formal data gathering and informal discussions) to develop theoretical understanding and applied solutions. Upon gaining access, I usually encountered no issues or challenges of being able to develop and maintain a longitudinal relationship with the studied community, though that does require work in itself (presence at industry events, supporting various stakeholders' projects, etc.). In addition, as the continuous physical presence (a desk at the research partners office, attendance and participation in events, visiting firms I interviewed instead of telephone or electronic communication) and virtual/social engagement (writing and publishing information, social media interactions, being "known") made me a true insider, so, when data collection was taking place, it became even more vital to assume, enact and internalize uninformedness.

As discussed previously, the uninformedness as part of this approach is conceptualized as both an acknowledgment of the true state of the knowledge asymmetry, as well as an epistemological necessity to remove as much bias and pre-conceptions as possible. The latter is particularly important for two reasons: on the one hand, to establish academic impartiality and associated integrity (an important condition of participatory action research), and on the other hand, to enable a holistic perspective on the field in order to successfully deploy abductivist research design. The considerations related to the role of uninformedness within the established epistemic regime are particularly interesting, as this is a deliberate act of the researcher's positioning.

The activities behind the exposition and enactment of the uninformedness of the insider researcher are structured in two ways - the upfront positioning through a personal introduction and the design and deployment of interview/data collection questions. In the positioning part, these are simple substantive moves of identification as a non-expert or acknowledgment of knowledge asymmetry in some other way. In the data collection part, however, the action of the researcher is more complex within the suggested framework of "role-play," which includes co-option and cooperation of the informant/interviewee. In the case of my project, the co-option and cooperation were obtained by an up-front explanation of the epistemic requirement for such procedure and behavior, which was well-received by all informants/interviewees. Similarly, my uninformedness was overall well-received in all informal interactions with the field and was harmonized well with my developed insider status.

Conclusion

Overall, I argue that the PERIpatetic Approach, based on participatory strategic ethnography of innovation and the "uninformed insider" position, provides an epistemologically optimal context for a social science researcher. As noted across this chapter, this type of positionality for close-quarters research within professional elites have already de facto been established in many fields of social science, but here I argue that it is equally applicable to STS research on innovation (and business communities) of practice. This was illustrated with an example of researching the Space Sector in Scotland, outlining facets of research design, experience in the field, and reflexive analysis.

In addition, I argue that the adoption of such an approach leads to a greater scientific quality of our empirical work, through strengthening our reflexivity and allowing greater attention to nuances and details. This has been recognized in many settings and has been labeled by many researchers in different ways; it needs more coherent and firmer systematic analysis in order to establish its scientific validity, hence the acute need for the type of analysis such as the one presented in this paper.

Hence, I would like to conclude with a few observations (and possible limitations) and put forward some suggestions for the future development of similar reflexive accounts and analyses. First and foremost, I have to highlight my appreciation for the degree of acceptance and support of my research ideas and, eventually, data collection by my research partners and gatekeepers, whose extraordinary readiness to assist me and engage with my work may not be universally applicable.

Secondly, I have to note the crucial and outstanding intellectual openness and academic support within my department, which accepted and enabled this somewhat exotic mode of research. I believe it is possible that, in this respect, I have been one of the lucky ones.

However, it is important to note the limitations of the past and present understanding of the PERIpatetic Approach and how it relates to other contemporary observations. First, its deployment so far has been limited to a single comprehensive study, and though its component elements are well articulated and documented elsewhere, further deployment and analysis are sorely needed. In addition, due to the longer time-scaled impact of innovation (eco)system development, the particular impact and success of the participatory and abductive research are yet to be fully captured. Though anecdotal evidence of the critical contribution my PERIpapetic project made to the consolidation of the Scottish Space Industry (for instance, through the use of strategic documents, intermediary's program development and planning, and the reach of awareness-raising), the longer-term effects of a PERIpatetic researcher's input into development policy has not been looked at. These are specifically interesting questions, as seen by the work on the "challenge function" of social scientists providing tacit and gradual knowledge transfer in the policy-making process (Kattirtzi 2016) to subordinate (even instrumental) use of social scientist within emerging science fields (Szerszynski and Galarraga 2013). It is hoped that novel paradigms of more participatory and collaborative integration of knowledge-making, such as Responsible Research and Innovation (RRI), will:

> allow for new and more fruitful science–society relations [..] and for the sake of collectively imagining better sociotechnical futures and directing action towards achieving them. [...] RRI requires an integration of the natural sciences and SSH [Social Science and Humanities] disciplines and that the nature of that integration must be researched as an empirical question. (Delgado and Åm 2018: 8)

Hence, in terms of further reflexive engagement with this theme, I would like to encourage other descriptive and analytical accounts of research philosophy and its relationship to positionality, which are currently, by and large, still a rarity in the STS literature. I hope that my honest and as complete as possible documentation of my experience in this respect attracts constructive feedback and (re)opens the discussion regarding the often falsely dichotomized and ignored issues of our presence in the arenas we study. I think of particular interest would be accounts of contention and rejection of the researcher's position, which have been seen as "negative outcomes" and, hence, shared a similar fate to inconclusive or negative scientific findings. However, it is precisely these instances of inability to form these research relationships or their breakdowns that might start to illuminate which of the many elements involved in constructing one's research approach and position with respect to the studied environment are critical for successful trust-building and acceptance and which are (merely) individual preferences.

Acknowledgments

This chapter is based on a part of my doctoral thesis, *Scottish Space Sector and Innovation: A PERIpatetic study of an emerging innovation system and the roles of innovation intermediaries.* This project was funded through a (1+3) Pathway Studentship, bestowed by the Scottish Graduate School of Social Science (SGSSS) [grant number: ES/J500136/1] on behalf of the Economic and Social Research Council (ESRC) and was conducted in collaboration with ATC Innovation team at the UK Astronomy Technology Centre (UK ATC), part of Science and Technology Facilities Council (STFC), UK Research and Innovation (UKRI).

This work greatly benefited from constructive feedback and suggestions from many colleagues, in particular, my doctoral mentors: Dr. Alessandro Rosiello, Dr. Niki Vermeulen, Prof Robin Williams, and Dr Julian Dines. Last but not least, I am very grateful to all participants in the study for their valuable time and expertise and to many others who have assisted me with my inquiries, as well as the editors Dr. Emine Öncüler Yayalar and Dr. Melike Şahinol, the Vernon Press editorial team; and the thorough and helpful anonymous reviewers.

References

Adlen, Sam. 2011. *Innovation in the Global Space Industry.* London: Imperial College London.

Berger, Roni. 2015. "Now I see it, now I don't: researcher's position and reflexivity in qualitative research," *Qualitative Research,* 15(2):219–234. Available at: https://doi.org/10.1177/1468794112468475.

Bhuiyan, Nadia. 2011. "A Framework for successful new product development," *Journal of Industrial Engineering and Management,* 4(4):746–770. Available at: https://doi.org/10.3926/jiem.334.

Blaikie, Norman. 2004. "Abduction" Pp. 1-2 in *The SAGE Encyclopedia of Social Science Research Methods* edited by M.S. Lewis-Beck, A. Bryman, and T.F. and Liao. London: Sage.

Bloor, David. 1991 *Knowledge and Social Imagery*. 2nd ed. Chicago: University of Chicago Press.

Bryman, Alan. 2016 *Social Research Methods*. 5th ed. Oxford: Oxford University Press.

CEN 2014 *SIST-TS CEN/TS 16555-1:2014 - Innovation Management - Part 1: Innovation Management System*. Available at: https://standards.iteh.ai/catalog/standards/sist/630f6191-8f1e-4597-ad54-8e16e0e90834/sist-ts-cen-ts-16555-1-2014 (Accessed: July 20, 2022).

Darrouzet, Christopher, Helga Wild and Susann Wilkinson. 2009. "Participatory ethnography at work." Pp. 61-94 in *Ethnography and the Corporate Encounter: Reflections on Research in and of Corporations*, edited by M. Cefkin. New York: Berghahn Books.

Delamont, S. 2013. "SAGE Qualitative Research Methods Collecting Data from Elites and Ultra Elites: Telephone and Face-to-Face Interviews with Macroeconomists," *Qualitative Research*, 7(2), pp. 203–216.

Delgado, Ana, and Heidrun Åm. 2018. "Experiments in Interdisciplinarity: Responsible Research and Innovation and the Public Good." *PLoS Biology* 16(3):e2003921. doi: 10.1371/journal.pbio.2003921.

Edwards, Tim, Rick Delbridge, and Max Munday. 2005. "Understanding Innovation in Small and Medium-Sized Enterprises: A Process Manifest." *Technovation* 25(10):1119–27. doi: 10.1016/j.technovation.2004.04.005.

Gieryn, Thomas F. 1983. "Boundary-Work and the Demarcation of Science from Non-Science: Strains and Interests in Professional Ideologies of Scientists." *American Sociological Review* 48(6):781. doi: 10.2307/2095325.

Green, Ken, Richard Hull, Andrew McMeekin, and Vivien Walsh. 1999. "The Construction of the Techno-Economic: Networks vs. Paradigms." *Research Policy* 28(7):777–92. doi: 10.1016/s0048-7333(99)00021-9.

Haidt, Jonathan. 2012. *The Righteous Mind: Why Good People Are Divided by Politics and Religion*. London: Penguin Books.

Harvey, Lee 2014. "Social Research Glossary: Interview". *Quality Research International*.

Hyysalo, Sampsa. 2009. "Review: How Packaged Software Conquers the Organization: Neil Pollock and Robin Williams, Software and Organizations: The Biography of the Packaged Enterprise System or How SAP Conquered the World", *Social Studies of Science* 39(4):635–42. doi: 10.1177/0306312709335847.

Hyysalo, Sampsa, Neil Pollock and Robin Williams. 2016. "Biographies of Artifacts and Practices Approach: Overview and Assessment," in *EASST/4S 2016*. Barcelona.

Hyysalo, Sampsa, Neil Pollock, and Robin Alun Williams. 2019. "Method Matters in the Social Study of Technology: Investigating the Biographies of Artifacts and Practices." *Science & Technology Studies* 32(3):2–25. doi: 10.23987/sts.65532.

Kattirtzi, Michael. 2016. "Providing a 'Challenge Function': Government Social Researchers in the UK's Department of Energy and Climate Change (2010–2015)." *Palgrave Communications* 2(1):1–9. doi: 10.1057/palcomms.2016.64.

Knapp, Mark L., Roderick P. Hart, and Harry S. Dennis. 1974. "An Exploration of Deception as a Communication Construct." *Human Communication Research* 1(1):15–29. doi: 10.1111/j.1468-2958.1974.tb00250.x.

London Economics. 2019. *Size and Health of the UK Space Industry 2018.* London.

Lundy, Patricia, and Mark McGovern. 2006. "Participation,Truth and Partiality: Participatory Action Research, Community-Based Truth-Telling and Post-Conflict Transition in Northern Ireland." *Sociology* 40(1):71–88. doi: 10.1177/0038038506058431.

Markman, Gideon D., Donald S. Siegel, and Mike Wright. 2008. "Research and Technology Commercialization." *The Journal of Management Studies* 45(8):1401–23. doi: 10.1111/j.1467-6486.2008.00803.x.

McIntyre, Alice. 2007. *Participatory Action Research.* Christchurch, New Zealand: Sage Publications.

Mikecz, Robert. 2012. "Interviewing Elites: Addressing Methodological Issues." *Qualitative Inquiry: QI* 18(6):482–93. doi: 10.1177/1077800412442818.

Morrisey, C.T. and Morrissey, C. 1998. "On Oral History Interviewing." Pp. 107-113 in *The oral history reader edited by* L.A. Dexter (ed.). Evanston: Northwestern University Press

Nutley, Sandra, Isabel Walter, and Huw Davies. 2007. *Using Evidence: How Research Can Inform Public Services.* London: Policy Press.

Odendahl, Teresa and Aileen M. Shaw. 2002 "Interviewing Elites." in *Handbook of Interview Research: Context and Method* edited by J.F. Gubrium and J. A. Holstein. Thousand Oaks, CA: Sage.

OECD 2007 *Oslo Manual.* Available at: https://doi.org/10.1787/9789264065659-es.

Ong, Beng Kok. 2012. "Grounded Theory Method (GTM) and the Abductive Research Strategy (ARS): A Critical Analysis of Their Differences." *International Journal of Social Research Methodology* 15(5):417–32. doi: 10.1080/13645579.2011.607003.

Ostrander, Susan A. 1993. "'Surely You're Not in This Just to Be Helpful': Access, Rapport, and Interviews in Three Studies of Elites." *Journal of Contemporary Ethnography* 22(1):7–27. doi: 10.1177/089124193022001002.

Pickering, Andrew. 1995. *The Mangle of Practice: Time, Agency and Science.* Chicago, IL: University of Chicago Press.

Platt, Jennifer. 1981. "On Interviewing One's Peers." *The British Journal of Sociology* 32(1):75-91. doi: 10.2307/589764.

Pollock, Neil, and Robin Williams. 2010. "E-Infrastructures: How Do We Know and Understand Them? Strategic Ethnography and the Biography of Artefacts." *Computer Supported Cooperative Work: CSCW: An International Journal* 19(6):521–56. doi: 10.1007/s10606-010-9129-4.

Richards, David. 1996. "Elite Interviewing: Approaches and Pitfalls." *Politics* 16(3):199–204. doi: 10.1111/j.1467-9256.1996.tb00039.x.

Schurz, Gerhard. 2018. "Optimality Justifications: New Foundations for Foundation-Oriented Epistemology." *Synthese* 195(9):3877–97. doi: 10.1007/s11229-017-1363-6.

Sedlack, Richard G., and Jay Stanley. 1992 *Social Research: Theory and Methods.* Boston: Allyn and Bacon.

Shank, Gary. 2008 "Abduction," in Pp. 1-2 in *The SAGE Encyclopedia of Qualitative Research Methods* edited by L. M. Given. Thousand Oaks CA: SAGE Publications. Available at: https://doi.org/10.4135/9781412963909.n1.

Space IGS. 2011. *The Space Innovation and Growth Strategy: Main Report.*

Stephens, Neil. 2007. "Collecting Data from Elites and Ultra Elites: Telephone and Face-to-Face Interviews with Macroeconomists." *Qualitative Research: QR* 7(2):203–16. doi: 10.1177/1468794107076020.

Sun, Peter Y. T., and Marc H. Anderson. 2008. "An Examination of the Relationship between Absorptive Capacity and Organizational Learning, and a Proposed Integration: Absorptive Capacity and Organizational Learning." *International Journal of Management Reviews* 12(2):130–50. doi: 10.1111/j.1468-2370.2008.00256.x.

Szerszynski, Bronislaw, and Maialen Galarraga. 2013. "Geoengineering Knowledge: Interdisciplinarity and the Shaping of Climate Engineering Research." *Environment & Planning A* 45(12):2817–24. doi: 10.1068/a45647.

The University of Edinburgh. 2002. *Policy on Conflict of Interest.* Edinburgh.

Thornberg, Robert and Kathy Charmaz. 2012 "Grounded theory" Pp. 41-67 in *Qualitative Research: An Introduction to Methods and Designs* edited by S. D. Lapan, M. T. Quartaroli, and F. J. Riemer. San Francisco, CA: John Wiley/Jossey–Bass

Vidmar, Matjaz. 2019a. "Agile Space Living Lab – the Emergence of a New High-Tech Innovation Paradigm." *Space Policy* 49(101324):101324. doi: 10.1016/j.spacepol.2019.05.002.

Vidmar, Matjaz. 2019b. *Scottish Space Sector and Innovation A PERIpatetic study of an emerging innovation system and the roles of innovation intermediaries.* The University of Edinburgh. Available at: https://era.ed.ac.uk/handle/1842/37018

Vidmar, Matjaz. 2019c. *The "Living Lab" Method: Discovery > Challenge > Innovation.* Edinburgh. Available at: https://doi.org/10.13140/RG.2.2.21744.20484.

Vidmar, Matjaz. 2019d. "The Ten Million Euro Question: How Do Innovation Intermediaries Support Smart Specialization?" *Croatian Economic Survey* 21(2):37–84. doi: 10.15179/ces.21.2.2.

Vidmar, Matjaz. 2020a. "Innovation Intermediaries and the 'Final Frontier' of the New Space Sector in Scotland." Pp. 39-41 in *Innovation Intermediaries and (Final) Frontiers of High-tech.* Springer International Publishing. https://doi.org/10.1007/978-3-030-60642-8_3.

Vidmar, Matjaz. 2020b. "New Space and Innovation Policy: Scotland's Emerging 'Space Glen.'" *New Space* 8(1):31–51. doi: 10.1089/space.2019.0032.

Vidmar, Matjaz, Alessandro Rosiello, Niki Vermeulen, Robin Williams, and Julian Dines. 2020. "New Space and Agile Innovation: Understanding Transition to Open Innovation by Examining Innovation Networks and Moments." *Acta Astronautica* 167:122–34. doi: 10.1016/j.actaastro.2019.09.029.

Welch, Catherine, Rebecca Marschan-Piekkari, Heli Penttinen, and Marja Tahvanainen. 2002. "Corporate Elites as Informants in Qualitative International

Business Research." *International Business Review* 11(5):611–28. doi: 10.1016/s 0969-5931(02)00039-2.

Willets, David. 2013. *Eight Great Technologies.* London, England: Policy Exchange.

Williams, Robin, and Neil Pollock. 2012. "Research Commentary: Moving beyond the Single Site Implementation Study: How (and Why) We Should Study the Biography of Packaged Enterprise Solutions." *Information Systems Research: ISR* 23(1):1–22. https://doi.org/10.1287/isre.1110.0352

Williams, Robin, and Diana Velasco. 2016 "How Did We Grow Apart" paper presented in *SPRU 50th Anniversary Conference.* Brighton: University of Sussex.

Yin, Robert K. 2009. *Case study research, Applied Social Research Methods Series.* London: Sage Publishing.

Zeitz, Gerald, Vikas Mittal, and Brian McAulay. 1999. "Distinguishing Adoption and Entrenchment of Management Practices: A Framework for Analysis." *Organization Studies* 20(5):741–76. doi: 10.1177/0170840699205003.

PART 4.
Case studies

Chapter 10

Post-fake artivism: how activism and art can break reflexivity

Pedro Alves da Veiga

Abstract: This text explores the historical evolution and intertwinement of propaganda, black-boxing, and attention control over the last century and their combined roles in influencing the masses, more specifically through the spread of fake information in various media formats. In this scenario, artivism (art + activism), in its many forms, hacks the black boxes and exposes their inner workings, algorithms, and strategies. Thus, the author highlights key aspects of propaganda, as posited during the Second World War, and their subsequent spread into advertising, becoming an intrinsic part of global politics and businesses. The concept of black-boxing, as introduced by Latour to designate an opaque process that takes input A and transforms it into output B, while hiding its inner workings or hidden purposes, is then extended and applied to all modern content and media production – including propaganda and advertising – that occludes its sources and information transformation from the public. The reflexive use of these extended back-boxes, together with technological determinism, is fostering what is presented as a widespread phenomenon, turning fake into real, largely supported by a culture rooted in propaganda, thriving in attention-capturing mechanisms and in the conscious use of logical fallacies to maximize public impact, epitomized by the phrase "if faking it gets the job done, who cares?" Form has become more important than function in the pursuit of goals. The appeal to emotion – overriding reason and fact – is privileged in public communication. Recent advances in generative AI black-boxed systems have densified the scenario, making it nearly impossible to distinguish between highly polished, often hyper-realistic, machine-generated deep-fakes and actual, human-generated media. In a post-fake reality punctuated by a barrage of cognitively overlaid buzzwords – from the simpler like friend, or tag to the more complex Internet of behaviors, artificial intelligence, extended reality, or enhanced connectivity – created and controlled by dominant cultures to inculcate habits and norms, and to consolidate power, the dismantlement of this unprecedented curtain of clichés becomes urgent, and post-fake artivism (art+activism) may be a step in the right direction.

Keywords: activism, art, attention, black-boxing, media

***

Introduction

When discussing digital communication tools, images of computers and mobile devices, applications, text, and image editors are among the first to come to mind. Many of these tools behave like black boxes, about which we know nothing except their expected behavior. However, nowadays, most of these sophisticated tools are as complex from a technological point of view as they are from social perspectives, and this complexity is proportional to the number of new connections and relations they are designed to hold, explore, and exploit. For Latour (1988), society and technics are two sides of the same Machiavellian ingenuity, and the black box is a machine that occupies a position defined by trade-offs and negotiations – some of them rendered invisible or *implicit* by the black box's success. The contemporary plane of discourse within post-digital cultures may thus be characterized by the observation that singular media entities have dissolved into a techno-social ecosystem inhabited by smart things that intend to engage and cooperate with human users. This socio-technological environment is translated into a hyper-nervous system of existences, which addresses, affects, and appropriates the human on a preconscious level (Hansen 2011). Most people with smartphone access have already become familiar with the idea of providing input A to black-boxed mobile applications, knowing that they will obtain output B in return. But they know very little – close to nothing – about the algorithms and inner workings of the black box, what data is being retrieved, and for what purpose: "when given the option to sign their agreement without reading the policy, most participants skip the policy altogether" (Steinfeld 2016:992). For Latour, spotlighting these inner workings and contractual small print is "not trying to innocent the Prince, but simply to give the analyst at least as much intelligence and deviousness as the Prince has" (1988:29).

These tools also foster *cognitive overlaying*, which consists of the act of assigning new meanings to existing concepts, words, and expressions, imbuing them with biased perception. The process is straightforward: it starts with a generally positive or attractive concept, and then the original meaning is adapted to designate something substantially different but intended to be well (or badly) accepted, superimposing the new meaning over the former. Words like *cookie, friend, like, and tag*, among many others, had their original meaning superimposed and redefined for the digital world. A cookie, a small file created on the users' devices when navigating through webpages, could be deemed intrusive, but calling it a *cookie* makes it sound harmless and almost desirable.

This phenomenon of cognitive overlaying has extensively been used, for example, in the American political oratory of former President Trump, where the expression of *the Chinese virus* was used to designate SARS-CoV-2 (Viala-Gaudefroy and Lindaman 2020), aiming to attribute by proximity (or overlay) a negative connotation of threat to other subsequent uses of the term *Chinese*. The previous examples appear linear in construction and meaning, but what happens when the redefinition takes place over two concepts that had previously been mutually exclusive, such as *real* and *fake*?

The exhibition "Fake," held at the Science Gallery in Dublin, explored the concept of authenticity, questioning its value through a variety of points of view:

> From fake meat to fake emotions, if faking it gets the job done, who cares? From biomimicry to forged documents, from scandals to substitutes — when is authenticity essential, when is copying cool, and what is the boundary between a fakery faux-pas and a really fantastic FAKE? We have long been obsessed with authenticity, the genuine article, the real deal. But, in both the natural world and human society, faking, mimicking, and copying can be highly reliable strategies for success. When appearance is everything — could a fake be just as valuable as the real thing? (Catts et al. 2018)

The purpose of the next sections is to cover the historical evolution and intertwinement of propaganda, advertising, black-boxing, and attention control over the last century and their combined roles in influencing the masses, more specifically through the spread of *fake* information in various media formats. In this scenario, artivism (art + activism), ranging from performative satire to technological hijinks, hacks the established black boxes and exposes their algorithms, thus promoting a better understanding of their societal influences and fostering critical thinking.

From Propaganda to Black Boxes

Propaganda

The Second World War propaganda and rhetoric can be placed at the root of most modern-day phenomena of mass-media-based information and counter-information, used by governments, advertising agencies, activists, and artivists alike to influence the general public.

The Second World War undoubtedly marked the assumed use of mass media by the involved countries as a tool for shaping public perception and opinion. Joseph Goebbels (1934), the main enforcer responsible for the rhetoric and propaganda of the Nazi regime, presented the foundations of *propaganda*, a term previously instituted by the Catholic Church (Huskinson 1982), to designate the dissemination of the faith among non-Christians, in other words, to spread a message and pass it as true among an unreceptive, possibly hostile audience. Some of these principles include the statement that political propaganda is aimed at the broad masses. Therefore, it should speak the language of the people. Propaganda is also defined as a means to an end, being neither good nor evil, its moral value determined only by the goals it seeks.

Propaganda's task is a creative high-art, which transforms complicated events and facts – just like a black box – through creativity and "productive fantasy" (sic) into simpler messages. In the same text, Goebbels clearly states that "since it is almost invisible, it is effective and powerful," thus reinforcing the black-boxed nature of propaganda, whose ultimate goal is *influence*: "The effective propagandist must be a master of the art of speech, of writing, of journalism, of the poster, and of the leaflet. He must have the gift to use the major methods of influencing public opinion such as the press, film, and radio to serve his ideas and goals" (Goebbels 1934). Then he proceeds to tackle the concept of *truth*: "Good propaganda (...) has no reason to fear the truth." However, "a good cause will lose to a bad one if it depends only on its rightness, while the other side uses the methods of influencing the masses" (Goebbels 1934).

German propaganda was considered by the Allies to be a danger to public opinion, which is why Goebbels was presented as a master of lies and deceit. But at the level of political and military summits, German reports of the battlefield were considered to be synthetic and accurate (NSA 1950).

An example of Western media being used as a black-box, blurring the distinction between true and false, was the Katyn massacre[1], in which Soviet secret services decimated the elite of Polish officers, burying the corpses in a mass grave near the town of Katyn. The Germans discovered this grave and publicized their finding to instill a distrust of the Soviets among the Allies, whose representatives were invited to examine the mass grave. At a time when the Allies sought to join the Soviet Union to defeat Germany, Soviet responsibility was ultimately concealed from American public opinion by the Roosevelt administration. In Europe, the Allies decided to attribute the crime to the German Army. The American media echoed this point of view, and The

---

[1] More information on the Katyn Massacre can be found, for example, at the Encyclopaedia Britannica website: https://www.britannica.com/event/Katyn-Massacre

New York Times, among others, blamed the Germans for the massacre (NYT 1945). It was not until 1990, with the opening of the archives of the Soviet secret police, that it was proven beyond any doubt that it was indeed an act of Soviet responsibility (NYT 1990).

This was the dawn of the manipulation of public opinion through mass media as *a service to a greater cause*, which was not necessarily the truth but rather what propagandists deemed necessary.

Advertising

According to Herman and Chomsky's propaganda model (1988), the 20th century is characterized by three developments of great political importance:

1. the growth of democracy,
2. the growth of corporate power and
3. the growth of corporate advertising as a means of protecting corporate power against democracy.

This type of use of the media as propaganda (at the service of politics and power) paves the way for its use with any intent, be it consumerist, ideological, cultural, artistic, or artivist.

The fine line between (politics driven) propaganda and (consumerism driven) advertising gets blurred in the following years, as do the lines between power, politics, and business. More communication campaigns are created with a clear and firm goal: to influence the people through any means and methods deemed necessary. Williamson's central premise is that "it is part of the deceptive mythology of advertising to believe that an advertisement is simply a transparent vehicle for a 'message' behind it" (Williamson 1978:17). But when the signifier has been conceptually overlaid, does the signified remain intact? As defined by Goebbels, propaganda is neither good nor bad: its value is determined by the objective it sets out to achieve. In other words, the end justifies the means. Or, in more modern terms, *real* and *fake* are beyond *good* or *bad*.

Greenberg, who stressed the opposition between avant-garde and kitsch, personifying everything that was artificially produced through industrialization, noted this *artificialization* of reality in 1939. For him, the media are a direct product of capitalism. These industries are driven by profit alone, and the mechanical culture they spread is intended to look like more than it actually is (Greenberg 1939).

During the 1930s and 1940s, the advertising industry adopted many of the visual references of surrealism, but that did not mean that the images used in advertising at that time were actually surreal, as they conveyed style without substance, nor were they objective, since they relied on surrealism to promote their very un-surreal products. On one extreme of the spectrum resides avant-garde creativity; on the opposing end lie market-driven trends, appropriating clichés through superficial stylistic innovation and using them to claim privileged status for those innovative objects of desire when, in fact, they are merely products of the new consumer culture. This marks the rise of artistic movements, marked by the use of their works at the service of activist causes, such as Fluxus, *L'Internationale Situationniste*, and the French student movement of May'68, with its striking poster art. Artists began using or incorporating into their artworks the same media, tools, and principles as propaganda and advertising at the service of their causes to influence the public.

The Media as Black Boxes

Most modern media inherited the strategy of operating behind curtains for both ethically sound and questionable purposes, ranging from undisclosed sources of protection to receiving inputs (facts) and producing biased outputs (news) without allowing for an external glimpse of their internal mechanisms and algorithms, thus preventing an objective external observation of the transformation processes (of facts into the news). This also explains why unchecked claims can quickly enter global circulation as news (Stencel 2015), as the public deems certain outlets more reliable (even if they are not). Trust and distrust influence acceptance, and to establish trust, these operations are sophisticated, admitting logical fallacies[2], but seldom obvious falsehoods. The extent and variety of this kind of fallacy are used not only in populist oratory, as it is often present in multiple areas of society, from politics to advertising.

The lack of transparency in these processes was coined by Latour as *black-boxing*, referring to the process of making scientific and technical work invisible through its own success. "Thus, paradoxically, the more science and technology succeed, the opaquer and obscure they become" (Latour 1999: 304).

---

[2] The list of logical fallacies is extensive – Bennett (2017) proposes more than 300 – and includes dispersal fallacies (e.g.: false dilemmas, *argumentum ad ignorantiam*, multiple interlocking issues), appeal to motivation or emotion rather than reason (e.g.: *argumentum ad baculum*, *argumentum ad misercordiam*, *argumentum ad consequentiam*, *argumentum ad populum*), wrong target (e.g.: *argumentum ad hominem*, *argumentum ad verecundiam*, anonymous authority, style without substance, *petitio principii*, *ignoratio elenchi*), inductive fallacies (e.g.: hasty generalization, limited sample, false analogy, lazy induction, omission of data), causal fallacies (e.g.: *post hoc ergo propter hoc*, conjoint effect, genuine but insignificant cause, complex cause), fallacies of ambiguity (e.g.: misunderstanding, amphibology, emphasis), *non sequitur* (e.g.: assertion of consequent, denial of antecedent, inconsistency, explanation, irrefutable, *ad hoc*, superficiality), among others (Bennett 2017).

Even if Latour's observations were made regarding science and technology, the present text posits their abstract amplification to a broader scope: in short, if one process achieves one known outcome, the act of black-boxing it is defined by keeping its internal mechanisms or algorithms hidden from the public, so that it cannot be disputed and questioned. If, on the one hand, it is easy, elegant, and attractive to use black boxes to hide complex functionalities, purposes, and limitations, on the other hand, they represent a type of *knowledge without knowing*, already widely dissected by Stiegler (2010). Pasquale also posits a definition of the *black box* in his seminal work *The Black Box Society*:

> The term "black box" is a useful metaphor for doing so, given its own dual meaning. It can refer to a recording device, like the data-monitoring systems in planes, trains, and cars. Or it can mean a system whose workings are mysterious; we can observe its inputs and outputs, but we cannot tell how one becomes the other. We face these two meanings daily: tracked ever more closely by firms and government, we have no clear idea of just how far much of this information can travel, how it is used, or its consequences. (Pasquale 2016:3)

The possibility of using a certain type of tool, knowing its input and output, but not knowing (and not questioning) its internal workings, is behind many of the data encoding processes taking place nowadays in mobile devices and applications, and going beyond data and information collection, generation, and dissemination.

Black boxes allow individuals to grasp, consume, capture, generate, transmit, duplicate, reproduce, manipulate, store, and retrieve digitally encoded multi-sensory information. Under cover of technology's apparent neutrality and appeal, black boxes can introduce invisible biases in their operation, and if some of these biases may appear as favorable (such as the skin-softening and eye-highlighting filters included in most current digital cameras, which result in more polished and supposedly appealing, albeit less trustworthy, portraits), others may pose threats to security and privacy (Bergman, Frenkel, and Zhong 2020).

A group of academics at the University of Harrisburg developed automated facial recognition software, allegedly capable of predicting the likelihood of someone committing a crime. It is not fiction, despite immediately triggering associations with the film Minority Report[3]. The announcement of its acceptance

---

[3] More information on the film can be found, for example, at the Internet Movie Database website: https://www.imdb.com/title/tt0181689/

for publication was received with strong peer opposition, deserving an open letter of repudiation signed by several scientists) – including the author of this text – under the name of Coalition for Critical Technology (CCT) and supported by the following arguments:

> This upcoming publication warrants a collective response because it is emblematic of a larger body of computational research that claims to identify or predict "criminality" using biometric and/or criminal legal data. Such claims are based on unsound scientific premises, research, and methods, which numerous studies spanning our respective disciplines have debunked over the years. Nevertheless, these discredited claims continue to resurface, often under the veneer of new and purportedly neutral statistical methods such as machine learning, the primary method of the publication in question. [...] Part of the appeal of machine learning is that it is highly malleable — correlations useful for prediction or detection can be rationalized with any number of plausible causal mechanisms. Yet the way these studies are ultimately represented and interpreted is profoundly shaped by the political economy of data science and their contexts of use. Machine learning programs are not neutral; research agendas and the data sets they work with often inherit dominant cultural beliefs about the world. These research agendas reflect the incentives and perspectives of those in the privileged position of developing machine learning models and the data on which they rely. The uncritical acceptance of default assumptions inevitably leads to discriminatory design in algorithmic systems, reproducing ideas that normalize social hierarchies and legitimize violence against marginalized groups. Such research does not require intentional malice or racial prejudice on the part of the researcher. Rather, it is the expected by-product of any field that evaluates the quality of their research almost exclusively on the basis of "predictive performance." (CCT 2020)

The Power of Black Boxes

The global business system, including media and IT companies, is radical in that it does not respect traditions or habits if they interfere with profit (James 2009). But it is also politically conservative since the media (and IT giants) are significant beneficiaries of the prevailing global social structure. Any change in social or property relations – especially if they interfere with the power of business – is not in their interest (McChesney 1999). One of the ways to alleviate interference with corporate power is to expand power itself, and this is achieved through concentration. In the 1980s, there were about fifty corporations that dominated the media in the United States (Gershon 1993). That number has since been steadily dropping through mergers and other financial concentration operations, and many of these new giants have, in the meantime, started operating on a global scale (Lutz 2012). But their initial dependency on third-party generated content, that is, on the events they reported, meant that in practical terms, they did not have direct power over that same content – and this realization implied a turning point.

A considerable part of what is seen, read, and heard today is generated by the media themselves, from the jet-set press to live entertainment and reality shows that fabricate social media stars, which then catapult their own events to the status of news, opening and closing episodes, red carpet parades, festivals, and awards for film, television, music or cooking, reality television, political, sporting, and cultural debates or opinion-making talk shows. Their employees – especially TV hosts and news anchors – are given the same star status, and all the feedback they generate, such as interviews (with the very same who used to be interviewers), book launches, and music recordings by media personalities, just to name a few, are given the same treatment.

When IT corporations started attaining the same order of magnitude as media corporations, social media networks were regarded as serious competition by traditional media. Fostered by the democratization of access to digital communication black-boxed tools, a growing number of individuals, as well as formal and informal groups based on social networks, produce and share countless media pieces, eagerly seeking the viral effect that would catapult them onto the global stage. These individuals gather their own fanbase on social media networks and continue to exert their influence over a wide range of subjects, often unrelated to their area of expertise. As Allcott and Gentzkow posit, "An individual user with no track record or reputation can in some cases reach as many readers as Fox News, CNN, or the New York Times" (Allcott and Gentzkow 2017: 211). This marked the point when media and technology started to merge, so a more appropriate description today would be *media technologies corporations* (MTC).

Big Data may also offer a complementary angle – from MTCs' point of view – to the success of black boxes, as they significantly contribute to user data acquisition processes, both visible and invisible:

> Big Data not only refers to very large data sets and the tools and procedures used to manipulate and analyze them, but also to a computational turn in thought and research. Just as Ford changed the way we made cars – and then transformed work itself – Big Data has emerged a system of knowledge that is already changing the objects of knowledge, while also having the power to inform how we understand human networks and community. 'Change the instruments, and you will change the entire social theory that goes with them,' Latour (2009:9) reminds us. (Boyd and Crawford 2012: 665)

The use of black boxes causes pronounced impacts on human self-conception, interactions between individuals, the metaphysical concept of reality, and how humans relate to and interact with it. Several studies point to the rise of individuality, conformity, and hedonism (Gallego et al. 2020; Uhls and Greenfield 2011). Young people today are even more obsessed with fame, and research suggests that narcissism is on the rise, with each successive generation surpassing the vanity levels of the previous one (Seigel 2013). The highlighted values in media content aimed at pre-teens in the twenty-first century make a greater appeal to individualism and sensationalism than in the 1990s, facilitating a narcissistic pre-teen culture (Rosen 2016).

The claims and statements repeated in sales pitches, self-motivation manuals, and lectures on creativity also shape public opinion: the creative individual is idolized and advertised as hyper-empowered, directed towards fame, using black-boxed apps such as TikTok, Instagram, FaceApp, YouTube, or Snapchat, moving between the physical and virtual universes. If *truth* and *facts* do not ensure likes, comments, shares, and more followers, then they must be replaced by something else that does.

Often presented as *technological determinism* or *corporate determinism*, this phenomenon assumes technology as the main driver of social change, replacing – or consolidating – the power of MTCs such as Google, Apple, Facebook, and Amazon (Natale, Bory, and Balbi 2019).

> Gaps in knowledge, putative and real, have powerful implications, as do the uses that are made of them. Alan Greenspan, once the most powerful central banker in the world, claimed that today's markets are driven by an "unredeemably opaque" version of Adam Smith's "invisible hand", and that no one (including regulators) can ever get "more than a glimpse at the internal workings of the simplest of modern financial systems". (Pasquale 2016:2)

Let us define, then, the power of the black box as the result of the rearrangement of society's resources in order to give the organizations behind them the power of independent negotiation over – and against – the other key sectors: countries, political, financial, and cultural elites, among others. This growing power over what is produced and communicated is what currently enables and consolidates the actual power of the black boxes being used by MTCs, whose intricacies remain hidden from the public.

But one could argue that the present text should be more concerned with the quality of information – its truth or lack of it – than with the analysis of power. However, one does have close implications over the other: as MTCs contribute to the creation of vast amounts of data, are those data being used to produce and deliver reliable and sound information? Or is that also beyond *good* or *bad*, as with propaganda?

Too often, Big Data enables the practice of apophenia: seeing patterns where none actually exist, simply because enormous quantities of data can offer connections that radiate in all directions. In one notable example, Leinweber (2007) demonstrated that data mining techniques could show a strong but spurious correlation between the changes in the S&P 500 stock index and butter production in Bangladesh. Interpretation is at the center of data analysis. Regardless of the size of a data, it is subject to limitation and bias. Without those biases and limitations being understood and outlined, misinterpretation is the result. (Boyd and Crawford 2012: 668)

Attention: Here Comes Post-Fake Reality

The Control of Attention

Mobile devices are currently responsible for more than half of the Internet traffic (STATISTA 2021), which is generated through black-boxed applications, to the detriment of conventional browsers. Each of these black boxes brings its own arsenal of notifications and attention requests, transforming the average user experience into a constant switch from application to application, from black box to black box[4].

Concentration and distraction form opposite poles: the usual, the customary, and the familiar tend to be enjoyed uncritically, and only the other, the strange, the weird, and the misunderstood is isolated and pointed out, as noted by the Coalition for Critical Technology (CCT 2020). Embodying this *other* is a common exercise in the populist politics of our time, often transferring to this *stranger* the causes of society's ills (Dhawan 2016; Itagaki 2015; Kaya 2010).

The way through which individual attention can be manipulated had already been mentioned by Münsterberg (1894), although in a broad context of psychological studies, and has since been analyzed by several other authors as well (e.g., Kahneman 1973; Wickens 1984; Tipper et al. 1989; Lang 1995; Pashler 2016), including Guattari (2000). For him, an individual's attention is captured by their environment, ideas, tastes, models, and ways of being. The images are constantly injected into them, even by the refrains that creep in and settle in their heads. Guattari exemplifies this division of attention:

---

[4] In 2019, the app market witnessed a total of 204 billion apps app downloads. Normally, an average person has 40 apps installed on his phone. Out of that 40 apps, 89% of the time is split between 18 apps (SIMFORM 2021).

> When I watch television, I exist at the intersection: (1) of a perceptual fascination provoked by the screen's luminous animation which borders on the hypnotic; (2) of a captive relation with the narrative content of the programme, associated with a lateral awareness of surrounding events (water boiling on the stove, a child's cry, the telephone...); (3) of a world of phantasms occupying my daydreams. My feeling of personal identity is thus pulled indifferent directions. How can I maintain a relative sense of unicity, despite the diversity of components of subjectification that pass through me? (Guattari 2000:8)

To remain efficient, propaganda had to learn and adapt to capture the public's attention or risk failing to influence the (distracted) masses. Boyd claims that the media is now shaped by these attention-grabbing mechanisms, while decentralized groups – including artivists – take advantage of digital tools and networks to hack the *attention economy* (Boyd 2017). The ways of governing the capture of attention are linked to what Kelly (2008) designates as the generative characteristics of intangible value and generators of attention. These generative characteristics are qualities or attributes that must be cultivated and nurtured, and cannot be copied, cloned, falsified, replicated, or reproduced. They are uniquely generated in a place and time. In the digital field, Kelly understands that generative qualities add value to things and experiences that are often free, turning them into something that can be sold. For him, the generative values of attention are:

- Immediacy: here and now.
- Customization: personalization and exclusivity.
- Interpretation: translation and simplification.
- Authenticity: realism and uniqueness.
- Availability: permanent access.
- Appropriation: use for other contexts and remixing.
- Sponsorship: low value paid directly to the author/creator by means of direct links and individualized recognition.
- Discovery Potential: ease of being found.

According to Crawford (2014), if the term *economy* applies to what is scarce and therefore valuable, then the debate about an *information economy* should be replaced by a debate about an *attention economy,* based on the fact that our cognitive ability is a valuable but limited resource and that decisions must be made about its use, with bad decisions resulting in losses. This economic phenomenon manifests itself as a crisis of self-ownership: the individuals' attention is not exclusively theirs so they can direct it wherever they desire. The changing technological environment creates an ever-increasing need for stimuli, and the content of stimuli becomes almost irrelevant. Propaganda is now delivered through notifications. If notifications do not arrive on our mobile devices, as expected, we feel anxiety over their absence.

To better understand this phenomenon, let us consider that attention can be categorized – even if not exclusively – in its orientation by goals (executive attention), at the service of the individual's own will, or by stimuli (involuntary attention), regardless of the individual's own will (Crawford 2014). An artist who is focused on his work devotes his executive attention to it. He, therefore, channels his attention to a goal. In contrast, if there is a sudden loud sound outside, his attention is stimulated. He can go out to investigate what is happening or not, but the claim for attention is involuntary and uncontrollable. The answer, however, requires a combined effort of executive attention if he is to resist external stimuli, and his capacity for such resistance is finite. However, executive attention can be manipulated, creating dependence on certain stimuli, for example, those obtained through interactions on social networks. Even outdoors, the ability to appreciate nature is impacted by the mediation of smartphones, which encourages treating the experience as a sequence of actions: observation, engagement, image/video capture, publication/sharing and archival/dismissal.

The notifications produced by black-boxed mobile apps range from inherently relevant and useful (incoming calls, messages, e-mails) to advertising/propaganda, whose purpose is often limited to keeping the users *engaged*, reaching for their phones, and opening the apps[5]. Some of these strategies may include calls to attention about features or updates in the app itself, very secondary – and often irrelevant – notifications about group activities (especially on Facebook), or even messages from *fake* accounts (bots or otherwise). But depending on the amount of privacy protection enabled on their smart devices, individuals risk being constantly monitored and influenced: where to go, what to see, where to shop and eat. The ability to exert influence on such a constant basis, as well as a wealth of personal and geographic data, is truly a propagandist's dream.

The distraction of individuals, or the inability to stay focused on a single intellectual activity (for example, admiring the landscape), seems to indicate that individuals are agnostic about the question of what deserves their attention, about what is, in fact, interesting. Systematic distraction, as a result of an overstimulation of attention, may be compared to the mental equivalent of physical addiction (Reber, Schwarz, and Winkielman 2004). Thus, elements

---

[5] On average, we check our phones every four minutes (Wheelwright 2022).

of distraction (or *notifications*) can hamper[6] creative production by disrupting the ability to keep multiple representations in mind. Indeed, disconnecting distraction devices and immersing yourself in a controlled environment over a long period of time can increase the production of creative verbal associations by 50% (Atchley, Strayer, and Atchley 2012). Becoming aware of the mechanisms that are already used by third parties – mostly through black-boxed devices or apps – to manage and capture the attention of individuals is an important step towards understanding their purposes and effects and, at the same time, consciously choosing to either remain under their influence or not.

Latour (1999:214) states that we already live and act as a collective and increasingly large number of humans, mixed with an increasingly large number of nonhumans, actively engaged in the making of aesthetics, politics, law, and morality, regardless of existing regulations. This may imply that black boxes are already an intrinsic part of this collective, as well as their attention-grabbing mechanisms. Our lives in the so-called *developed societies* imply the coexistence with nonhumans, and this extends beyond our conscious choices. Therefore, conscious and unconscious acts, voluntary and involuntary choices always intermingle, producing hybrid states in which actors are dispersed in an ever-complexifying network.

Some of today's information systems behave like attention control black boxes, appealing to the fluency of their users in certain basic concepts and typical vocabulary, providing assertive – or even dogmatic – mechanisms along with easy-to-accept information (regardless of its validity or truthfulness) through user profiling. Black-boxing and attention control can thus be understood as current propaganda mechanisms, widely used in keeping the masses actively engaged in social media networks – while often disengaged from reality. And if the trick to maintaining this level of engagement is the widespread dissemination of (ethically, factually, or scientifically) questionable

---

[6] However, it is also suggested by other authors that distraction can improve disruptive creativity, making individuals more open to remote associations between concepts. In verbal creativity, not fixing the dominant meaning of a word allows subordinate meanings to remain active, which can promote atypical connections between concepts and therefore creative associations. This phenomenon can be interpreted as a distraction, preventing the individual from focusing on the most likely interpretation of a stimulus, thus creating unusual interpretations and connections. The discovery that the attention disorder or hyperactivity – which can be interpreted as a failure to control attention – may be linked to creativity, divergent thinking, the ability to focus too far, non-conformism, adventure, self-acceptance and sublimation is consistent with this view (White and Shah 2011; Sedgwick, Merwood and Asherson 2019).

information, institutional and corporate priorities are clear: influence and profit above all else.

Post-Fake Reality

Watching footage of actual death became commonplace through television coverage of bombings, shootings, war, or executions. Shocking and unusual footage is guaranteed to trigger virality more easily. To this effect, individuals mount cameras on cats and dogs seeking unique angles on their secret life while training them to behave as desired in simulated situations in controlled environments; drones are used to film pornography as if they were spontaneous flashes of reality: reality is being forged and shared as the genuine article. As early as 1922, in Robert J. Flaherty's documentary[7] "Nanook of the North," an Inuk actor is led by the director to perform staged actions, such as hunting seals with harpoons (instead of guns, as was the custom at the time), or the exaggeratedly inappropriate use of a gramophone to fallaciously illustrate ethnographic concepts (Kara and Reestorff 2015).

In May 2016, the suicide of young Océane (Williamson 2016), broadcast in real-time by herself through the Periscope social network to an audience of more than 1,000 users, launched the debate on the differentiation between fantasy and reality, when both are mediated by the same technology. Several people watching the live broadcast assumed it was a simulation, while others expected the worst, some fearing it and others urging its outcome. It was also at this time that the Blue Whale challenge (Adeane 2019), which supposedly started as a hoax, gained notoriety. Its universal projection and dissemination increased its impact among adolescents through a complex phenomenon, imbued with imitation, tribalism, and acceptance rituals, as well as the appeal of risk and transgression, leading several social networks to change their search mechanisms to try to stop the situation (Hempel 2017).

More recently, in 2020, the coronavirus challenge, which consisted of licking a toilet seat in a public space, was released on the TikTok network by a 21-year-old influencer. Ava Louise, whose accounts on social media networks have since been suspended, in addition to filming herself – allegedly in an airplane bathroom – also had the help of Larz, a friend who posed as an unknown follower, who was also filmed licking a toilet seat. Later, Larz claimed on social media that he had contracted the disease and even showed images of his hospital stay. Except that the hospitalization had occurred before the toilet-

---

[7] By definition of The Oxford Dictionary: "Either television, film, video, or radio programme dealing with factual material rather than fictional material, usually with some defined goal to create new insight or exposure to facts." retrieved from https://www.oxfordreference.com/display/10.1093/oi/authority.20110803095724407

seat-licking shoot and had no relationship with COVID-19. Ava Louise then dismantled the whole situation as fake, assuring she only did it with the – achieved – purpose of gaining more notoriety. When faced with the multitude of negative reactions on various social networks (and many positive ones, too), she claimed to be happy with the outcome, which she reinforced in the Dr. Phil TV show[8].

In the era of the *post* prefix (post-real, post-internet, post-modern), it is worth positing yet one more: *post-fake.* A *post-fake* event may be defined as follows: the outcome of any false or fallacious creation that one wishes to posit and communicate as true in order to cause (actual) impacts or changes in reality can be designated as *post-fake.*

Due to the wide dissemination that the term *fake* underwent, especially during the Trump administration, *fake* is thus used as the opposite of *real* since *fake,* in this context, stands for non-existence and, therefore, non-reality. Fake (fictitious) situations in cinema, theatre, or television have already led to (real) aggression to the actors involved in them, as people mistook the actors for the characters, hence resulting in a post-fake event, such as the attack on two cast members of an LGBT themed theatre play in Southampton, U.K. (Perraudin 2019). *Post-fake* is thus the outcome of *fake* affecting and changing the *real.*

In the controversial electoral campaign in Brazil in 2018, which culminated in the election of President Jair Bolsonaro, social networks were flooded with heavily politicized memes, based on various fallacies imbued with emotional appeal, ranging from the collage of the concept of modern Socialism to Nazi National Socialism, to the spreading of fake news about several candidates, including the elected President.

WhatsApp was the most used network to this effect, as there is no centralized or public monitoring of what is being shared. All communications are encrypted, and therefore the type of scrutiny exercised on Twitter or Facebook cannot be applied, making it a perfect black box. In addition, groups are created in an organic and free way, in which anyone can add friends, professional colleagues, and family members not needing their prior acceptance. The El País newspaper conducted an investigation (Gosálvez and del Ser 2018) into the topic and found that 66% of Brazilian voters shared political news using WhatsApp, and 90% of those users did so more than 30 times a day. To further understand this reality, a survey (IPSOS 2018) revealed that 62% of the Brazilian population admitted to having already believed in fake news. When asked for the main reason behind this, nearly half of the respondents blamed the politicians and the media, which, they felt, deliberately deceived the people.

---

[8] An excerpt can be watched at: https://www.youtube.com/watch?v=ayuS5W45LlE

However, 37% assumed that this could be due to their biased view of society and the world, to generalizations from personal experiences, or to the influence of social networks.

WhatsApp does work like a black box, obscuring the possibility of verifying the facts it relays, enabling the assembly of fallacies such as the "gay kit for six-year-old children" (GLOBO 2018a) attributed to candidate Haddad – a distortion of the initiative "School without Homophobia" – or, in the opposite direction, the rumors about Bolsonaro's terminal cancer (GLOBO 2018b), which would prevent him from finishing his term. It is also worth noting the fact that much of this information circulates in the form of a meme, some of which exhibit careful photographic editing and composition, context extraction, and adulteration – much like what happens in most modern advertisements. This phenomenon of communication through images, often complemented by the abundant use of emojis, further corroborates the emotional – rather than rational – nature of the message, which mere text cannot convey with comparable impact (Kaye, Malone, and Wall 2017).

The output of this black box was voter influence through a set of fake or fallacious stories about the various candidates, spread by their supporters with the intention of undermining the voting outcome of their opponents. The Brazilian presidential election could thus be deemed as *post-fake*.

One final example of what could be considered[9] as *post-fake authorship* of artistic work is Jason Allen's AI-generated piece, "Théâtre D'opéra Spatial," whose notoriety stems from the fact that it won a prize in 2022 at a fine arts competition, where all other competing pieces were non-AI generated artworks, being the first known example of this type of situation.

Art, Attention, and Black Boxes

Too Much Art, Too Little Time

The process of cognitively superimposing virtuality and materiality – or the human, machine, and nature (Floridi 2015) – is similar, in operation, to the cognitive overlay of authenticity and falsehood and can be assumed as a direct heir of Goebbels' propaganda: seeking acceptance without questioning, dismissing critical analysis, positioning itself as necessary, beyond good or bad.

In the context of popular music, Gauvin (2017) stated that the attention economy has introduced significant changes in the processes of creation and

---

[9] While the competitors claimed he didn't author the work, Jason Allen stated that it was submitted under the name "Jason M. Allen via Midjourney", and that he hadn't deceived anyone about its origins. (Roose 2022)

construction of musical pieces to deal with an online streaming market characterized by a constantly growing offer that can easily become overwhelming. This implies that the possibility of quickly ignoring any content that does not capture one's attention or interest during the first few seconds of contact is on the rise. Capturing attention by providing information with a strong appeal to emotionality and subjectivity while preventing its fixation is a viable strategy. Those initial few seconds play a crucial role since it is also when most users decide to share what they are experiencing before exerting any critical thinking and analysis, mostly based on emotions and immediacy.

Regarding how individuals exercise their choice to direct attention (toward art, music, news, or information sources), Pomerantsev suggests that "one has to look beyond just news and politics to also consider poetry, schools, and the language of bureaucracy and leisure to understand the formation of attitude" (2019: xv). This intricate and interconnected mesh delivers a constant flow of attention-capturing stimuli, designed to guarantee uncontested acceptance by the user: digital artworks become easily forgettable as they are time-lined by the hundreds on Instagram and Facebook, or streamed on Spotify, Google Music, Netflix, or Apple TV. Users often rely on word-of-mouth to focus on a particular song or series – much like they do with information and news, with uncontested and acritical acceptance. An observer whose attention is concentrated before an artwork is absorbed by it. By contrast, the distracted/abstracted masses absorb the work – often relegated to a secondary role as *background noise* or *decoration*. For the public, the greater the decline in the social meaning of an artwork – even an art form – the greater is the gap between a critical attitude and distraction, and in this case, literally, *everything* risks being reduced to mere amusement, playfulness, and entertainment.

In the context of visual art (including digital art), several strategies have emerged to counter this phenomenon and provide the audience with more immediate, meaningful connections. For example, a work of art with a title that relates to its content produces better responses (appreciation, subsequent interest, remembrance) than using unrelated titles, or the absence of a title (Belke et al. 2010). The semantic context improves the processing fluency, and therefore, it also improves the evaluation and the reaction. Aesthetic enhancement occurs as a function of the viewer's ability to mentally process the properties of the artifact in a significant way, with the viewer's attention being a critical part of the aesthetic experience. Seeing a work and saying it is interesting is a much more superficial aesthetic experience than seeing the same work and noticing that it is interesting and that it has additional value because of its author's creative process, its original intent, its communication goal, its place in history and the influences, techniques and production processes used; just as fluency for aesthetic experiences is enhanced by access to more information and deeper connections between one's experience and the audience's cognitive processes, so the ability to think creatively – creative literacy – is supported by access to a rich and broad knowledge network, which can also be interrupted by an attention deficit.

Roger (1997) proposed the concept of *artialization*[10] *in visu*, which designates the process of elevating the landscape to the status of a work of art, but the present author posits its generalization to all visual representations, culminating in what could be designated as the digital *artialization* of the mundane (Whitfield and Destefani 2011).

The artist's conventional identity, as one who develops projects or works that are intended for a merely receptive (i.e., non-creative) audience, is challenged by generations of individuals actively engaged in the use of black boxes, even before the dawn of modern generative AI systems, to produce their own *artialized* content for public viewing and sharing – including photographs of pets, food, personal styling, trendy locations, status granting objects, only to name a few. The power of the *like* on social media pressures them to become noticed, craving comments and reactions (Rosen 2012). Hundreds of millions of *artialized* multimedia content, either original, remixed, or simply appropriated, can be found, for example, on YouTube, Instagram, Tumblr, Twitter, Flickr or SoundCloud social networks. But... is it art?

> I can no longer think what I want to think. My thoughts were replaced by moving images. [...] And this song? Hear without listening. [...] It's fake music. Preserved music. It comes out of the music slaughterhouse like breakfast sausages out of the pig slaughterhouse. [...] It's music in cans. (Duhamel 1934: 24-52)

The belief that individual creativity and expression are unlimited, while at the same time it increasingly depends on the use of (limiting) black boxes filled with machine learning and artificial intelligence, is widespread and encouraged. The field of digital art provides relevant examples of this dependence on black boxes, both in terms of creation as well as dissemination. Adobe Photoshop was the ultimate example of a creative black box before the advent of generative AI systems. The term *photoshop* and its derivative verb and adjective have entered dictionaries, loosely used to designate an alteration to a photographic image, even though the low-level operations performed on the individual pixels remain a mystery for most of its regular users. But when it comes to disseminating or exhibiting the artworks on social media for a larger audience, the artists that are promoted are the same that the system

---

[10] *Artialisation*, in the original French.

commoditized, arbitrarily defining the value of their works according to their participation level in platform service subscriptions, paid advertisements, or digital marketing, among others, while ignoring everything else that is being created outside its sphere of influence. Even the Non-Fungible Tokens recent phenomenon happened through a similar type of manipulation, being heralded as a break from the art-market chains – while trying to deploy a new set of similar, black-boxed attention control mechanisms (Nadini et al. 2021).

Think Outside the (Black) Box

Artivism is a term used to reflect the use of art to the service of activism, using several forms, such as *subvertising, brandalism,* memes, *hijinks,* street performances, spoof-websites, and electronic disobedience events, among many others. To this effect, several art collectives have been using propaganda, advertising, attention-grabbing mechanisms, and post-fake reality to actually expose the black-boxed inner workings of events, organizations, corporations, and governments, forcing them into – mostly unwanted or unsought – action and accountability. The fact that they often resort to what is later perceived as camouflaged parody or satire, resulting in entertaining and humorous outcomes, is a merit, not a flaw. These collectives can thus be called *post-fake artivists.*

> REFF was born in Italy in 2008. Since then it continuously operated using *fake, remix, reinvention, recontextualization, plagiarism and reenactment as tools for the systematic reinvention of reality.* Defining what is real is an act of power. Being able to reinvent reality is an act of freedom. REFF promotes the dissemination and reappropriation of all technologies, theories and practices that can be used to freely and autonomously reinvent reality. To do this, REFF established an international competition on digital arts, a worldwide education program in which ubiquitous technologies are used to create additional layers of reality for critical practices and freedoms, a series of open source software platforms and an Augmented Reality Drug. (REFF 2011)

REFF is a paradigmatic example: an Italian fake cultural institution that has its origin in a protest of the organization of an art competition – RomaEuropa WebFactory – which strictly prohibited artistic remixes, *mashups,* and other artistic combinations for fear of copyright-related problems. But that same organization reserved the right to be able to use and remix the submitted works for further exploration. Then came the idea of occupying the event by registering a similar name and web domain (RomaEuropa FakeFactory), and a parallel competition was launched, in which it was mandatory to use all kinds of remixes, highlighting the importance of the new open licensing models: commons, public domain, copyleft, among others. In short, the fake organization's competition not only outranked the original competition's success, but it also received official recognition around the world, being received by the Cultural Commission of the Italian Senate and considered an official initiative of the European Community's Year of Creativity in 2009, surpassing in media coverage the original event. When doing an online survey about the original competition, most of the organic results pointed, in fact, to the fake one, which may lead to the conclusion that *fake* surpassed and transformed *reality*, giving rise to a *post-fake* event and organization.

Another relevant example is "The Yes Men," a culture-jamming artivist duo that deploys a satirical yet carefully crafted aesthetic approach, including communication materials and media, posing as powerful entities and making shocking statements that caricature the ideological position of the actual entity, either a person or an institution. The Yes Men use spoofed – yet credible – websites and media outlets to disseminate their ideas, which have included the prospect of selling one's vote or a proposition that the poor should consume recycled human waste. This has resulted in fake news reports, such as those covering the demise of the World Trade Organization or Dow Chemical paying compensation to the victims of the Bhopal disaster – two of the many institutions that The Yes Men have posed for as spokespeople (TYM 2021). An early example of their satirical actions was the 2004 group called "Yes, Bush Can!" that went on tour, encouraging supporters to sign a *Patriot Pledge,* agreeing to keep nuclear waste in their backyards and send their children off to war (TYM 2004). The Yes Men are Jacques Servin, an author of experimental fiction, and Igor Vamos, an associate professor of media arts at Rensselaer Polytechnic Institute, New York. They encourage activist and student groups into unconventional – yet highly impactful – actions, such as crashing conferences, creating fake websites, and getting arrested. The post-fake outcome of their actions is the actual damage control the real entities impersonated by them then need to make due to the raised public awareness of the issues they tackle, as such public shaming seeks to assert social demands through their social impact.

One final example, among many other possibilities, is the 2016 action art *Bundeserpressungskonferenz*[11], which was held at Berlin's Maxim Gorki Theatre, launching the project "Flüchtlinge Fressen – Not und Spielen"[12] (Gorki 2016) by artists Phillip Ruch, Cesy Leonard, Stefan Pelzer, and André Leipold – under the moniker "Center for Political Beauty" (CPB 2020). Their claim was to feed volunteer refugees to tigers if their political demands were not met, and it relied on the shock value and word-of-mouth support to gain enough traction

---

[11] Federal blackmail press conference, a take on the German expression for federal press conference.

[12] Devour Refugees – Misery and Games.

in order to force the authorities to control the public damage. All these projects have the *post-fake effect* in common, as they all rely on a spontaneous response of public disobedience or protest from their audiences, which is "the dream of every artivist: a passing of the baton to civil society" (Banks 2016).

Conclusion

The interconnected networks of digital communication seem to lead to conformism and self-indulgence in real life, in contrast to online petitions, written confrontations in the comment sections, and memes. *Propaganda* set the rules over which *true* and *fake* could be "justifiably" combined to influence the masses, placing it at the root of *post-fake* genealogy. The fragmentation of attention and a growing set of logical fallacies disseminated by black-boxed applications contribute to *form* becoming more important than *function*. There is little transformative and collaborative action, given that the appeal to emotion, overriding reason, is widely used by the media. We are witnessing the global aestheticization of all areas (Lipovetsky and Serroy 2014; Groys 2009), from the dishes served in gourmet restaurants to political events, from landscape interventions in the best cities to visit, to disposable and polluting packaging and COVID-19 fashion facemasks.

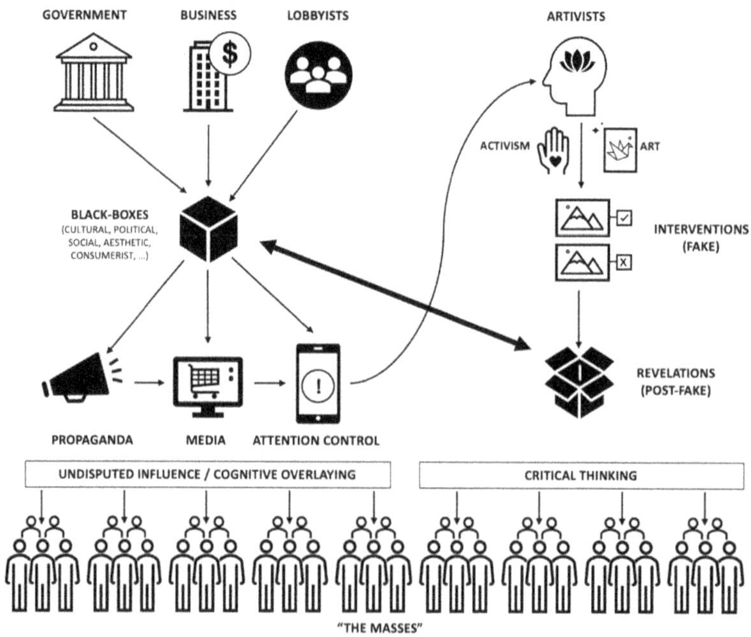

Figure 10.1. A summarizing visual representation of the presented argument
Source: the author

But the study of fallacies also gives individuals the tools to explore the truth and expose the falsity of ideas and, as such, to participate in society and politics (Rosen 2006). More than just living uncritically with global aesthetics and contributing to the perpetuation of fallacies, post-fake artivism can help bring the public closer to a critical, logical – and entertaining – spirit, sustained by philosophical debate. These areas of knowledge have somewhat been pushed to a secondary role by the dominant STEM and entrepreneurship culture, which – unlike them – seem not only to coexist rather well with the present reality but also strongly contribute to maintaining it, as the novel success of generative AI tools, such as ChatGPT, DALL-E or MidJourney, attests. Despite the hype surrounding ChatGPT and its ability to produce elaborate, human-like answers to user prompts, its FAQ section clearly states that "ChatGPT (...) can occasionally produce incorrect answers. It has limited knowledge of world and events after 2021 and may also occasionally produce harmful instructions or biased content" (ChatGPT n.d.). However, using an artificially intelligent black box that may or may not deliver reliable, fact-based answers has in no way hampered its public success.

In a *reality* guided by the cognitive overlaying of a barrier of buzzwords, produced and controlled by the dominant culture to inculcate habits and consolidate power, it is necessary to dismantle this "curtain of clichés, which has become trivial and selfish, in an unprecedented way" (Berger 1983: 279).

Art and activism can help understand the structures that sustain modernity and colonialism by creatively exposing the underlying relationships and mechanisms that structure social fabric, the logical fallacies, and assumptions that underpin the unquestioned acceptance of knowledge-without-knowing, the individually and collectively manifested *otherphobia*. Post-fake artivism can contribute to a wider understanding of the complexity of influence and attention-grabbing mechanisms through their exposure. By rendering less opaque the dispersed and unpredictable algorithms that lead to manipulation and bias in information and communication, post-fake artivism can also foster critical thinking and analysis, as opposed to the undisputed acceptance of influence and cognitive overlaying.

References

Adeane, Ant. 2019. "Blue Whale: What is the truth behind an online 'suicide challenge'?". BBC News. https://www.bbc.com/news/blogs-trending-46505722

Allcott, Hunt, and Matthew Gentzkow. 2017. "Social Media and Fake News in the 2016 Election". *Journal of Economic Perspectives* 31, no 2: 211-236.

Atchley, Ruth Ann, David L. Strayer, and Paul Atchley. 2012. "Creativity in the Wild: Improving Creative Reasoning through Immersion in Natural Settings." *PloS One* 7(12):e51474. doi: 10.1371/journal.pone.0051474.

Banks, Summer. 2016. "Feeding on Refugees: The Center for Political Beauty." *Exberliner*. https://www.exberliner.com/stage/andre-leipold-interview/.

Belke, Benno, Helmut Leder, Tilo Strobach, and Claus Carbon. 2010. "Cognitive Fluency: High-Level Processing Dynamics in Art Appreciation". *Psychology of Aesthetics, Creativity, and the Arts* 4(4): 214 –222.

Bennett, Bo. 2017. *Logically Fallacious: The Ultimate Collection of over 300 Logical Fallacies (Academic Edition)*. Sudbury, MA: Ebookit.com.

Berger, John. 1983. "The production of the world". In *Selected Essays* edited by G. Dyer, London: Bloomsbury.

Bergman, Ronen, Sheera Frenkel, and Raymond Zhong. 2020. "Major TikTok Security Flaws Found." *The New York Times*, January 8. . https://www.nytimes.com/2020/01/08/technology/tiktok-security-flaws.html

Boyd, Danah. 2017. "Hacking the Attention Economy". Medium.com. https://points.datasociety.net/hacking-the-attention-economy-9fa1daca7a37

Boyd, Danah, and Kate Crawford. 2012. "Critical Questions for Big Data". *Information, Communication & Society* 15(5): 662-679. https://doi.og/10.1080/1369118X.2012.678878

Catts, Oron, Nicola Marples, Fiona Newell, and Lynn Scarff. 2018. "Fake". Dublin Science Gallery. https://dublin.sciencegallery.com/fake

CCT. 2020. "Abolish the #TechToPrisonPipeline. Crime prediction technology reproduces injustices and causes real harm". Medium.com. https://medium.com/@CoalitionForCriticalTechnology/abolish-the-techtoprisonpipeline-9b5b14366b16

ChatGPT. n.d. "ChatGPT General FAQ". OpenAI.com. https://help.openai.com/en/articles/6783457-chatgpt-general-faq

CPB. 2020. "Political resistance in the 21st century: weaponizing the power of history". Center for Political Beauty. https://politicalbeauty.com/about.html

Crawford, Mathew. 2014. *The World Beyond Your Head*. New York: Farrar, Straus and Giroux.

Dhawan, Nikita. 2016. "Homonationalism and state-phobia: The postcolonial predicament of queering modernities". Pp. 51-68In *Queering paradigms V: Queering narratives of modernity*, edited by M. A. Viteri and M. Lavinas Picq. Bristol: Peter Lang.

Duhamel, Georges. 1934. *Scènes de la vie future*. Paris: Éditions Fayard.

Floridi, Luciano. 2015. *The Onlife Manifesto*. Cham, Heidelberg, New York, Dordrecht, London: Springer Open.

Gallego, Javier Paez, Ángel De-Juanas Oliva, Francisco Javier García Castilla, and María José Díaz Santiago. 2020. "A Study about Social Values and Psychological Well-Being in Young People: Implications for Educational Social Work." *Social Work Education* 39(6):721–36. doi: 10.1080/02615479.2020.1793935.

Gauvin, Hubert Léveillé. 2017. "Drawing listener attention in popular music: Testing five musical features arising from the theory of attention economy". *Musicae Scientiae* 22(3): 1-14. https://doi.org/10.1177/1029864917698010

Gershon, Richard A. 1993. "International Deregulation and the Rise of Transnational Media Corporations." *Journal of Media Economics* 6(2):3–22. doi: 10.1080/08997769309358235.

GLOBO. 2018a. "É #FAKE que Haddad criou 'kit gay' para crianças de seis anos". Globo News Network. https://g1.globo.com/fato-ou-fake/noticia/2018/10/16/e-fake-que-haddad-criou-kit-gay-para-criancas-de-seis-anos.ghtml

GLOBO. 2018b. "É #FAKE mensagem que diz que Bolsonaro simulou ser vítima de facada para disfarçar câncer". Globo News Network. https://g1.globo.com/fato-ou-fake/noticia/2018/10/17/e-fake-mensagem-que-diz-que-bolsonaro-simulou-ser-vitima-de-facada-para-disfarcar-cancer.ghtml

Goebbels, Joseph. 1934. "Der Kongress zur Nürnberg 1934". *Frz. Eher Nachf.* Munich: Zentralverlag der NSDAP: 130-141. https://research.calvin.edu/german-propaganda-archive/goeb59.htm

Gorki. 2016. "Flüchtlinge Fressen – Not und Spiele. Devour Refugees – Misery and Games". Gorki Theatre. https://www.gorki.de/en/fluchtlinge-fressen-not-und-spiele

Gosálvez, Patrícia, and Guiomar del Ser (Coord.). 2018. "Os 'whatsapps' de uma campanha envenenada". El País. https://brasil.elpais.com/especiais/2018/eleicoes-brasil/conversacoes-whatsapp/

Greenberg, Clement. 1939. "Avant-Garde and Kitsch". In *The Partisan Review,* 34-49. New York: The John Reed Club of New York.

Groys, Boris. 2009. "Self-design and aesthetic responsibility". E-flux Journal. http://www.e-flux.com/journal/07/61386/self-design-and-aesthetic-responsibility/

Guattari, Félix. 2000. *The Three Ecologies.* London: The Athlone Press.

Han, Ruixia, Jian Xu, and David Pan. 2022. "How Media Exposure, Media Trust, and Media Bias Perception Influence Public Evaluation of COVID-19 Pandemic in International Metropolises." *International Journal of Environmental Research and Public Health* 19(7):3942. doi: 10.3390/ijerph19073942.

Hansen, Mark B. 2011. "Medien des 21. Jahrhunderts, technisches Empfinden und unsere originäre Umweltbedingung". Pp. 365-409 in *Die technologische Bedingung. Beiträge zur Beschreibung der technischen Welt,* edited by Erich Hörl. Berlin: Suhrkamp.

Hempel, Jessi. 2017. "Killing the Blue Whale Challenge". Wired Magazine. https://www.wired.com/story/killing-the-blue-whale-challenge/

Herman, Edward, and Noam Chomsky. 1988. *Manufacturing Consent.* New York: Pantheon Books.

Huskinson, Janet M. 1982. *Concordia apostolorum: Christian Propaganda at Rome in the Fourth and Fifth Centuries - A Study in Early Christian Iconography and Iconology.* Oxford: British Archaeological Reports.

IPSOS. 2018. "Fake news, filter bubbles, post-truth and trust". IPSOS Mori, Social Research Institute. https://www.ipsos.com/en/fake-news-filter-bubbles-and-post-truth-are-other-peoples-problems

Itagaki, Ryuta. 2015." The anatomy of Korea-phobia in Japan". *Japanese Studies* 35(1): 49-66.

Kahneman, Daniel. 1973. *Attention and Effort.* New Jersey: Prentice-Hall.

James, Kyle. "Catholic bank admits investing in contraception and armaments". Deutsche Welle. 2009. https://www.dw.com/en/catholic-bank-admits-investing-in-contraception-and-armaments/a-4535186

Kara, Selmin, and Camilla Møhring Reestorff. 2015. "Introduction: Unruly Documentary Artivism." *Studies in Documentary Film* 9(1):1–9. doi: 10.1080/17503280.2014.1002665.

Kaya, Ayhan. 2010. "Migration debates in Europe: migrants as anti-citizens". *Turkish Policy Quarterly* 10(1): 79-91.

Kaye, Linda K., Stephanie A. Malone, and Helen J. Wall. 2017. "Emojis: Insights, Affordances, and Possibilities for Psychological Science." *Trends in Cognitive Sciences* 21(2):66–68. doi: 10.1016/j.tics.2016.10.007.

Kelly, Kevin. 2008. "Better than Free". *Edge.* https://www.edge.org/conversation/better-than-free

Lang, Peter J. 1995. "The Emotion Probe: Studies of Motivation and Attention." *The American Psychologist* 50(5):372–85. doi: 10.1037/0003-066x.50.5.372.

Latour, Bruno. 1999. *Pandora's Hope: Essays on the Reality of Science Studies.* London, England: Harvard University Press.

Latour, Bruno. 1988. "How to write 'the Prince' for machines as well as for machinations". Pp. 20-43 in *Technology and Social Process,* edited by Brian Elliott. Edinburgh University Press.

Lipovetsky, Gilles, and Jean Serroy. 2014. *O capitalismo estético na era da globalização.* Lisboa:Edições 70.

Lutz, Ashley. 2012. "These 6 Corporations Control 90% Of The Media In America." *Business Insider,* June 14. https://www.businessinsider.com/these-6-corporations-control-90-of-the-media-in-america-2012-6

McChesney, Robert. 1999. "Noam Chomsky and the struggle against neoliberalism". *Monthly Review* 50(11): 40-47.

Münsterberg, Hugo. 1894. "The intensifying effect of attention". *Psychological Review,* 1: 39-44.

Nadini, Matthieu, Laura Alessandretti, Flavio Di Giacinto, Mauro Martino, Luca Maria Aiello, and Andrea Baronchelli. 2021. "Mapping the NFT Revolution: Market Trends, Trade Networks, and Visual Features." *Scientific Reports* 11(1):20902. doi: 10.1038/s41598-021-00053-8.

Natale, Simone, Paolo Bory, and Gabriele Balbi. 2019. "The Rise of Corporational Determinism: Digital Media Corporations and Narratives of Media Change." *Critical Studies in Media Communication* 36(4):323–38. doi: 10.1080/15295036.2019.1632469.

NSA. 1950. "German Radio Intelligence by Albert Praun; Department of the Army Office of the Chief of Military History. Includes NSA Memo Comments on Report". National Security Agency: 3. https://www.nsa.gov/news-features/declassified-documents/friedman-documents/assets/files/publications/FOLDER_240/41748999078819.pdf

NYT. 1990. "Upheaval in the East; Gorbachev Hands Over Katyn Papers". *The New York Times,* April 14. https://www.nytimes.com/1990/04/14/world/upheaval-in-the-east-gorbachev-hands-over-katyn-papers.html

NYT. 1945. "Germans Indicted in Massacre Of 11,000 Poles in Katyn Forest; With Russians Scheduled to Prosecute Case, Accused Are Expected to Recall Former Charge That Red Army Killed Victims". The New York Times, October 19. https://www.nytimes.com/1945/10/19/archives/germans-indicted-in-massacre-of-11000-poles-in-katyn-forest-with.html

Pashler, Harold. 2016. *Attention*. New York: Psychology Press.

Pasquale, Frank. 2016. *The Black Box Society: The Secret Algorithms That Control Money and Information*. Cambridge: Harvard University Press.

Perraudin, Frances. 2019. "Southampton theatre cancels shows after actors in LGBT play attacked". The Guardian, June 9. https://www.theguardian.com/uk-news/2019/jun/09/southampton-theatre-cancels-lgbt-play-performances-after-actors-attacked

Pomerantsev, Peter. 2019. *This Is Not Propaganda: Adventures in the War against Reality*. London: Faber & Faber.

Reber, Rolf, Norbert Schwarz, and Piotr Winkielman. 2004. "Processing Fluency and Aesthetic Pleasure: Is Beauty in the Perceiver's Processing Experience?" *Personality and Social Psychology Review* 8(4):364–82. doi: 10.1207/s15327957pspr0804_3.

REFF. 2011. "REFF is a fake cultural institution enacting real policies for arts, creativity and freedoms of expression all over the world". Roma Europe Fake Factory. http://www.romaeuropa.org/

Roger, Alain. 1997. *Court Traité du Paysage*. Paris: Gallimard.

Roose, Kevin. 2022. "An A.I.-Generated Picture Won an Art Prize. Artists Aren't Happy." *The New York Times*, September 2. https://www.nytimes.com/2022/09/02/technology/ai-artificial-intelligence-artists.html

Rosen, Frederick. 2006. "The Philosophy of Error and Liberty of Thought: J.S. Mill on Logical Fallacies." *Informal Logic* 26(2): 121-147.

Rosen, Johanna. 2016. "Examining Preadolescent Television Programming and the Rise of Generation Me." *Elon Journal of Undergraduate Research in Communications* 7(1). http://www.inquiriesjournal.com/a?id=1476

Rosen, Larry. 2012. *iDisorder: Understanding Our Obsession with Technology and Overcoming its Hold on Us*. New York: Palgrave Macmillan.

Sedgwick, Jane Ann, Andrew Merwood, and Philip Asherson. 2019. "The Positive Aspects of Attention Deficit Hyperactivity Disorder: A Qualitative Investigation of Successful Adults with ADHD." *Attention Deficit and Hyperactivity Disorders* 11(3):241–53. doi: 10.1007/s12402-018-0277-6.

Seigel, J. 2013. "The Hannah Montana Hypothesis." *Nautilus*, September 16. http://nautil.us/issue/5/fame/thehannahmontanasis

SIMFORM. 2021. "App Usage Statistics 2021 that'll Surprise You (Updated)". SIMFORM. https://www.simform.com/blog/the-state-of-mobile-app-usage/

STATISTA. 2021. "Percentage of mobile device website traffic worldwide from 1st quarter 2015 to 1st quarter 2021". Statista. https://www.statista.com/statistics/277125/share-of-website-traffic-coming-from-mobile-devices/

Steinfeld, Nili. 2016. "'I Agree to the Terms and Conditions': (How) Do Users Read Privacy Policies Online? An Eye-Tracking Experiment." *Computers in Human Behavior* 55:992–1000. doi: 10.1016/j.chb.2015.09.038.

Stencel, Mark. 2015. "Politicians Keep Pushing Claims despite What Fact Checks Say." *American Press Institute*. (https://www.americanpressinstitute.org/publications/reports/survey-research/politicians-pushing-claims-despite-fact-checks/).

Stiegler, Bernard. 2010. *For a New Critique of Political Economy*. Oxford, England: Polity Press.

Tipper, S. P., T. A. Bourque, S. H. Anderson, and J. C. Brehaut. 1989. "Mechanisms of Attention: A Developmental Study." *Journal of Experimental Child Psychology* 48(3):353–78. doi: 10.1016/0022-0965(89)90047-7.

TYM. 2021. "Hijinks/Behind the Curtain". The Yes Men. https://theyesmen.org/hijinks-all?view=btc_embed_block

TYM. 2004. "Yes Bush Can". The Yes Men. https://theyesmen.org/content/yes-bush-can

Uhls, Yalda, and Patricia Greenfield. 2011. "The Rise of Fame: An Historical Content Analysis". *Cyberpsychology: Journal of Psychosocial Research on Cyberspace* 5(1). https://cyberpsychology.eu/article/view/4243/3289

Viala-Gaudefroy, Jerome, and Dana Lindaman. 2020. "Donald Trump's 'Chinese virus': the politics of naming". The Conversation, April 21. https://theconversation.com/donald-trumps-chinese-virus-the-politics-of-naming-136796

Wheelwright, Trevor. 2022. "2022 Cell Phone Usage Statistics: How Obsessed Are We?". *Reviews.org*. https://www.reviews.org/mobile/cell-phone-addiction/

White, Holly A., and Priti Shah. 2011. "Uninhibited Imaginations: Creativity in Adults with Attention-Deficit/Hyperactivity Disorder." *Personality and Individual Differences* 40(6):1121–31. doi: 10.1016/j.paid.2005.11.007.

Whitfield, T. W. Allan, and Lucila R. de Destefani. 2011. "Mundane Aesthetics." *Psychology of Aesthetics, Creativity, and the Arts* 5(3):291–99. doi: 10.1037/a0023038.

Wickens, Christopher D. 1984. "Processing resources in attention". Pp.63-101 in *Varieties of Attention* edited by R. Parasuraman, and R. Davies. New York: Academic Press.

Williamson, Judith. 1978. *Decoding Advertisements: Ideology and Meaning in Advertising*. London: Marion Boyars.

Williamson, Lucy. 2016. "French Periscope death stirs social media safety fears". BBC News. https://www.bbc.com/news/world-europe-36274056274051

Chapter 11

Comparative dialogues on reflexivity: what do pre-service teachers need to understand about 'objectivity' and 'evidence' in their emergent workplace during a post-truth age?

Janine Aldous Arantes

Abstract: Science, Technology and Society can be reimagined to address the challenges pre-service teachers face in this 'post-truth' age. With the increasing complexities of artificial intelligence evident via commercial apps and platforms now part of educational practice, the notion of a simple, linear model of knowledge creation and diffusion must be challenged. This chapter shifts the focus from the student to the classroom as a workplace laden with digital infrastructures that modulate and interconnect information for our emerging pre-service teachers. Concentrating on the reciprocal sharing of language across diverse groups associated with capturing and using socio-emotional datasets in education, this chapter offers a comparative perspective of the word 'bias' in K-12 educational settings across Australia. It provides both a theoretical and an empirical approach, drawing on the findings of the author's doctoral thesis, the Apps in Australian Classrooms Project. The intended contribution is to provide a detailed account of how concepts of truth, evidence and objectivity are entwined within the notion of bias in Australian classrooms. Firstly, it does so by considering 'commercial bias', an idea governed by the Australian Competition and Consumer Commission. Secondly, it considers 'algorithmic bias', a complex computational concept, increasingly of interest to the Australian Human Rights Commission in a teacher's workplace and associated with employment conditions. An interdisciplinary approach to comparing dialogues on reflexivity in digital and infrastructural materiality of the post-truth age is provided through these lenses. The following question guides the chapter's development, "What do pre-service teachers need to understand about 'objectivity' and 'evidence' in their emergent workplace during a post-truth age?"

Keywords: Post-truth education, Digital infrastructures in classrooms, Algorithmic bias in education, Socio-emotional datasets, Objectivity and evidence in the classroom

***

Introduction

A decade ago, concerns about commercialization in the classroom in Australia were based on advertising and private partnerships. People joked about having MacDonald's stickers on their whiteboards. Moving forward, commercialization is hidden and masked in data and the analytics used to develop artificial intelligence (AI) solutions in education (edtech). As a result, it is data that is the price teachers pay for using commercial edtech as part of their educational practice. With claims of technological neutralism rampant, it is profitable to promote the notion that commercial edtech solutions with AI perform better than teachers, often discussed within the frame of fallible human beings. I suggest that data allows for the mediation of reality in a post-truth era. Commercial edtech provides a powerful but hidden voice in shaping this reality in Australian education. Secondly, I suggest that such topics could be institutionalized into pre-service teacher training programs to flag concerns raised by the Australian Competition and Consumer Commission (ACCC) and the Australian Human Rights Commission (AHRC). Training into the notion of market power and free choice and a relationship between AI and human rights could shift training programs away from edtech as a learning tool and towards a more rights-based assessment *before* it enters the classroom.

Whether it is data, big data, personal data, datafication, data science, algorithms, learning analytics, talent analytics, artificial intelligence, or machine learning, edtech is working to monetize it in educational settings across Australia. My focus is not on assessment data. I am focusing on data collected by commercial edtech companies. This focus is significant, as teachers act as a conduit to the provision of commercial edtech in education. As an Australian teacher once said to me: "the way that they do it is through educators as the conduit… you provide them with free stuff and questionable legality things, so that then you have the teachers, so then you have the teachers as loyal users of apps who then share it with other teachers via Twitter and so forth". (Dimble, as referenced in Arantes, 2020)

Therefore, I flag the need to shift focus when talking about data in Australian education. It needs to stop being fundamentally about students and assessment and instead focus on the teacher in a commercialized post-truth era. Teachers use a variety of apps, whether for communication, feedback, audiovisual media, homework, or to enable ubiquity for those not otherwise able to attend class. Due to their prevalence, the apps and platforms participate in the socio-technical context. Teachers are targeted explicitly through their social media, coupled with rewards and badges, and once established as part of the pedagogy, making their withdrawal difficult.

We need a discourse calling for in an epoch of post-truth discourse about teachers' deficits. For example, the prevalence of predictive analytics that perpetuates algorithmic bias is being promoted, and yet a teacher's capacity to predict outcomes is widely (and rightly so) criticized. The truth is lost in a commercialized classroom. How is it possible to describe the classroom as it is? It can be argued that data and platforms offer interpretations under the guise of objectivity. However, their reliance on the financial support of commercial organizations is, from a philosophical point of view, somewhat naïve. There are different situations within the classroom during a supposed 'truth age' where the four walls were the boundaries to education. Now, within a post-truth age, we see that globalization, analytics, and data trails can be used to moderate and modulate the actions and behaviors of teachers. The post-truth age sees teacher behavior equalized with artificial intelligence and automation development.

To assume that the classroom can be described as it is without the complexities well discussed and researched within educational systems should raise doubts about such sponsored initiatives. Their ways of moderating what is true and false in the classroom stand symptomatically for their rise as part of the postmodern abandonment of truth. As Nietzsche has stated, "All things are subject to interpretation whichever interpretation prevails at a given time is a function of power, not truth." These new conditions created by AI contribute to this fallacy of interpretation through seeming objectivity. And we remain yet to understand and unpack how sociotechnical changes are a part of this.

When discussing data and the commercialization of education, we have to consider the classroom as a conduit for commercialization and a workplace with digital infrastructures in a post-truth era. We need to acknowledge that the boundaries of workplaces and personal spaces are blurred and that this is profitable. We need to recognize (or is it admit?) that teachers are being commercially tracked, monitored, and profiled for profit as part of their working conditions and that by encouraging or enforcing the trial and use of edtech platforms, more training is required for teachers to understand their human, consumer, and workplace rights. Throughout this chapter, I draw mainly on teachers' voices, as taken from the interviews completed with Australian teachers in 2019.

In a post-truth era of education, teachers' reality is modulated by commercial data, engagement metrics, and digital profiles built from interconnected datasets both in and beyond the classroom. We have to enter into this discussion acknowledging that commercial data in educational settings is not objective. Nor is it neutral. It is political, has power, and has implications for teachers' consumer, human, and workplace rights, increasingly needing institutionalized programs to address. In this chapter, I have taken into consideration teachers' views. I use their voice in part throughout, weaving within their narrative works from an interdisciplinary approach.

This chapter draws on my Ph.D. project, the Apps in Australian Classrooms Project (Arantes, 2020). I talked with twenty-four Australian educators over 12 months about using and understanding the data and algorithms that underpinned edtech and how they related to their educational practice. The participants were educators from across Australia and included support staff, librarians, principals, classroom teachers, and ex-teachers working in government roles. My Ph.D. project explored educator negotiation of commercial apps and platforms offering personalization. The dissertation drew on close to forty hours of semi-structured interviews with twenty-four Australian K-12 educators, an online survey (n=129), and a platform analysis. This chapter, predominantly based on the interview data, discusses these findings, focusing on how teachers negotiate the algorithmic systems that underpin commercial platforms.

This study builds on the voice of educators and is divided into two parts. The first part establishes the importance of understanding truth in terms of the data used in the commercialization of education in Australia. It begins with a story about my own experiences of understanding the size and scope of correlations made for profit through targeted advertising. It then details how the current practice relates to contemporary considerations of consumer rights and commercial bias. In the second part of the chapter, I amplify teachers' lack of interest in predictive analytics, about teachers thinking more scientifically about educational practice informed by predictive recommendations. I examine how values and beliefs relating to 'evidence' are inherent in how edtech can be profitable and how hidden forms of discrimination resulting from algorithmic bias may be embedded in educational settings. Throughout, I suggest that pre-service teachers need to be encouraged to think more scientifically about data and its infrastructures, including ways that may be interpreted as controversial due to data's perceived objectivity. By considering truth, evidence, and objectivity in commercialized classroom settings across Australia, I communicate uncertainty in scientific evidence and address the problematic and complex processes needed to equip our emergent pre-service teachers to work in this post-truth era of commercialized educational settings.

Propaganda or Pedagogy? The commercialization of education in Australia

And so, the conversation about the commercialization of education begins. Many moons ago, I worked as a Marketing Manager in the Australian not-for-profit sector. Targeted advertising provided access to people I had not met but were able to be algorithmically predicted to engage with the not-for-profit potentially. Their data enabled a relatively accessible way to align their interests with our strategic goals. The ethics of reaching thousands for only ten responses through their data was not considered, as the focus was on a means of advertising that was new and innovative. Move forward ten years, and people's social media is swamped with targeted advertising, and socially mediated perceptions of reality are hidden in data infrastructures. Targeted advertising now does not have the same return on investment. The way data collection occurs is more intangible and interconnected, resulting in intangible forms of commercial bias in the classroom. Commercial bias refers to content integrity or how we're/they're monetizing teachers' the teachers understand data.

Part one focuses on how commercial bias is intangibly entering the pedagogy in Australian classrooms and how teachers understand commercial data flows.

Examining the range of cookies and tracking devices used in contemporary Australia reveals how edtech can track and monitor teacher behavior for profit. Commercial cookies and other tracking devices are considered part of a teacher's workplace, particularly if they are encouraged or forced to use technology as part of their employment. One teacher stated

> Yeah, I definitely feel like we've got that freedom (autonomy to choose free apps).... In terms of trialing new things, there's a lot of encouragement to do that... I'm just always researching new stuff, so I download apps to my phone, it's absolutely full of everything I've ever tried, and I try them out, and then I suggest them to others. (Lisa T, as referenced in Arantes, 2020)

Although trialed, many of the edtech apps were not being used in the classroom and all but one reported that they were not removing cookies from their devices after trialing. In this context, do we value the trial process or the product of data collection? **Table 11.1** provides a list of tracking technologies discussed by the Australian Competition and Consumer Commission to which teachers in the study may have been exposed.

Technology	Description of the Technology
Cookies	First-party cookies: used by an app chosen by the teacher to recall information about the teacher.
Web beacons or pixel tags	Third-party cookies are set by companies other than the teacher's app and are often used for advertising / to track the teacher across different sites.
Device or browser fingerprinting	An embedded technology on the app or email is deliberately not visible to the teacher. "This technology can be used to recognize the same user across multiple online sessions even if cookies are deleted, user log-in changes or IP addresses are hidden or changed." (ACCC, 2019: 388)
Biometric Software	The collection of data related to patterns of how teachers use a device, which enables the identification of a specific teacher.
Mobile Device Tracking	The teacher, as a prosumer, can be tracked on their personal mobile device, including the location that tracks the movement of the device, and that can then use radio signals to communicate with mobile devices passing nearby.
Cross-device Tracking	Multiple methods are combined to identify an individual teacher across different devices. This may be the teacher's log-ins, combined with de-identified data to create connections between separate devices.
Audio Beaconing	Audio beaconing drops a cookie onto a teacher's personal device, enabling the playing of inaudible ultrasonic code through the teacher's personal device's speakers. This code "can be picked up by other smart devices with appropriate software installed and used to link the devices being used by the same person." (ACCC, 2019: 389)

Table 11.1. Online Tracking Technologies

Modified from the ACCC Digital Platforms Final Report (ACCC 2019: 387-389)

An extensive array of tracking devices is apparent. However, "teachers" were not aware of a systematic databasing of such devices that could be apparent due to them trialing edtech. Secondary school teacher Dimble stated, "There's no recording of it. There's no policy on it, no anything", and Linsimma stated, "You can bring any free app that you recommend, yes." Linsimma acknowledged that the mapping or documentation of free commercial apps as part of educational practice provided an ongoing challenge for schools and teachers

(Communications referenced in Arantes, 2020). What happens when commercial data collection from the teachers' devices is not systematically acknowledged? Will this lead to our best educational knowledge to contain commercial bias? Arguably, edtech is shamelessly flexing power dynamics through governmental support and challenging trust in teachers' professional competencies by spruiking a *need* for technology in the classroom. Leveraging the linear model of knowledge creation at universities, educators were left with reflexivity to understand how commercial bias may be modulating their pedagogy and independently blamed if they did not 'clean' the cookies from their devices. Therefore, reflexive dialogue in this context in the post-truth age emphasizes the importance of reciprocal sharing of knowledge through diverse group dynamics. By doing so, institutions may allow the infrastructural materiality of the post-truth age to be made real when a teacher is impacted by commercial bias as a consumer. Thus, new ways of thinking are needed in terms of how such groups may be formed and interacted with so that commercial bias does not become evident in the classroom.

Commercial bias can become evident as edtech monetizes data collected about teachers. Intimate data, including emails, phone, location, and device type, is shared repeatedly when teachers try edtech and are used further to target them on their social media accounts with advertising or to nudge them towards commercial products, conferences, and groups. Commercial bias benefits from the internet being seeded with commercial dynamics. Internet's hyper-connectivity has manufactured rivers of digital data that flow from teachers' interactions with commercial edtech. Where a teacher uses a platform for pedagogical reasons, the data collected and curated by the platform enables forms of modulatory control through predictive and prescriptive analytics. As such, commercial digital platforms in education can modulate what is considered 'truth'. Although a myriad of apps and platforms are used within classrooms, modulatory control in education could be likened to the free commercial platform YouTube.

Teachers do not lock themselves in an enclosure to watch videos; instead, they can leave at any time. However, as every video they choose to watch is continuously tracked, YouTube can predict when they are likely to leave and tempt them with an offer to stay. The videos they ultimately (or only partially) watched in the past are not forgotten, and YouTube uses this information to control what is offered to them in the future. Like being drugged, YouTube controls to what extent teachers can be awakened to different content and keeps them constantly needing the platform. Pragmatically, modulatory control may recommend content or insights for the teacher to use when developing curriculum materials or Google search results. Therefore, what passes for 'truth' in these circumstances is a commercial contingency. Thus, teachers become encapsulated in commercial rhetoric from their social media to interactions with brand ambassadors and the collective use of edtech at school. However, the technology does not need to justify itself according to established learning theory to enter the pedagogical practice. This remains in the realm of teachers exposed to commercial bias. Apps and platforms are considered to participate in the sociotechnical context, resulting in commercial bias. Arguably, commercial bias through apps could concern literally any profession apart from teachers. However, teachers actively try multiple apps and platforms, are mandated to use commercial apps, and feel the 'need' to use free apps – all of which deposit cookies on their personal devices at arguably such a high level that the role of commercial bias in education, is of particular concern.

In trying to understand the size and scope of how teachers employ edtech apps and how these may shape their pedagogical approaches, it became clear that truth and causality were built on teachers' digital profiles. That is, impartiality is profitable in terms of targeted advertising. Cookies collect data, and targeted advertising remains a correlation only. This is because the correlation can be profitable. Should predictions not be accurate, they remain a lucrative venue for Facebook, Twitter, and others for both advertising and the ongoing development of new technology. With data collection comes analytics, and with analytics comes artificial intelligence, and AI is profitable. It is plausible to assume that there is commercial bias in the educational process due to commercial forms of educational technology. Such commercial bias challenges objectivity and truth potentials within the classroom, arguably shadowing established learning theory that underpins evidence-informed pedagogy taught to pre-service teachers in higher education. The teachers I spoke with were part of multiple online communities to increase knowledge and skills, share resources, and collaborate with teachers in Australia and internationally. Many were also 'employed' (working for free) by edtech companies or had been commercially trained.

For example, Jay was a Seesaw Ambassador, and Mr. D, Dimble, Moo, and Jonesy had Google Certifications. Mrs. Jackson and KL were Classdojo Ambassadors. Lisa T and various other educators were Microsoft Innovators (Arantes, 2020). This process has been described as a *classification situation* (Fourcade and Healy 2013*)*. When an educator obtains a commercial badge, the platform categorizes them through their data. It initiates a process of ongoing monitoring of their engagement that can be associated with others within that category (Williamson 2014).

Moreover, 81% were also part of an online commercial group or had received push notifications to further engage with a commercial platform. This results in commercial marketing collateral or propaganda embedded in the classroom but often unfelt and unseen. Commercial bias relates to content integrity (the alignment of the content to understood intent by the teachers). Commercial conflict of interest influences the role of the teacher more for-profit than for educational outcomes. The intangibility of the marketing collateral is considered to be evidentiary of commercial bias. But where else can we see evidence of commercial bias? It is everywhere but deliberately tricky to pinpoint. This research demonstrates that most teachers do not erase cookies from their devices. Secondly, they are mostly unaware of how free apps work concerning collecting their data. Commercial platforms are not for profit – although we cannot prove that this leads to commercial bias, the reasoning concerning the role of commercial bias in Australian schools is valid.

Laws concerning the management of personal data through AI in Australia are in the development stage. For example, although the General Data Protection Regulation is more established in the EU, such discourse is only beginning within Australia. Within the context of Australian classrooms, what the GDPR refers to as 'personal data' and what is collected by companies continues to be revealed to third parties for commercial purposes. These practices further trigger concerns over technology, market power, and free choice in Australian educational systems.

Of course, these concerns suggest that teachers can't be critical. I propose a different stance than this. Although pragmatically employed in a role that encourages the use of edtech, educators are made commercially comatose to such discussions through targeted advertising, badges, certificates, and online groups. No teacher would readily provide their intimate 'data' for a for-profit edtech platform. No teacher would knowingly engage in commercial forms of surveillance. When asked how commercial platforms offer their services for free, M stated: "Honestly, I cannot answer that question...I honestly do not know." Another participant, Adele, stated, "I don't think that I've ever really ... as I said, I often wonder how you can get some of these things for free.... I haven't had a good look at their terms and conditions, but I would assume that they don't sell your stuff." However, some educators provided a clear understanding of sales strategies. For example, Dimble says, "I think the sum is usually tech companies offer a teacher convenience, and we provide the student and teacher data." Alluding to the commercial value of keeping commercial intent obfuscated, Dimble was aware and understood that a sales strategy underpinned edtech, mainly if it was free (Communications referenced in Arantes, 2020). The normalization of commercial bias in education smothers the recognition of the significant market power edtech has, conflates what is meant by *need,* and discounts professional knowledge.

Secondly, education and teacher training in this post-truth context often spruik how edtech can support teaching and learning. Amiss of the ways edtech can accelerate forms of knowledge being circulated for profit and how it may be at odds with their pedagogy, training programs actively encourage critical engagement. How does this criticality not translate into the classroom? Acknowledging that teachers are professionals in their fields, capable of critically evaluating edtech, why do they not question how an app can be free? The answer lies in commercial bias's intangibility, and the enormous market power commercial platforms hold. Teachers are coerced or forced by commercial edtech, the government, and their employment conditions to use technology. This has resulted in this criticality being seen as complaining.

Linsimma stated, "I know that if I raised this at a staff meeting, I would be seen as someone who is not in favor of the school philosophy of engaging in technology, and there are not the forums within schools to stop and think about, well, what are we doing policy-wise for that." Dimble also stated similar concerns, "Yeah. If I were to go to my leadership team and say, 'We need to develop a policy on AI,' that would raise many questions. Or data, even data maybe is so far removed from most people's experience" (Communications referenced in Arantes, 2020). The post-truth appeal of knowledge is now commercially biased in educational settings across Australia. Although the Australian Competition and Consumer Commission has flagged the impacts of such power, it is yet to translate into the accepted criticality of edtech in training programs and the classroom. "We're just looking at the positive education of technology, and so if you raise some of these questions, then you're not a team player in implementing the digital revolution, as such." (Linsimma) There is a need to draw on the Australian Competition and Consumer Commission (ACCC 2019), which makes clear that the systems used by edtech, such as cookies and other tracking devices, are capable of modulating behavior so that teachers are no longer exposed to governmental and school-based mandating of edtech use.

We need to consider commercial data and the consumer rights of teachers and listen to our educators. For example, Technologica asked, "But who's collecting that data? How's that data being used? Where does that data exist?" And, "Can you see that you're desensitiz[ed]... to data collection and surveillance?" (Communications referenced in Arantes, 2020) While teachers use edtech for pedagogical purposes, we need to train our emerging teachers regarding the consequences of commercial surveillance in a teachers' workplace. It is no longer intangible and normalized (Manolev, Sullivan and Slee 2019). We may shift this context from being openly 'asymmetrical' due to its pejorative nature by dismantling the universality of commercialized post-truth rhetoric. The exposition of the power relations buttresses altered conceptualizations of truth associated with commercial bias that the teacher and their 'truths' may challenge the assumption that Big Tech was not negotiable to resist edtech's disruption.

The following section will demonstrate how these tensions between commercial in Australian classrooms and their potential impacts on teachers. These inconsistencies will be explored from the teachers' perspective as comparatively new users of AI in the classroom. There will be an emphasis on how they understand and perceive the consequences of use. In particular, the following section highlights the potential human rights violations that teachers must grapple with, guided by the Australian Human Rights Commission's report on Human Rights and Technology.

Cookies and what constitutes evidence: It's a matter of discrimination

As the conversation about commercialization considers what constitutes truth, we also need to consider what constitutes 'evidence.' Commercial asymmetry leverages 'evidence' claims in knowledge creation to promote a perceived pedagogical need. However, if personalized insights drawn on predictive analytics are not valid but encouraged by commercialized evidence claims, what may the impacts be? With an increased commercial promotion that technology is *needed* in educational settings, the educators were aware of sales strategies and concerned about commercial bias. Secondly, they were also very clear that predictive technologies could not be valid. To explore evidence and predictive analytics used by AI in education, the language to describe 'bias' associated with predictive analytics from a computational perspective now shifts towards a complex analytical term called 'algorithmic bias.'

Algorithmic bias refers to the interactions between aggregated de-identified data and personalized data that make up the decision matrix or pipeline for insights and recommendations that result in erroneous predictions. Where commercial bias refers to the conversion of the teachers' digital interactions into data that can be monetized, stored, and repurposed, algorithmic bias refers to the process of analyzing the data and the inherent trade-off between accuracy and bias. Notably, algorithmic bias is a fundamental component of AI that uses predictive analytics, and the removal of algorithmic bias cannot currently be guaranteed. Further, teachers largely under-negotiate algorithmic bias (Arantes 2019), and the causes of algorithmic bias are varied. For example, algorithmic bias can occur due to historical biases in the datasets, missing data, algorithmic objectives, and proxies where aggregated de-identified data is used to infer an identity about an individual (Pessach and Schumeli 2020). Algorithmic bias has arisen in AI-informed decision-making across multiple contexts in Australia and beyond.

Data and technology are often presented as being objective. A recent report from the OECD states, "While technology is ethically neutral, it will always be in the hands of educators who are not neutral. The real risks do not come from AI but the consequences of its application" (OECD Education 2021). This notion, however, obscures the socio-emotional and socio-cultural context in which data was collected and the policy or legislation that enabled or promoted the algorithmic design of the edtech. There is a trade-off between the accuracy of insights and bias that manifests across a broad range of educational contexts, leading to "a danger of false positives" (Bulger 2016). False positives can emerge due to dominant social norms represented in the data used by edtech companies. It is well proven in STS research that different interpretations can emerge from the same data; hence, paradigmatic considerations are rampant in educational research. Such considerations are lacking in largely positivist approaches. By examining the evidence used to make claims of objective or ethically neutral AI, it is clear that algorithmic bias is increasing and perpetuating the presence of personal and social harms (Calo 2011). Therefore, data and technology cannot be seen as neutral or objective. Instead, there is a need to acknowledge that stereotyping and intangible forms of discrimination can occur at all stages of the datafication process, reproduced as predictive insights and recommendations. But is technology any less biased than the teachers themselves? Some argue that these ways of thinking occur as algorithmically informed 'solutions,' are perceived to integrate more data than humans, can perform more complex computations, and that human decisions are ultimately subjective (Pessach and Shumeli 2020).

However, they clarify that it is an incorrect belief that edtech solutions with AI perform better than teachers as human beings. Secondly, corporate marketing machines continuously advertise that they are better than humans. Thus, humans are perceived as less capable and more biased than algorithmic systems due to huge corporations' intangible data infrastructures and marketing collateral. This, in turn, encourages a way of thinking that sees edtech based on predictive analytics as more objective than human teachers, providing 'evidence' that produces more accurate results. This is further entrenched through highly financed and marketed claims. For example, Google's description of its adaptive technology is likened to some form of 'magic.' However, this is simply not the case, and our teachers know it.

Clearly articulating why predictive analytics were problematic, Linsimma provided various examples whereby individual complexities could not be quantified and expressed a feeling of danger concerning predictive analytics within the classroom. Linsimma stated:

> And that's why [there is] the danger of data prediction...because there are so many variable factors... there are a lot of things that impact student success that's got nothing to do with the student's individual data...I can think of about 20 situations where students have been thrown out in different cohorts for different events and different year groups. I have a teaching experience that one student I taught, it would have been many years ago, totally lost it because it was the year that her brother was sent to East Timor with the Australian Army and she just totally lost it because she was scared that he was going to get killed. I mean - you can't quantify that. (Communications referenced in Arantes, 2020)

Dimble imagined a similar concern, focusing on Google and how its perceived influence over educational practice might play out in the future, and stated, "I could foresee a future in five-ten years where there is a range of massive court cases where a [parent] says, 'Google stole my child's learning'" (Communications referenced in Arantes, 2020). Explaining why such concerns were justified, Linsimma stated, "there are so many variables about issues that would also concern me about how valid the data is and how to construct it. What's its baseline?" (Communications referenced in Arantes, 2020). As such, pre-service teachers require training beyond pedagogical practice and educational outcomes.

Australian Human Rights Commission has flagged how algorithmic bias has obscured and entrenched unfairness or even unlawful discrimination in contexts ranging from the criminal justice system, advertising and recruitment healthcare, and policing. The commission has made the following recommendation (Farthing et al. 2021):

> **Recommendation 18:** The Australian Government should resource the Australian Human Rights Commission to produce guidelines for government and non-government bodies on complying with federal anti-discrimination laws in the use of AI-informed decision making.

The profiling of teachers based on data collected by cookies and other tracking devices to draw inferences about their behavior for profit flags the risk of unfair and discriminatory treatment. Since edtech utilizes predictive and prescriptive analytics as either a targeted advertising mechanism or as part of a teaching and learning tool, there is a practice of constant comparison between individuals and large de-identified datasets, resulting in algorithmic bias that may be legally questionable. Australia's federal anti-discrimination laws prohibit discrimination based on protected attributes, such as age and gender identity. However, these are common categories used in targeted advertising and the design of edtech.

Thus, it is a question of compliance with anti-discrimination laws. As 'evidence' is determined by which commercially selected proxies are used, and commercially derived proxies may be curated through the use of protected attributes, should a commercial proxy drawing on a protected attribute be profitable but erroneous, issues of commercial biasassociated with anti-discrimination laws become problematic but remain intangible. With commercially selected proxies used as 'evidence,' commercial bias may also promote claims based on proxies that are profitable over their 'truths.' Predictive insights and recommendations contain commercial and algorithmic bias in this educational context. As protected attributes may have been used in AI decision-making, edtech may not comply with federal anti-discrimination laws. Where it could be argued that commercial bias is not infringed on consumer rights, as commercial edtech and teachers receive reciprocal benefits, the impacts of algorithmic bias take the discourse into legalities. No teacher knowingly would want predictions made about them based on a potentially discriminatory data set, as there are federal laws to protect them against such activity.

Secondly, algorithmic bias is a less considered topic in education. Jonesy stated, "I wouldn't say [algorithmic bias] is such a big talking point amongst people who are just looking for a solution to get something done in the classroom. They might not consider the implications of the data use and the way it's collected." Thus, the commercial asymmetries may be modulating any impact of algorithmic bias. In this context, where algorithmic bias is intangible in data-driven insights and recommendations, true and false claims are not just beliefs; they become manifested in practice (Hempel 1945) and policy (Jasanoff 1999).

When considering the impacts and implications of algorithmic bias in educational settings in Australia, the notion of 'reciprocal benefits' flags that evidence in a post-truth era is related to discrimination and arguably human rights violations; that is, data is categorized to make predictions, which from a human rights perspective aligns very closely with stereotyping. To understand the complex and dialectical nature of how technology and social data are made real through edtech's datafication of the classroom, algorithmic bias claims of being grounded in neutrality are challenged.

This new way of thinking is essential, as it further impacts teachers' employment. As commercial edtech, data collection, predictive analytics, and AI are now significant aspects of what teachers experience as part of their conditions. With notions of class, race, and gender represented through commercially derived proxies, teachers' membership and group classifications via badges, certificates, and ambassador titles arguably provide fodder for discrimination due to confirmation bias. Acknowledging commercial and algorithmic bias remains fundamentally intangible to the teachers' negotiating it, but their employers can garner insights from commercial talent analytics.

Further impacts to consider are related to employment. Talent analytics and Human Resource analytics provide data about who to recruit, promote and justify the termination of employees. These tools use predictive analytics to link various data streams based on commercial proxies to obtain a picture of teacher behavior. Employers can forecast teachers' cultural fit and predict employee mental health. Other indicators, such as behavior and performance, are used within workplaces across Australia and the globe in terms of workplace intervention (He et al. 2015). Teachers arguably have larger datasets due to the ongoing trial of edtech platforms that leave cookies on their devices. Commercial platforms claim that they can use metrics from their data aggregated with risk management metrics to act as proxies for mental health and well-being. Other platforms claim to offer a means to screen toxic workplace behavior via monitoring social media. Data objectivity in terms of confirmation bias and workplace rights impacts education. It affects educational practice, as intangibly, some pre-service teachers may or may not be witness to advertisements, promotions, and feedback based on the proxies used to interpret their data commercially. With the digital and infrastructural materiality of the post-truth age flagging the blurring of boundaries between learning analytics and talent analytics, we must re-think how these boundaries became blurred to better address the issues of confirmation bias in the classroom as a workplace. Recognizing that boundaries have become blurred is the start.

Pre-service teacher training needs to flag that the unique complexities of teachers, teaching, and education can never be fully quantified and datafied. The Australian Human Rights Commission suggests vulnerabilities built into decisions made with inferred identities created (Guihot and Rimmer 2019). As such, edtech's datafication of the classroom may be seen to present "issues of inclusion and fairness... including around race, gender, sexual orientation, socio-economic status" (Standards Australia 2019).

There is a growing recognition that where past datasets about teachers and teaching are used as proxies, the social and historical biases hidden in the data may result in "unlawful discrimination... to appear in new forms," and teacher training programs need to shift towards the inclusion of human rights as part of the discussion around technology use in the classroom (Guihot and Rimmer 2019). Teachers can explore the technology development that produces proxies and explain that it does not follow a linear synthesis of knowledge creation. The pedagogy they create is built on the knowledge that commercial proxies that can result in erroneous insights, predictions, and recommendations can then address impacts associated with algorithmic bias. This may result in bias in adaptive language models whereby offensive associations related to gender and race through publicly available word embeddings are provided as recommendations (Hovy and Spruit 2016) being challenged. Or, where computer vision training through "off-the-shelf pre-trained models" claim to offer useful insights as they are representative of "lighter male subjects than darker female ones" are not purchased (Ntoutsi et al. 2020:5). As algorithmic bias results in products that could perpetuate racial forms of discrimination, a policy shift towards forcing edtech to explain how they have addressed algorithmic bias may become enacted. Bias can manifest in all forms of data through sensitive features and the resultant causal influences, even through the under or over-representation of certain groups. Our emerging teachers require training far beyond current pre-service teacher training programs in Australia that focus on improving teaching and learning. Thus, this new way of thinking flags the essential nature of the reciprocal sharing of notions of 'evidence' across diverse groups to enrichen the training teachers receive.

Conclusion

Using the voice of Australian educators, this chapter has allowed me to communicate a contextually relevant snapshot of how a group of Australian educators understands data, analytics, commercialization, and its impacts on their educational practice. From their role as consumers engaging with marketing and trial apps to their physical use of predictive insights as educators in a classroom to their workplace experience as employees, the word 'bias' has entered education in intangible forms. This chapter explored how commercial data allows for the mediation of reality in a post-truth era in Australian classroom contexts and responded to the question: "What do pre-service teachers need to understand about 'objectivity' and 'evidence' in their emergent workplace, during a post-truth age?" The discussion and analysis present the following two conclusions:

Commercial edtech provides a powerful but hidden voice that is shaping Australian education, and we need teacher training to challenge the notion of data objectivity by associating it with their consumer rights. Predictive analytics used by edtech has obscured a relationship between AI and human rights. We need teacher training to challenge the notion of data as valid evidence by associating it with human rights.

The topics covered in this chapter have crucial importance for the future of education from societal, legal, technical, and political points of view. The post-truth age demonstrates the power that pre-service teacher educators have to respond to these conclusions in Australian education. Teachers can challenge the nature of 'truth' and 'evidence' when discussing edtech supporting educational practice and any perceived *need.* Such educators can enable emerging teachers to challenge commercialized 'facts' even in the face of policy and curriculum by flagging research by large Australian commissions, such as the Australian Competition and Consumer Commission and the Australian Human Rights Commission. We can understand whether our emergent pre-service teachers understand what commercial bias and algorithmic bias really mean when artificial intelligence is digitally profiling them. We can then encourage others to explore: How can we begin to educate pre-service teachers about the scope and scale of these new data-rich classrooms so that they can truly grasp what is at stake?

References

Arantes, Janine Aldous. 2019. "Equity Implications of Predictive Analytics in K-12 Classrooms." *Ubiquitous Learning: An International Journal* 12(2). https://doi.org/10.18848/1835-9795/CGP/v12i02/63-84

Arantes, Janine Aldous. 2020. "Big Data, Black Boxes and Bias" (Doctoral dissertation, The University of Newcastle, Australia).

Australian Competition and Consumer Commission (ACCC). 2019. *Digital Platforms Final Report.* Canberra, ACT, Australia: ACCC.

Guihot, Michael, and Matthew Rimmer. 2019. *Artificial Intelligence: Governance and Leadership - A submission to the Australian Human Rights Commission and World Economic Forum.* Australian Human Rights Commission and World Economic Forum, Australia.

Farthing et al., 2021. *Human Rights and Technology Final Report 2021*, Australian Human Rights Commission. https://tech.humanrights.gov.au/sites/default/files/2021-05/AHRC_RightsTech_2021_Final_Report.pdf

Bulger, Monica. 2016. "Personalized learning: The conversations we're not having." *Data and Society* 22 (1): 1-29.

Calo, Ryan. 2011. "The Boundaries of Privacy Harm." *Indiana Law Journal* 86(3): 1131.

Fourcade, Marion and Kieran Healy. 2013 "Classification situations: Life-chances in the neoliberal era." *Accounting, Organizations and Society*, 38(8): 559-572. https://doi.org/10.1016/j.aos.2013.11.002.

He, Wu, Jiancheng Shen, Xin Tian, Yaohang Li, Vasudeva Akula, Gongjun Yan, and Ran Tao. 2015. "Gaining competitive intelligence from social media data: Evidence from two largest retail chains in the world." *Industrial Management & Data Systems.* 115(9): 1622-1636. https://doi.org/10.1108/IMDS-03-2015-0098

Hempel, Carl G. 1945. "Studies in the Logic of Confirmation (I.)." *Mind* 54 (213): 1-26. https://doi.org/10.1093/mind/LIV.213.1

Hovy, Dirk and Shannon L. Spruit. 2016. "The social impact of natural language processing." In *Proceedings of the 54th Annual Meeting of the Association for Computational Linguistics (Volume 2: Short Papers)*:591-598. https://doi.org/10.18653/v1/P16-2096

Jasanoff, Sheila. 1999. "STS and public policy: Getting beyond deconstruction." *Science, Technology and Society* 4(1): 59-72. https://doi.org/10.1177/097172189900400104

Manolev, Jamie, Anna Sullivan, and Roger Slee. 2019. "The datafication of discipline: ClassDojo, surveillance and a performative classroom culture." *Learning, Media and Technology* 44(1): 36-51.

Ntoutsi, Eirini, Pavlos Fafalios, Ujwal Gadiraju, Vasileios Iosifidis, Wolfgang Nejdl, Maria-Esther Vidal, Salvatore Ruggieri et al. 2020. "Bias in data-driven artificial intelligence systems—An introductory survey." *Wiley Interdisciplinary Reviews: Data Mining and Knowledge Discovery* 10(3): 5 https://doi.org/10.1002/widm.1356

OECD Education, and Skills Today. 2021. "How Can We Radically Reimagine What Teaching and Learning Can Be When Powered by Digital Technology?" *OECD Education and Skills Today.* https://oecdedutoday.com/how-radically-reimagine-teaching-learning-digital-technology/.

Pessach, Dana, and Erez Shmueli. 2020. "Algorithmic Fairness." *arXiv preprint arXiv:2001.09784.*

Standards Australia. 2019. *An Artificial Intelligence Standards Roadmap: Making Australia's Voice Heard - Final Report.*

Chapter 12

Masculinity of the Scientific Tradition in the Post-truth era: a comparative analysis of Sweden and Turkey

Ezgi Pehlivanli

Abstract: STS has received contributions from feminist scholars who have identified men's monopoly of technology as an important source of their place in the social and academic hierarchy. This approach takes gender relations as an integral constituent of the higher education institutions and projects from which technologies and scientific knowledge emerge. In the post-truth era, knowledge claims about gender tend to be different versions of debates of anti-feminism, which seek to delegitimize the knowledge produced and disseminated by feminist scholars and movements while also trying to establish an alternative set of purposefully selected legitimized truth about gender. This paper situates current discussions in STS, of women's position in STEM departments in (Lund University and Middle East Technical University (METU) in the wider context of post-truth rhetoric on gender. It aims to explore how gender-specific debates are conceived, questioned, and justified in two contexts while at the same time trying to understand the material and symbolic manifestations of gender dynamics of two institutions in relation to deep-rooted traditions in STEM departments. Findings highlight that organizational structures within faculties need to be reinterpreted by looking at post-truth claims, especially for STEM departments where the ratio of female participation and LGBTIQ+ representation is relatively low.

Keywords: Post-Truth, Masculinity, Scientific Tradition, STEM, Gender, Sweden, Turkey.

***

Introduction

The interdisciplinary community of Science and Technology Studies (STS) contains resistance to an uncritical acceptance of scientific authority. As STS developed, scholars sought to understand how scientific research and production of knowledge were planned, situated, and practiced (Latour and Woolgar 1979; Harding 1986; Fox-Keller 1987; Clarke and Fujimura 1992; Fujimura 1996). The epistemologies and methodologies have been questioned, and possibilities emerged for an inclusive science and technology. In this tradition, facts are not simply waiting to be discovered in nature. They are negotiated within society by different groups, such as consumers, producers, and experts. They are also mutually engaged with values and norms because knowledge producers are human, and they are made up of social narratives composed of values and norms (Haraway 2020: 305).

Social understandings of science and technology were analyzed by feminist contributions in the twentieth century. Feminist theorists criticized the claims of neutrality of science; they argued that science had been granted 'cognitive authority' (Laslett and Brenner 1989:1) because of its claim of objectivity (Harding 1986: 4). Such privilege to science is mistaken because the practice of science, like any other branch of human endeavor, cannot be disembodied from the value systems and implicit biases and ideologies of its practitioners (Harding 1991:89). Science historically has been produced and conducted mainly by middle-class, white, European men who are largely thought to be heterosexual (Cech and Waidzunas 2021:2). Thus, it cannot be neutral from the values of such situated representation. Mentioned critical contributions created a certain degree of diversity awareness in STS and in other disciplines, which triggered attention to underrepresentation in areas of science, especially in natural sciences, technology, engineering, and mathematics, as in the widely used acronym STEM.

The post-truth era is defined as "denoting circumstances in which objective facts are less influential in shaping public opinion than appeals to emotion and personal belief." (Tsipursky 2017:13). Not exactly an opposition to "science," the post-truth narrative highlights the selective use of emblems and idioms of scientific authority and political administration of selective treatment of science advice. (Fujimura and Holmes 2019). Oppositional voices are often expressed through the rhetoric of science, voiced by credentialed experts who claim to present "alternative facts" (Cann and DeMeulenaere 2020:25). Given the dynamics of the criticisms, STS and the notion of gender in science have been dragged into a debate on scientific validity.

The current politics of knowledge in the world, the mobilization against "gender," is de-democratizing knowledge production in academia and making it difficult to sustain. Academic`s changing research dynamics due to Post-truth claims, including anti-gender movements and conservatism (Paternotte and Verloo 2021; Mede and Schafer 2020; Korolczuk and Graff 2017; Kovats and Petö 2017; Kuhar and Paternotte 2018) affect academics and institutions. Despite the efforts for diversity and equal representation in science, current scientific knowledge has been under criticism, especially after the COVID-19 pandemic (Fujimura and Holmes, 2019:1253; Eslen-Ziya and Pehlivanli 2022: 200).

In the post-truth era, the norms governing claims-making are being significantly reconstructed, with significant consequences for critiques of gender inequality at the institutional level (Budgeon 2021:248, Drew and Canavan 2020:167). It is argued that the post-truth rhetoric used to dismiss women's experiences provides a challenge that feminism must confront (Ringrose 2018). Attempts to discredit gender claims with the rise of conservative and authoritarian populist attacks on gender as 'ideological' (Mihăilă et al. 2018: 160) make for a troubling context for the main feminist argument that cognition is a human practice mediated by theoretical propositions which illuminate the question of what can be known. Taking feminist argument as the basis for my study, I defend the role of personal experience in the post-truth era; I believe an awareness of the shift in knowledge production and perception would create possibilities of transformation that would benefit and protect feminist claims, in return.

In the framework of this study, I take STEM departments as one of the organizations mentioned by Bijker and Law (1994), which are populated mainly by men, are constructed by the masculine tradition of science, repeating the old stories I mentioned above by building academic networks, researching masculine questions with their own teams within the university environment. Thus, this study asks whether the corporate culture of STEM exhibits similar or differentiating aspects for the two countries under analysis, namely Sweden and Turkey, and addresses to what extent STEM organizational culture displays aspects of masculine culture. The goal of this article is to outline new directions for gender in STEM by comparing how supposedly two different countries experience gender equality in STEM academics. In the so-called "post-truth" world, there exists widespread confusion and disagreement over what is known, how to know, and whom to trust. Current higher education has largely failed to meet the challenges of this world that are needed to promote gender inclusion in the "post-truth" landscape.

Based on these aims, this paper situates current discussions of gender in STEM departments at Lund University and Middle East Technical University in the broader context of post-truth debates on gender in STS. To do this, it explores the material and symbolic roots of masculine scientific tradition via the work-life experiences of full-time academicians. Findings highlight that organizational structures within faculties need to be reinterpreted by the "gender as doing" approach, especially for STEM departments where the ratio of female participation and LGBTIQ+ representation is relatively low.

Methods

A growing body of literature exists on gender and STEM academia (Schmader 2022:226; Grossman and Porche 2014: 700). Across North America and Western Europe, the dominant model for understanding the scarcity of women in STEM has been the 'pipeline.' The pipeline model was introduced in the USA in the 1970s and exported to a range of European countries, including Sweden (Lövheim 2014:1768) and the United Kingdom (Archer et al. 2013). It is now a part of what Ball (Ball 2008:1) calls education 'global policyspeak.' The model describes a linear sequence of steps defined as necessary to become a scientist or engineer and sets out the corresponding numbers that are 'needed' to maintain national competitiveness. Thus, the pipeline model has been the basis of the quantitative modeling of labor market supply, and policy interventions have been based on these models.

There are several problems with the pipeline model. First, its engagement with social justice is limited to adding women, LGBTQ+, and people of color to the existing science workforce. It does not consider the intersecting forms of privilege that enable some women and people of color to progress through the pipeline or what impact these may have on science. Second, the pipeline model assumes science itself is unproblematic. The model does not engage with debates around science as an elite occupation or one implicated in neoliberalism. It does not address and implicitly dismisses questions about the relationship to the science of those who eventually leave/leak.

Hence, within the mainstream approaches such as the pipeline, an analysis of how issues regarding gender equality have become subject to regulatory mechanisms within purposes related to the impact of the mentioned practices on the daily life experiences of scholars is lacking. According to studies, many women in STEM report inequities in their career outcomes. 2020 OECD survey of academic scientists[1] suggests that women earned 5–6% less than their male peers, controlling for individual and job characteristics. This pay gap was particularly wide in more lucrative careers in engineering, computer science, and senior management. Thus, although progress toward gender equality has been made in the life and social sciences, the physical sciences continue to be marked by persistent gender disparities in representation, feelings of inclusion, and pay equity.

A comparison of two countries through an analysis of two cases, namely, Lund University and METU, provides an example of how gender can operate in daily practices (Martin and Fischer-Ari 2021:1080). That is to say that, on the individual level, academics engage in the career practices by which the gender

---

[1] OECD International Survey of Science - OECD Accessed on 06.02.2023.

order is constituted in STEM in universities (Martin 2003:354). For instance, this includes acquiring visibility, cultivating networks, and reconciling time to acquire academic capital (O'Hagan et al. 2019:75). These practices are gendered and provide an indication of the way gendered privilege interacts with the global research economy to ensure that minorities are kept away (Morley 2016:35). Moreover, gender discourses in higher education usually operate over the distribution of genders in academic departments.

Thereby, this study compares the two countries, which appear to be different from each other according to the UN[2] Gender Equality Index (2018): Sweden (UNDP GII 7th) and Turkey (UNDP GII 48th). Within the scope of this research, gender-oriented developments in academia are handled in two ways:

1. Policies that guide and link universities to support the gender equality facility for academicians in STEM areas
2. Ensuring gender-based equality in addition to numerical data; re-interpretation of the ongoing dynamics in access to permanent positions in higher education, the rise in duty and access to scientific research resources

Within the framework of this study, it is accepted that there are different experiences of gender and that all genders, including men, are affected by gender-based structures in institutions. It is aimed to discuss the experiences of participants through in-depth interviews. I reached the participants via snowballing sampling technique through personal relationships and participants' roles in gender-related committees such as WINGS[3]. Overall, for the profile of participants, please see Table 12.1 below.

Interviews conducted at Lund University			**Interviews conducted at METU**	
	Woman	**Man**	**Woman**	**Man**
Mechanical Engineering	3	2	3	2
Electrical/Computer Engineering	2	3	2	3
Mathematics	3	2	2	3
Physics	2	3	3	2
Total	20		20	

Table 12.1. Profile of Participants

[2] HDR21-22_Statistical_Annex_GII_Table.xlsx (live.com) Accessed on 06.02.2023.

[3] https://www.wings.lu.se/

Two sets of questions were prepared, one for male and one for female participants. Following two pilot studies, the semi-structured questionnaire was finalized. The interviews were carried out in Lund and in Ankara, usually in the participants' offices. Interviews started with demographic questions, followed by questions about their experiences. The transcriptions and the fieldwork notes constituted the main data set of this study. For the secrecy of participant identities, I named participants from P1 to P40, referring to her/his order in interviews.

In this context, the participants are asked semi-structured questions about their academic life experiences, organizational culture, career selection backgrounds, coping mechanisms for academic workload, and access to research resources. Grounded theory was deployed as a research method, aiming at an iterative process in which I defined core theoretical concepts and themes through the interpretation of the data. In this process, I tended to create links between the theoretical core and the data I collected to create a meaningful abstraction level without losing the systematic conceptualization of the data. After reaching a consensus on the categories of existing data, I prepared interpretation material to decide under which narrative the emerging themes and categories fell.

In addition, the case analysis method was used. The main reason for this is that case studies are a useful method for discovering new variables, producing new hypotheses, and clarifying causal relationships (Yin 2006:44). The case study method allows a unit to be analyzed closely and in detail, and consequently allows new approaches and hypotheses that are not visible beforehand (Yin 2006:47). Therefore, this research characterizes a research process in which explanatory variables and causal relationships will be clarified.

Results

Among many policy instruments, gender diversity strategy is one possible proposal, and it is widely adopted as a part of new organizational fabrics. The findings of this study showed that structural gender inequality is evident in the field, and it results from an interplay of factors at the individual, institutional, and structural levels. Findings also confirmed that there is no gender neutrality in academia regarding organizational culture (Morley 2005:420), and institutions of Lund and METU have highly gendered practices as well as the culture of the organization.

Statistics from Lund University and METU show that STEM departments have been populated mostly by male academicians. However, it is also observed that Lund University achieved an increase in the ratio of women in

STEM from 19% to 37% throughout the 2000s, but the ratios for METU seem to stay stagnant between 35% and 37%. The academic person—the "scientific persona" (Daston and Sibum 2003:5)— in both institutions is constructed as masculine (Beaufaÿs and Krais 2005:60), often with a dependent spouse. Other practices continue to rely on patronage and the old boys' networks (Rees 2001:245), which are made visible, for instance, in the fact that men obtain research funding more successfully than their female colleagues. Additionally, the two institutions have regulatory mechanisms based on policy statements published in Lund on 8 September 2011 and in METU on 24 May 2016 to reach gender equality within every field of academic life. There is less blatant sexism than in earlier years, but a subtle bias persists. This results in minorities dropping out of scientific careers, which leads to a lack of role models in the upper divisions and administrative units of both institutions.

This part will proceed into four subtitles to examine the findings of this study: a masculinist restoration after the post-truth, departmental culture, diversity policies, and real-life experiences, and finally, LBGTQ+ and non-mainstream masculinities.

A masculinist restoration after Post-Truth?

We live in a time of massive disinformation and post-truth (Özdemir and Springer 2022). Epistemic climates, especially higher education institutions, are the ones mostly affected by the weakening of public trust; public disagreements about established empirical findings are rampant, and opinion often takes precedence over evidence (Barzilai and Chinn 2020; Buckingham 2019). Science has a deep-rooted tradition. It is based on the traditional epistemic endeavor, mainly relying on old boys' networks. The changing dynamics of the academy due to post-truth claims, including anti-gender movements and illiberalism (Micheale, Bitzan and Petö 2017:178), point to a societal crisis beyond gender equality and LGBTQ rights, but these are symptoms of deeper crises of democracy and academic freedom.

Feminist scholars have argued that the most fundamental aspects of systematic theory in the natural sciences have been pervaded by masculine perspectives deriving from masculine experiences (Siann and Callaghan 2010: 90). According to Fox-Keller, the ideology of modern science provided men with a new basis for masculine self-esteem and male ideology over natural processes. The scientists, technologists, and managers of capitalist societies found opportunities to show their "superior masculinities" (Easlea, cited in Fox-Keller 1987: 64). In addition, as time passed, definitions of male and female were polarized in ways that were suited to the growing division between paid work and homework required by growing capitalism (Fox-Keller 1987, 61).

The collected data also confirms these statements. The interpretation of participants' narratives shows the impact of mathematical and technical ability and the influence of relatives and teachers in their career selection. Prestige and income were other major factors that played a role in their choice. They also reported that the image of a scientist is male in both countries.

> I chose Physics because I was good at math and sciences. Also, Physics was very prestigious. More prestigious than Biology for instance. I have always wondered why I did not choose biology, because the image around biology was so female. It is soft. I recognized that after I think about gender issues. It did not have the prestige that Physics had. Perhaps, that is the reason I did not choose biology.[4]

> Did you see the big poster on the department hall? It says 'The Men of Mathematics' [5] Every day I look at that picture and I tell myself that the poster should be my trigger in my academic life. It is sad, though. [6]

The quotations above reveal how gender stereotypes are embedded in professional perceptions. As mentioned by P3, hard and soft sciences are ideological and based on a complex web of general discourses formed around traditional gender roles, family, technique, technical know-how, masculine hardness, and feminine softness. The embeddedness of gender in scientific inquiry and academia cannot be uncovered simply from statistics. So, the question of underrepresentation is not only about numerical scarcity. This problem has other dimensions that are concealed in historical formations, daily expressions, prejudices, and in interaction styles. It originates from the way gender roles are learned; it is because of how gendered social structure is internalized and related to the capitalist relations that maintain and reinforce those gendered bondages.

> This is a faculty of engineering, its culture historically, it's a masculine thing. Patriarchy comes sort of naturally, this is how I take it. [7]

Meanwhile, in the universities of the world, by the 2020s, the focus on gender had left its place to alternative knowledge claims and distrust in science. Diversity defenders were trying to carve out a small space of maneuver for themselves by relying on the UN`s inclusive agendas, and then COVID-19

---

[4] P3, interview with author, Lund, February 12, 2019.

[5] A chronological poster showing important figures in the history of Mathematics.

[6] P21, interview with author, Ankara, June 03, 2020.

[7] P15, interview with author, Lund, February 9, 2019.

arrived and smashed out the achievements. Already debated post-truth claims articulated with anti-gender, anti-science (Eslen-Ziya and Giorgi 2021:50) mobilization and indirectly supported a new type of conservatism. The findings of this study revealed that there is an ongoing tendency to produce and reproduce the naturalized character of patriarchy which confirms Deniz Kandiyoti`s (2016: 110) argument on masculinist restoration as being the reproduction of masculinity when its naturalized character is called into question.

> There is serious pressure on us. When we said that we are getting our rights as full women, we lost most of them with the pandemic. Now it's back to the offices, and we see that we lost some of the areas we won to men again.[8]

I argue that the types of resistance and subversion are a consequence of post-truth claims articulated with patriarchal pretensions of power and become the occasion for new forms of restoration of masculinity where patriarchy is deep-rooted, as it is in STEM departments. Thus, the situation calls for accountable governance for higher education institutions and democratic representation to create new transformation axes that need to start with understanding the departmental cultures.

Departmental culture

Participants of this study told various stories about their department lives. Some took the gender aspect seriously and tried to fight against it. Some participants did not have an awareness of this departmental culture, while others intentionally ignored it as a survival strategy. Gendered jokes, and phrases were mentioned as covert forms of discrimination.

> We experienced it many times. For instance, the professor comes to class and says 'good morning, gentlemen.'.[9]

P20's statement shows how professors' daily language is determined by gender; observing the presence of female students and not addressing them might be an unconscious act. However, the act unintentionally also shows the gendered image of the engineer in the professors' minds. Feeling left out was only mentioned by women participants in this study. Similar to Robinson and McIlwee's research, women academics argued that they do not share the same

---

[8] P3, interview with author, Ankara, June 20, 2020.

[9] P20, interview with author, Lund, February 20, 2019.

burdens as their male colleagues. Some women respondents suggested they must cope with feelings of being left out and decreased self-esteem.

> When I first came to this department, I could not make any friends. Only women and there were two women. One was so blind, or I do not know maybe she chose to act as if everything is awesome. The other one helped me so much, when I had my first child. She was head of the department at that time, and you know women are usually the head of the department. Because men do not want to do the job. Men simply ignored me as a colleague. If women professors do not encourage you, you stay alone. [10]

Male professors emphasized that they believe they treat all colleagues equally. Some women participants, on the other hand, were sensitive about the issue. They insisted that, as professors, they also experienced silence or neglect in their departments.

> Departmental culture is my favorite (she laughs). I mean it is masculine. Especially at our department. It is like male professors are our fathers and we are conforming daughters. (laughs) [11]

Silence about gender is one of the ways to maintain hierarchies. Professors might think they behave positively by remaining silent about the issue of gender o. They might even think they do it in the name of equity. However, by being silent, professors not only ignore women but also sustain the existing status quo between genders. Social relations in this department were described as a combination of education-related and leisure-time activities. As mentioned previously, women were subjected to gendered jokes and implications. One choice for women was to adopt one of the categorized identities. Some chose to remain reserved, while others tried to fit in, whether they really liked it or not.

> They say things like 'wow, she is really good, we didn't expect this' which is kind of insulting. Although she is a woman and she get to hear things like 'what do you do in Alex's office, do you make coffee?'[12]

Participants from Lund University mentioned gender committees that were created to overcome gender inequality and raise awareness of problems both at the faculty and at the university level. They also mentioned that this

---

[10] P18, interview with author, Lund, February 17, 2019.
[11] P32, interview with author, Ankara, June 01, 2020.
[12] P11, interview with author, Lund, February 24, 2019.

institutionalization positively impacted recruitment and promotion processes and that the mechanism facilitated changes in the departmental cultures.

> When I began here 8 years ago, the departmental culture was masculine. There has been a large shift since then. There's been some people retiring and some people changing their attitude at least. [13]

> People are willing to change. It is an opportunity to be a selected member of a group representing the science faculty in doing gender work. We were supposed to create an action plan against sexual harassment and harassment in general. We will do something operational, and I believe we will make a difference.[14]

Since METU is a state university, it is bound by statutory regulations. Therefore, no specific gender equality and/or equality measures, mechanisms, or provisions exist in the University's regulations. This creates difficulties for female academic and administrative staff, as well as for students, since no positive action measures can be applied. However, since 2016, there has been a strong willingness to improve recruitment, work, and promotion conditions for women and develop women-friendly policies among all parties consulted.

Diversity policies and real-life experiences

Europe shows great diversity in both national research landscapes and gender settings. This diversity is important to keep in mind when exploring and comparing gender dynamics in available research funding across Europe (Husu and Cheveigné 2010:45). The question of gender and research funding has recently attracted substantial attention and is still less often addressed in the literature than gendered structures and career dynamics.

The Gender Challenge in Research Funding report (Severin and Egger 2021:649) drew on and continued these efforts with a specific focus on research funding. One of the most consistent findings in the literature on research productivity is that women tend to have somewhat lower publication rates than men (Ceci et al. 2014:100). This may partly be because women collaborate less than men and have less-developed collaboration networks that would help them achieve external grant awards (Treviño 2018:975).

---

[13] P11, interview with author, Lund, February 09, 2019.
[14] P17, interview with author, Lund, February 11, 2019.

Based on this literature, I wanted the participants to discuss the available research opportunities and asked if gender could be a factor in receiving grants. Despite the lack of direct causal explanations, several elements have been identified from the interviews, such as division of labor at the department, time efficiency, intellectual stimulus, intellectual renewal or new skills learned from previous collaborators, companionship, and a sounding board to the discussion of research, access to equipment, communication of new information, and new publishing opportunities.

> No, it's not so easy to find grants. I mean if you go for the large calls, it's really competitive and the teams are already being built. There have been quite large grants where there has been, like, people are in the same team for years. [15]

Modern science's increasingly interdisciplinary, complex, and costly characteristics encourage scientists to get involved in collaborative research. Funding agencies, particularly government agencies, facilitate active research collaboration as part of their funding conditions. Participants also mention that most grants are for teams of researchers, and those who are working on grants often have commitments to devoting a certain percentage of their time to collaborative or team-based goals, projects, and publications.

> So, it could be like, if women are more isolated in the research, they do not have the same networks that men have. People would think that she chose not to be a part of the collaboration but she has never been given the opportunities. You didn't give her a few extra months or pay for some time to write an application or this kind of support. [16]

Both universities also asserted that research opportunities are not restricted to women but simply cannot be a part of informal social networks within the department. These social networks contain conversations and discussions about work; women remain isolated from such informal work-related topics.

> I think.., we don't usually attend the Fika. Do you know what fika is? It's the coffee break In Sweden it's very important. It has an impact. If you do not attend, you lose contact. If you attend you can make connections, you can create network, you can be a bit more confident in your own. [17]

---

[15] P17, interview with author, Lund, February 15, 2019.
[16] P4, interview with author, Lund, February 18, 2019.
[17] P1, interview with author, Lund, February 01, 2019.

Fika was mentioned as a social networking mechanism in the case of Lund University. Similarly, participants from METU indicated that smoking is a crucial factor in building networks. Another factor reported was the relationship between football and women academics' isolation.

> Men are networking through football. Our department has a football team composed of male colleagues. They organize football games outside working hours and on weekends. I never attended these games, but I am sure they talk about work, research, co-authorship. I am sure of that. [18]

The composition of wages is quite different for Lund University and METU. At Lund University, faculty members get a certain percentage of their salary from the state; for the rest, they need to find research grants. Thus, there is harsh competition. As for METU, %100 of the wages of faculty members are paid by the state. If they secure research grants, they receive an additional wage. When we look at the distribution of research projects, Lund University has a 50 %distribution ratio between male and female project directors, yet at METU, 60 %of the grant holders are male.

> In our faculty, male professors are in a higher place in the hierarchy. Getting to that level as a researcher and as a teacher requires a lot of work socially as well as doing one's job. The way you do your work is under the spotlight, you're under the spotlight. [19]

Academic research undertaken by young researchers can be criticized for its methodology. But in an era of 'post-truth' and 'alternative facts,' when witnessing the incremental but unmistakable rise of forms of totalitarian political discourse, it seems imperative to be careful and accurate in distinguishing rigorous academic research from dogma and ideology (Suissa and Sullivan 2021:75). The language of harm, safety, and neglect within a faculty must be treated critically and seriously. While it should all be vigilant in addressing the disadvantage and discrimination faced by various minority groups, it should be able to distinguish between the expression of dissenting views and actions and speech that constitute overt forms of harassment, intimidation, and threats towards individuals.

---

[18] P27, interview with author, Ankara, June 09, 2020.

[19] P7, interview with author, Ankara, July 1, 2020.

LGBTIQ+ and non-mainstream masculinities

While little is known about the experiences of women and racial/ethnic minorities in male-dominated fields such as STEM, the experiences of lesbian, gay, and bisexual (LGBTIQ+) identifying individuals remain unstudied. This article cannot break this silence but provides an exploratory perspective for further analysis. The existing literature approaches gender in STEM from the perspective of women because there is a common tendency to assume that we know all about masculinity. On the other hand, studies that analyze the masculine culture among STEM fields assert that the common type of masculinity in STEM might also be oppressive over some men (Cech and Waidzunas 2021:4). Within the frame of this study, I accept that there are several masculinities that are affected by the operations of gender in STEM fields.

I asked the participants about the departmental climate towards LGB students and faculty members. Participants from Lund and METU mainly asserted the same type of attitude; they argued that although the climate is heteronormative, they personally think it should change, and some argued that it has changed since the 1970s.

> I mean when I was a student in the 1970's and 80's it would have been impossible for two men to hug or kiss each other in the corridor, or show openness about sexuality, but that is, that is okay. [20]

> I think we have many men feels a little bit estranged by the general culture of Physics, I think it is heteronormative, people are very careful of saying that they are open-minded about LGBT. [21]

The mapping of a gender binary onto a technical/social dualism is consequential for the construction of heteronormativity, which relies on a strong distinction between genders as a basis for the construction of heterosexuality. However, the ways in which a technical/social dualism may map onto a heterosexual/homosexual binary are not so straightforward. The social stereotypes pervasive in STEM fields of gay men as more 'feminine' than straight men and lesbian women as more 'masculine' than straight women become complicated.

A compelling dimension of this disadvantage is mapping the 'masculine/feminine' binary onto the technical/ social dualism. This dualism

---

[20] P16 , interview with author, Lund, February 23, 2019.
[21] P11, interview with author, Lund, February 03, 2019.

has consequences for women, who face a cultural non-congruence between their gender and their scientist identity, threatening their stance as 'real' scientists and their reception as competent, skilled, and valuable technical workers. We needed to observe that at both universities, participants argued that "there should be LGB academics, but they do not know any." When we asked the reason for this silence, the heteronormative climate was the answer given.

> Even though they are gay engineers, they probably never talked about that, they pretended they were not around, I guess.[22]

According to previous studies, heteronormativity is fostered by general expressions of sexual prejudice, while norms specific to STEM silence the concerns of LGB academics. A key mechanism that reproduces heterosexism is 'heteronormativity,' the interpersonal and institutional enforcement of norms that create a strict male/female sex binary and designate heterosexuality as the only acceptable relationship form.

> So that means that if it's a masculine culture if you introduce any kind of minority into any kind of majority, there is stereotype threats, there are social processes pushing out to the margins and everything that goes on. And as I said you know I hate that. I don't think it's fair, I think it's bad for the organization but it's also bad for the people doing it and for those who experience it. It's not this is not how people should behave. [23]

Within this professional culture, broadly held stereotypes about social groups can be built into or 'co-constructed' with professional norms and perceptions of competence.

> For sure I mean we still have the Herculean idea of a successful scientist The brain so whatever, and, and being very forceful and competitive and all that, that is still what we describe as a success if you listen to people.[24]

> I usually say that I mean, women are choosing not to be here and a certain group of men too. [25]

---

[22] P35, interview with author, Ankara June 30, 2020.
[23] P03, interview with author, Lund, January 18, 2019.
[24] P12, interview with author, Lund, January 15, 2019.
[25] P19, interview with author, Lund, January 21, 2019.

One ideology within the culture of engineering that is particularly salient to the treatment of LGB students is the so-called 'technical/social dualism.' Following Hacker, Faulkner (2001) theorizes the 'technical/social dualism' as an ideological separation between 'technical' activities and skills (such as design, science, and math-related activities) and 'social' tasks and skills (such as management, communication with other employees and clients, etc.). This dualism is central to the scientist's identity: to be considered a competent scientist requires 'throwing oneself' into technical activities. Importantly, this ideological dualism is a false distinction in practice: technical and social dimensions are components of all academic STEM work.

Broadly, the technical/social dualism casts issues like the experiences of LGB students/academics as 'social' or 'political' and thus irrelevant to serious discussions about the profession in classrooms, office hours, or study groups. The rendering of science as an 'apolitical' and 'technical' space, combined with the relegation of equality issues to the 'social,' may marginalize LGB people and make them feel as though discussions of their circumstances are silent.

However, we know that gender is political. Doing gender and experiencing the outcome of gender is political. LGBT students at METU have struggled to be officially accepted as a student club since 2007[26]. Denial of the applications shows that METU disregards the right of the freedom of association with its institutionalized homophobic actions. Participants from METU declared they do not share the institutional standpoint, yet they also added that they do not take action to ease the heteronormative climate in their departments.

Conclusion

Current feminist debates on this topic highlight the tensions between the desire to maintain the critical edge and the need to be recognized as a source of legitimate knowledge and have an impact on STEM policy in the post-truth era. This article examines new directions for gender in STEM by comparing how supposedly different two countries experience gender equality in STEM academics. In the "post-truth" world, current higher education has largely failed to meet the challenges of this world that are needed to promote gender inclusion in the "post-truth" landscape. Institutions need to seriously consider whether some of the organizations they work with operate in a way that is incompatible with the core value of academic freedom. This article seeks to examine the dialectical relationship between different types of gender knowledge based on experience. Of the major findings from this study, the unequal gender distribution of STEM faculty has deeper roots than is shown in

[26] http://www.kaosgl.org/page.php?id=28073

the statistics. Masculine features in the science tradition, the male image of the scientist created certain ideal types whose definitions are not inclusive enough to comfort the underrepresented gender.

The abovementioned definitions live in the departmental culture, manifest through small talk, jokes and prejudices. They also reinforce hierarchies in the very concealment of how they structure organizations. If leadership is distributed, rather than being 'in' the body of the leader, it is also *unevenly* distributed. *Uneven,* as one participant suggested, women are usually thought to be suitable department chairs because they are good at tasks such as organization and administration. On the other hand, METU, having a comparatively younger science tradition, attributed masculine features to leadership positions in contrast to Lund University but parallel to the paternal structure of Turkish society. Thus, the commitment to diversity and gender equality of those in leadership positions at STEM faculty does not influence what happens more than the commitment of staff who inhabit seemingly 'lower' positions within the hierarchies of departments.

Departments with stricter scientific traditions tend to rely more on ideal images of masculine scientists. Another finding of this study is that older scientific traditions, such as Lund University, have stricter departmental cultures, and it takes more effort to create mechanisms to recruit and promote minorities in STEM. This study shows that even when universities allocate resources to diversity and equity initiatives, that allocation seems to depend on individual persistence. Although Lund University has created these mechanisms, and statistically, the institution is becoming more diverse on some levels, participants of this study highlighted individual persistence as a form of resistance.

Yet, the allocation of research resources shows that both Lund University and METU have unequal fund distribution among women and men. Participants in this study argued that social networking is a significant factor in getting large research grants. Fika and football matches are reported as the main ways of creating social networks, especially between male colleagues.

The most repeated assertion in the interviews was that gender matters in STEM. It is important because such an attitude shows the binary ideology against the idea that science and scientists are not affected by social values and that gender dynamics are fading away. Such a perspective can also become a strategy to use anti-discrimination to focus on relations of advantage – to recognize that one party may have a significant advantage over the other, thus turning the gaze away from discourses about women's disadvantages, often constructed as only women's problems. Furthermore, moving in this direction also requires us to consider not only how gender works but also how whiteness and class impart privilege, how that is marked internally by differentiation along class, race and ethnicity, age, ability, and sexual orientation (Bacchi 2000:50).

As mentioned, it was crucial for us to observe that at both universities, LGB academics navigate a heteronormative STEM climate by covering or downplaying cultural characteristics associated with LGB identities. According to participants of this study, the unspoken category of heterosexuality is legitimated on the ideal level. Despite the resembling heteronormativity in STEM culture, the two institutions are incomparably different in terms of institutional attitude towards LGB identities which creates a more egalitarian environment at Lund University and insecurity for students and academics at METU.

Further research on inequalities in STEM should consider how sex and race categories encompass people with often very different experiences depending on their location within interwoven dimensions of oppression. I also hope that further research on gender and sexuality in STEM will look closely at people's experiences in the profession and explore the issue of gender nonconformity. Even in contexts where feminist knowledge and expertise have been under severe attack, there are still possibilities for intervention.

References

Archer, Margaret, Roy Bhaskar, Andrew Collier, Tony Lawson, and Alan Norrie, eds. 2013. Critical Realism: Essential Readings. Routledge.

Bacchi, Carol. 2000. "Policy as discourse: What does it mean? Where does it get us?." *Discourse: studies in the cultural politics of education* 21(1): 45-57.

Ball, Stephen J. 2008. "New philanthropy, new networks and new governance in education." *Political studies 56* (4): 747-765.

Barzilai, Sarit, and Clark A. Chinn. 2020. "A Review of Educational Responses to the 'Post-Truth' Condition: Four Lenses on 'Post-Truth' Problems." *Educational Psychologist* 55 (3): 107-119.

Beaufaÿs, Sandra, and Beate Krais. 2005. "Women in Scientific Careers in Germany: The Hidden Mechanisms of Power." *Travail, Genre et Societes* 2: 49-68.

Bijker, Wiebe E., and John Law. 1994. *Shaping technology/building society: Studies in sociotechnical change.* MIT Press.

Buckingham, David. 2019. *The Media Education Manifesto.* John Wiley & Sons.

Budgeon, Shelley. 2021. "Making feminist claims in the post-truth era: the authority of personal experience." *Feminist Theory* 22(2):248-267.

Cann, Celeste, and Elizabeth DeMeulenaere. 2020. *The Activist Academic: Engaged Scholarship for Resistance, Hope and Social Change.* Myers Education Press.

Cech, E. A., and T. J. Waidzunas. 2021. "Systemic Inequalities for LGBTQ Professionals in STEM." Science Advances 7(3). doi: 10.1126/sciadv.abe0933.

Ceci, Stephen J., Donna K. Ginther, Shulamit Kahn, and Wendy M. Williams. 2014. "Women in academic science: A changing landscape." *Psychological science in the public interest* 15(3): 75-141.

Clarke, Adele E., and Joan H. Fujimura. 1992. "What Tools? Which Jobs? Why Right." In *The Right Tools for the Job: At Work in Twentieth-Century Life Sciences,* edited by Adele E. Clarke and Joan H. Fujimura, 3-44. Princeton University Press.

Daston, Lorraine, and H. Otto Sibum. 2003. "Introduction: Scientific personae and their histories." *Science in context* 16(1/2).

Drew, Eileen, and Siobhán Canavan. 2020. *The Gender-sensitive University: A Contradiction in Terms?.* Taylor & Francis.

Eslen-Ziya, Hande, and Ezgi Pehlivanli. 2022. "The search for alternative knowledge in the post-truth era: Anti-vaccine mobilization during the COVID-19 pandemic in Turkey." *Cogent Social Sciences* 8(1).

European Institute for Gender Equality. 2018. *Gender Equality Index 2018: Measuring Gender Equality in the European Union 2008–2018.* Retrieved from http://eige.europa.eu/rdc/eige-publications/gender-equality-index-2018-measuring-gender-equality-european-union-2018-report

Faulkner, Wendy. 2001. "The technology question in feminism: A view from feminist technology studies." In *Women's studies international forum,* 24(1): 79-95. Pergamon.

Fujimura, Joan H. 1996. Crafting Science: A Sociohistory of the Quest for the Genetics of Cancer. Harvard University Press.

Fujimura, Joan H., and Christopher J. Holmes. 2019. "Staying the Course: On the Value of Social Studies of Science in Resistance to the "Post-Truth" Movement." *In Sociological Forum,* 34:1251-1263.

Graff, Agnieszka, and Elżbieta Korolczuk. 2017. "Towards an Illiberal Future: Anti-Genderism and Anti-Globalization." *Global Dialogue* 7 (1): 27-30.

Grossman, Jennifer M., and Michelle V. Porche. 2014. "Perceived gender and racial/ethnic barriers to STEM success." *Urban Education* 49(6): 698-727.

Haraway, Donna. 2020. "Situated knowledges: The science question in feminism and the privilege of partial perspective." Pp. 303-310 in *Feminist Theory Reader.* Routledge.

Harding, Sandra G. 1986. The science question in feminism. Cornell University Press.

Harding, Sandra. 1991. *Whose science? Whose knowledge? Thinking from women's lives.* Cornell University Press.

Husu, Liisa, and Suzanne De Cheveigné. 2010. "Gender and gatekeeping of excellence in research funding: European perspectives." Pp. 43-59 in *GenderChange in Academia.* VS Verlag für Sozialwissenschaften.

KAOSGL, Retrieved from https://kaosgl.org/en/single-news/quotthe-prevention-of-metu-lgbtrsquos-request-for-being-a-student-club-at-the-universityquot.

Kandiyoti, Deniz. 2016. "Locating the politics of gender: Patriarchy, neo-liberal governance and violence in Turkey." Research and Policy on Turkey 1(2): 103-118.

Kováts, Eszter, and Andrea Pető. 2017. "Anti-Gender Discourse in Hungary: A Discourse without a Movement." In *Gender and Far Right Politics in Europe,* edited by Mieke Köttig, Renate Bitzan, and Andrea Pető, 117-131. Palgrave Macmillan.

Köttig, Michael, Renate Bitzan, and Andrea Pető, eds. 2017. *Gender and Far Right Politics in Europe.* Palgrave Macmillan.

Fox-Keller, Evelyn, and Gertrude Scharff-Goldhaber. "Reflections on gender and science." (1987): 284-286.

Laslett, Barbara, and Johanna Brenner. 1989. "Gender and social reproduction: Historical perspectives." *Annual Review of Sociology* 15 (1): 381-404.

Latour, Bruno, and Steve Woolgar. 1987. *Laboratory Life: The Construction of Scientific Facts.* Princeton University Press.

Lövheim, Daniel. 2014. "Scientists, engineers and the society of free choice: enrollment as policy and practice in Swedish science and technology education 1960–1990." *Science and Education* 23 (9) :1763-1784.

Martin, Ian. 2003. "Adult Education, Lifelong Learning and Citizenship: Some Ifs and Buts." International Journal of Lifelong Education 22 (6): 566-579.

Martin, Anne E., and Teresa R. Fisher-Ari. 2021. ""If We Don't Have Diversity, There's No Future to See": High-school students' perceptions of race and gender representation in STEM." *Science Education* 105(6): 1076-1099.

Mede, Nikola G., and Mike S. Schäfer. 2020. "Science-Related Populism: Conceptualizing Populist Demands toward Science." *Public Understanding of Science* 29 (5): 473-491.

Mihăilă, R., Kovacova, M., Kliestikova, J., Kliestik, T., and Kubala, P. 2018. "Deconstructing masculinist power politics in society: Oppression, control, and domination." *Contemporary Readings in Law and Social Justice*, 10(1):158-164.

Morley, Louise. 2005. "Opportunity or exploitation? Women and quality assurance in higher education." *Gender and Education* 17(4): 411-429.

Morley, Louise. 2016. "Troubling intra-actions: gender, neo-liberalism and research in the global academy." *Journal of Education Policy* 31(1): 28-45.

McIlwee, Judith S., and J. Gregg Robinson. 1992. *Women in Engineering: Gender, Power, and Workplace Culture.* Albany, NY: State University of New York Press.

O'Hagan, Anthony. 2019. "Expert knowledge elicitation: subjective but scientific." *The American Statistician* 73, no. sup1: 69-81.

Paternotte, David, and Roman Kuhar. 2018. "Disentangling and Locating the 'Global Right': Anti-Gender Campaigns in Europe." *Politics and Governance* 6 (3): 6-19.

Paternotte, David, and Monique Verloo. 2021. "De-democratization and the Politics of Knowledge: Unpacking the Cultural Marxism Narrative." *Social Politics: International Studies in Gender, State & Society* 28 (3): 556-578.

Rees, Teresa. 2001. "Mainstreaming gender equality in science in the European Union: The 'ETAN report'." *Gender and education* 13(3): 243-260.

Ringrose, J. 2018. Digital feminist pedagogy and post-truth misogyny. *Teaching in Higher Education*, 23(5): 647-656.

Schmader, Toni. 2023. "Gender Inclusion and Fit in STEM." *Annual Review of Psychology* 74(1):219–43. doi: 10.1146/annurev-psych-032720-043052.

Severin, Anna, and Matthias Egger. 2021. "Research on Research Funding: An Imperative for Science and Society." *British Journal of Sports Medicine* 55(12):648–49. doi:10.1136/bjsports-2020-103340.

Siann, G. and Callaghan, M. 2010. Choices and Barriers: Factors Influencing Women's Choice of Higher Education in Science, Engineering and Technology. *Journal of Further and Higher Education*, 25, 85-95. http://www.tandfonline.com/doi/abs/10.1080/03098770020030524

Suissa, Judith, and Alice Sullivan. 2021. "The gender wars, academic freedom and education." *Journal of Philosophy of Education* 55(1): 55-82.

Springer, Simon, and Veysel Özdemir. 2022. "Disinformation as COVID-19's Twin Pandemic: False Equivalences, Entrenched Epistemologies, and Causes-of-Causes." *OMICS: A Journal of Integrative Biology* 26 (2): 82-87.

Treviño, Len J., Luis R. Gomez-Mejia, David B. Balkin, and Franklin G. Mixon Jr. 2018. "Meritocracies or masculinities? The differential allocation of named professorships by gender in the academy." *Journal of Management* 44(3): 972-1000.

Tsipursky, Gleb. 2017. "Towards a post-lies future: Fighting 'alternative facts' and 'post-truth' politics." *The Humanist* 77(2).

Yin, Robert K. 2006. "Mixed methods research: Are the methods genuinely integrated or merely parallel." *Research in the Schools* 13(1): 41-47.

Chapter 13

Sociotechnical equalities and inequalities in the post-truth age: The case of remote work in Greece during the COVID-19 pandemic

Alexia-Sofia Papazafeiropoulou

Abstract: The chapter examines the case study of remote work in Greece during the COVID-19 pandemic. The study's central hypothesis is that both the remote work and the 'post-truth' phenomena are co-constructed with the same socio-technical infrastructure of the internet. They share common structural features, such as the expansion of virtual communication, the digitalization of the identities of socio-technical agents, and the disconnection of their interaction from specific spatial and temporal contexts. Therefore, the socio-technical networks of remote work are the perfect example to examine the multiple interactions between technology and society. The chapter first discusses how the STS helps us understand the post-truth phenomenon. Then, it examines the multiple socio-technical changes caused by remote work, focusing on the case of remote work in Greece during the COVID-19 pandemic. The empirical research has been conducted through a questionnaire available online from May to June 2021. It was addressed to remote work employees who answered the questions anonymously. It aimed to shed the spotlight on the voices of those immediately involved in socio-technical changes in order to understand the heterogeneous structure and the inequalities of the networks in which they participate. Based on the research findings, this chapter argues that the inequalities built within the remote work networks contribute to understanding post-truth aspects. Finally, the Conclusion discusses how remote work inequalities relate to the post-truth phenomenon.

Keywords: Post-truth age, STS, Remote work, COVID-19 Pandemic, Sociotechnical Inequalities

***

Introduction

At the dawn of the twenty-first century, a series of phenomena, including the global economic crisis and, more recently, the COVID-19 pandemic, have restructured sociotechnical labor networks. Additionally, the intensification of mobilities and the explosion of social media have created new conditions for knowledge evaluation and sharing. These conditions have contributed to challenging the core values of modernity, such as democracy or scientific expertise. Hence, the optimistic narrative of science and technology as drivers of social progress has been unsettled, setting the framework of the 'post-truth' age (Iyengar and Massey 2019:7656-7661).

Regarding the COVID pandemic, one of the most notable sociotechnical changes has been the immediate increase in remote work. Due to the urgent health protection measures in many countries, millions of employees worldwide were obliged to change how they worked within a few weeks (Brynjolfsson et al. 2020: 1-26). This caused rapid sociotechnical changes, such as the modification of work methods, the challenge of labor conditions and rights, the reorganization of work and domestic space, and the transformation of family and social relations. In the long term, work delocalization might include restructuring urban life and mobility, redistributing work and communication networks, resigning the role of metropoles, and reconstructing what is local and global. Such conditions alter the geopolitical relations between the so-called centers and peripheries (Castells 2004:3-45). These issues have been broadly discussed within the mobility paradigm (see Cresswell 2006; Cresswell 2015; Fari and Moraglio 2016: 10-17; Papazafeiropoulou 2016: 28-50).

The sociotechnical networks of remote work are the perfect example for examining the multiple interactions between technology and society. The remote work networks are heterogeneous systems comprising employers and employees, technological equipment and infrastructure, labor market, and legislation. Therefore, this case is worth researching through the scope of Science and Technology Studies (STS). At the same time, this subject urges the revisiting of the STS paradigm. The STS methodology examines the interaction of the sociotechnical factors by usually emphasizing the users' perspective; therefore, the case study of remote work is helpful since it focuses on the employees' aspect. Secondly, remote work modifies space and time as interaction parameters. That is a subject thoroughly examined within the mobilities bibliography; therefore, an STS analysis should take into consideration the approach of this paradigm because it can also be helpful for an in-depth understanding of the entanglement between technology and society. Finally, a reason for the possible renegotiation of the theoretical approach and the boundary of STS concerns the new paradigm of knowledge sharing during the post-truth age.

This chapter discusses the above-described matters through the case study of remote work in Greece during the COVID-19 pandemic. The starting point of this study is the following questions: How can the methodology of STS contribute to understanding the post-truth conditions? Do these conditions challenge the STS paradigm, and if so, how does this happen? How is remote work connected to post-truth conditions? Furthermore, how does the case study of remote work help us understand the challenges of the 'post-truth' conditions for the STS?

This study's central argument is that examining remote work sociotechnical networks can contribute to understanding aspects of the post-truth context and the questions it brings to the STS research. Concerning the two last of the abovementioned questions, the study's central hypothesis is that both the remote work and the 'post-truth' phenomena are co-constructed with the same sociotechnical infrastructure of the internet. Their heterogeneous systems share common sociotechnical factors and, therefore, common structural features, such as the expansion of virtual communication, the digitalization of the identities of sociotechnical agents, and the disconnection of sociotechnical interaction from specific spatial and temporal contexts. Hence, the sociotechnical inequalities constructed within remote work networks, as discussed below, partly relate to social and political pathogeneses of the post-truth age.

The chapter unfolds in six parts. The second part discusses the challenges of the post-truth age for STS. In doing so, it seeks to respond to the first of the abovementioned research questions, namely, how STS helps us understand the post-truth phenomenon and how the latter challenges the theoretical paradigm of the former. The third part examines multiple sociotechnical changes caused by remote work, while the fourth part focuses on the case of remote work in Greece during the COVID-19 pandemic. This case study originates from the hypothesis that shedding the spotlight on the voices of those immediately involved in sociotechnical changes is crucial because it contributes to understanding the heterogeneous structure and the inequalities of the networks in which they participate. The fifth part discusses research findings and the final discusses how sociotechnical inequalities of remote work relate to the post-truth phenomenon.

Post-truth and STS

How can the intellectual tools of STS contribute to understanding the post-truth conditions, and how do these conditions challenge the STS paradigm? To seek the answer to those questions, one can examine the genealogy of the term 'post-truth' in the public discourse. It was introduced in the Oxford Dictionary in 2016 as 'the word of the year,' aiming to describe how knowledge is circulated, valued, and applied during the age of the so-called digital revolution. The rapid growth of information sources was anticipated to spread knowledge and rationality to societies. However, skepticism towards modern values and institutions, such as democracy or scientific expertise, along with the rise of alt-right nationalism, has been perceived as the growing domination of irrationality in public discourse (Iyengar and Massey 2019: 7656-766; Hoffman 2018: 444-452). These phenomena are testified by incidents, such as the democratic election of politicians or policies that have stood against democratic institutions and values.

These incidents have been attributed to poor communication among scientists, policymakers, and the public or the fragmented and polarized media environment. The social media explosion has created multiple alternative media sources that claim to have equal credibility as journalistic or academic sources. Voters can access diverse discourses of bloggers, influencers, or even 'trolls' or internet bots, which are not subjected to documentation based on evidence and journalistic or scientific ethics. Many of these sources use sensationalist methods, promoting populist argumentation; however, they all claim they talk about the truth. Many analysts were aghast by the fact that public decisions of significant political impact, such as the election of the US president or Brexit in 2016, can be massively influenced by subjective, partisan, and erratic opinions, which count more than facts. Hence, the inauguration of the term 'post-truth age' indicates pessimism –if not moral panic- about a digital society where core values of modernism, such as reason and the scientific method, are challenged because opinions of both experts and the public are perceived to be equally valid (Jasanoff and Simmet 2017:751-770; Bhambra 2017: 214-232; Bonikowski 2017:181-213).

From that point of view, many theories –including STS- have been accused of introducing intellectual and moral relativism. The reason for that is that such approaches provide a critical perspective on the positivist Enlightenment narrative about science's absolute authority and impartiality. Central theses of STS scholarship challenge the concept of universally valid scientific truths; consequently, they could be responsible for sustaining epistemological relativism. Such ideas include, for example, the concept of symmetry. According to this concept, scientific arguments have both epistemological and social explanations. Another 'relativist' suggestion of STS is deconstructing the historiographical perspective of science and technology as significant factors of linear progress and successive innovations. Finally, a central STS argument is that scientific facts are co-constructed with political norms and social meanings. These allegations in the STS bibliography have caused a recent discussion on whether STS should be accused of paving the road to the emergence of a 'post-truth' age (Jasanoff and Simmet 2017: 751-770; Hoffman 2018, 444-452; Lynch 2017: 593-599; Sismondo 2017: 3-6; Marres 2018).

It is a matter of fact that STS has challenged epistemological certainties concerning the impartiality of science of its self-evident independence from other domains, like politics or society. Nevertheless, before discussing the potential impact of STS on wider audiences and its possible relation with the rise of alt-right ideologies and racism, one should first question the validity of 'post-truth' as a concept. Such a term seems to presume a mythical era when the public discourse was fully attuned to the scientific signification of facts. Moreover, the belief in a single scientific truth is teleological. It presupposes that scientific truth is a diachronic concept and, therefore, independent from the social context, the institutional norms, and the scientific and technological methods with which it is co-produced. (Jasanoff and Simmet 2017: 751-770; Marres 2018). Such an approach cannot explain the asymmetrically different scientific and technological paradigms deemed valid in different historical, cultural, and institutional contexts, as Kuhn has shown (1962). On the other hand, the STS paradigm aims to show how science and technology are co-constructed with society. Therefore, through the STS scope, the post-truth concept is grounded on ahistorical presuppositions. Hence, the argumentation of the two paradigms is oppositional.

Still, what remains as a problem is the chaotic situations within the public discourse caused by the simplistic argument that scientific and political claims are equally valid. From that point of view, the 'post-truth' concept is useful because it describes a context in which the persuasion method has a more significant influence than facts on public decisions with massive impact. The authority of science is based on its impartiality; unlike politics, which uses opinions, science is expected -within the public discourse at least- to ground its argumentation on facts. The illustration of the institutional alignment of science with politics and the description of scientific facts as 'constructed' within STS could potentially be misinterpreted as this would imply they are either 'subjective' or even 'fabricated.' In that case, the validity of scientific expertise is undermined. However, such a conclusion is not within the scope of the STS paradigm.

On the contrary, STS aims to provide the intellectual tools for an in-depth understanding of the interaction of science, technology, and society. If we acknowledge the potential limitations of truth, that does not mean we abandon its seeking; it means we should seek a deeper grounding of its argumentation and validity. Hence, according to STS, scientists should not claim their allegations are universally true; on the contrary, they should consider the diversity of the sociotechnical groups co-constructed with the networks in which they participate. Within STS, there is a strong emphasis on the reflexivity of science, contributing to the production of socially robust knowledge (Nowotny, Scott, and Gibbons 2001). From that point of view, scientists should focus on the multiple impacts of the sociotechnical changes on different groups, giving voice to those who benefit from such changes and those subjected to potential inequalities.

Consequently, while the term 'truth' is common in both the STS bibliography and the 'post-truth' discourse, their meaning is asymmetrical. On the one hand, the STS bibliography aims to reach the meaning of scientific truth dialogically by illustrating the heterogeneous factors that define this truth. By doing so, it does not aim at diminishing the worth of scientific knowledge; on the contrary, it aims to demonstrate the structure of the scientific institutions and methods to call for strategies that enforce their validity (Sismondo 2017: 3-6). It also aims to democratize both its dissemination and impact. On the other hand, within the post-truth discourse, different versions of truth are denounced, so a single and subjective version of the truth is promoted as undoubtedly valid. Such a process does not contribute to the development of knowledge. As historically demonstrated, undermining the worth of knowledge is the first step towards authoritarianism and reinforcement of inequalities (see, for example, Herf 1984).

The STS approach can contribute to understanding the 'post-truth' phenomenon without imposing moralistic or ahistorical views. More specifically, the STS scope can show that the 'post-truth' discourse does not derive from a kind of ideological 'irrationality' or 'illusion of the masses'; on the contrary, it originates from sociotechnical inequalities, such as the unequal distribution of flows of capital and information, as well as the unequal access to education, health, work, or even natural resources. As it is argued, many people cannot feel sympathy with abstract problems in their everyday lives, such as respect for differences or climate change, while their elementary labor rights and quality of life are renegotiated, challenged, or even threatened (Jasanoff and Simmet 2017: 751-770). The rapid sociotechnical changes due to remote work could justify the idea that this might risk elementary labor rights, causing new inequalities. These inequalities have also been discussed within the mobilities paradigm, and they are termed 'unequal mobilities' or 'restricted motility' (Sheller 2018: 45-74; Kaufmann, Bergman, and Joye 2004: 745-756).

In conclusion, the STS argumentation aims to reach the most remarkable objectivity possible and ensure that science and technology work inclusively. Therefore, it illustrates the sociotechnical inequalities upon which the 'post-truth' discourse is rooted. The following parts of the chapter discuss these inequalities to show their connection with the 'post-truth' phenomenon.

Reorganizing urban sociotechnical networks and mobilities through remote work

After discussing the challenges that the post-truth conditions bring to the STS, we can proceed to the following questions we already set in the Introduction: How is remote work connected to the post-truth conditions? Furthermore, how does the case study of remote work help us understand the challenges of the 'post-truth' conditions for the STS? To seek the answer to these questions, one should first discuss the new conditions related to remote work.

The digitalization of public discourse in the post-truth era has coincided with the digitalization of remote work. Even though remote work has been technologically feasible for a couple of decades, it is only during the COVID-19 pandemic that it was massively applied worldwide. This change caused rapid as well as structural changes in multiple sociotechnical networks. The diverse and multiple effects of remote work have been discussed in several academic papers recently (Mollino et al. 2020; Manzo and Minello 2020; Hardill and Green 2003, 212-222; Nash and Churchill 2020, 1-14; Keeling, Clemens-Croome, and Rosch 2015, 880-898). Two main categories of scales concern these effects; the macroscale level concerns the environment and urban space or the macroeconomy, while the microscale level focuses on work and domestic space.

On a macroscale level, the digitalization of work has caused the reorganization of urban space and mobilities, bringing the discussion about the relation of remote work with sustainability to the forefront. The diminished physical mobilities due to the application of remote work might eliminate severe problems in contemporary cities, such as traffic and gas emissions. Therefore, the massive application of remote work could be part of larger-scale technological changes, which include new forms of mobilities, communications, and economic interaction aiming at the reconstruction of cities as 'green' and 'eco-friendly' (Prasad et al. 2020; Kylili et al. 2020). Such a change would bring structural changes to the urban scape. From the 18th century onwards, the spine of modern cities has been the sociotechnical infrastructure constructed according to a particular economic model. This model's structure was built on a distinction between work and domestic spatial and temporal contexts. Its infrastructure concerned urban planning, transportation, energy systems, public services, and health services (Hård and Misa 2008). Hence, the fact that a great range of work and transactions can potentially take place in cyberspace would result in reduced physical mobilities, as well as reduced fuel transmission and traffic congestion. At the same time, it would trigger the development of innovative city technologies, which mainly concern Internet of Things (IoT) functionalities. These IoT functionalities would enable the decentralized control of the best possible management and use of networks and natural resources (e.g., gas, electricity, data, transportation) through smart buildings (Kylili et al. 2020). To sum up, due to the massive application of remote work, cities can be reconstructed both as urban machinery and as mobility flows' junctions (Hård and Misa 2008).

The delocalization of work can modify the metropoles' role at a global level. The historical role of international trade centers has been crystallized around localized work economy; however, the disconnection of work from locality brings the 'death of geography' as a step towards what Manuel Castells has termed 'network society.' This term describes the social, economic, and political networking developed through internet communications at the beginning of the twenty-first century (Castells 2010). Although the emphasis is usually on the global instead of the local level of interaction, the case of remote work shows that networking diminishes the role of geographical space to the extent that the local and the global levels are mutually transformative. This paradigm has been described as a net locality, aiming to show how virtual mobilities give a chance to simultaneously interact on a local and global level, creating hybrid spaces (De Sousa e Silva and Gordon 2014). In conclusion, the sociotechnical reconstruction of economic and geographical hierarchies of 'centers' and 'peripheries,' along with the restructuring of the sociotechnical networks of cities as financial centers, is a macroscale process that also relates to the doubt and renegotiation of conceptual categories of the modernist narrative, such as democracy or science.

The massive application of remote work has also brought structural changes on a microscale level regarding the sociotechnical organization of domestic space and time. The boundaries between work and home life are confounded, changing the households' technological equipment and spatial planning. From a historical aspect, this shift is critical: modern houses equipped with home appliances have been central to the twentieth-century consumer culture (See, for example, Oldenziel 2009). Because of remote work, houses become multi-spaces where the symbolic terrains of private and public life merge. Additionally, the family roles are rearranged according to diverse working patterns and daily routines, bringing a great range of family types (Manzo and Minello 2020; Hardill and Green 2003; Nash and Churchill 2020).

The recasting of work and life boundaries has advantages and liabilities for balancing different aspects of daily routines. The distinction between home and work life has been the basis for claiming labor rights' legal protection, such as the eight-hour workday, sick leaves, parental leaves, and access to social insurance services. Within remote work, these issues are considered private matters for which the employees must find solutions independently. Moreover, their cooperation with work unions becomes more difficult. This situation entails increased unpaid overtime work and the lack of social policies about health problems or childcare issues. As pointed out in different case studies, during the pandemic, employees have claimed they had to be available for their employers, colleagues, or clients practically at any time, while they also had to struggle to balance their professional and private or family life (Manzo and Minello 2020; Hardill and Green 2003; Nash and Churchill 2020).

Moreover, there is a gender aspect concerning remote work. As relevant research shows, most domestic tasks usually burden female employees, who simultaneously juggle many things. This issue is interesting from the aspect of gender and technology. Technology is usually approached through a positivist and optimistic narrative within public discourse connected to social progress. However, as shown in the relevant literature, technology was constructed as a professional domain during the nineteenth and twentieth centuries in a way that reproduced the institutional mechanisms of patriarchy. Therefore, the stereotypical representation of women as passive receptors of technology has restrained their ability to participate in the shaping of technology as designers or engineers (Oldenziel 2009). Some feminist theorists describe technology through a dystopic lens, arguing that it reproduces power relations of gender subordination. On the other hand, more recent theorists argue that digital technologies have been connected to what is termed within the STS as 'interpretive flexibility,' allowing women to promote their views of technology through alternative participation networks; therefore, they can reconstruct their professional identity freed from gender stereotypes (Wajcman 2004: 102-115; 2009:1-10). Still, in the case of remote work, it seems that the digitalization of sociotechnical networks does not contribute to the reconstruction of the existing forms of gender discrimination.

Regarding the above, the massive application of remote work has caused new forms of sociotechnical inequalities on a microscale level, including doubt about labor rights, the renegotiation of the domestic and professional sphere, and the rise of new forms of gender discrimination. As discussed in the second section of this chapter, such inequalities are also connected to the unsettlement of core values of modernity, contributing to the rise of the 'post-truth' discourse.

In summary, the rapid changes in the sociotechnical labor networks caused by remote work, both on the macroscale and microscale level, are also connected to the conditions that contribute to the post-truth phenomenon, such as the expansion of virtual communication, the digitalization of the identities of sociotechnical agents, and the disconnection of sociotechnical interaction from specific spatial and temporal context. The following section examines these issues through a specific case study.

The case of remote work in Greece during the COVID-19 pandemic: empirical research

The theoretical context described above has been used for the questions of the empirical case study of remote work in Greece during the COVID-19 pandemic. More specifically, the research examined the sociotechnical networks by focusing on the views of the participating agents. Therefore, the study was conducted through an online questionnaire. The survey addressed people who either worked or had worked remotely at some point during the pandemic since March 2020, when the first lockdown was imposed in Greece until June 2021. (Papazafeiropoulou 2021).

During the data collection period, there were several total or partial lockdown periods. Different measures were applied once every few weeks, announced in several issues of the *Journal of the Government of the Greek Republic* in 2021. These measures included, among others, the enforcement of obligatory remote work either for all the employees in companies and public services or for a percentage of them through a rotation schedule and obligatory remote education for students and teachers. They also included the obligatory digital conduction of the public services' exchange with citizens, like the application for state documents and their issuance exclusively online. Other businesses, such as shops, bars, restaurants, or gyms, were either fully suspended for specific periods or operated with heavy restrictions (*Journal of the Government of the Greek Republic* 2020; 2021).

The online questionnaire was available online as a Google Form Questionnaire from May 13th to June 14th, 2021, and 342 respondents filled it. It was circulated through social media on pages or forums for professional teams, which mainly targeted university teachers, schoolteachers, public and private sector administrators, and engineers. The questionnaire was addressed to remote work employees who answered the questions anonymously. It did not categorize different groups of interviewees according to their age, educational, or professional level. This choice was made because the research aimed to record a multi-vocal aspect of the working conditions since remote work was massively applied. For similar reasons, the research did not focus on a specific area in Greece. Since the sociotechnical labor networks were delocalized, the research sample included respondents from different areas of this country. Besides, many employees moved from big cities to their home villages and worked remotely during the lockdowns.

The questions of the questionnaire were categorized into six groups. The first group aimed to outline a general profile of the interviewees and their households, as shown in the tables below.

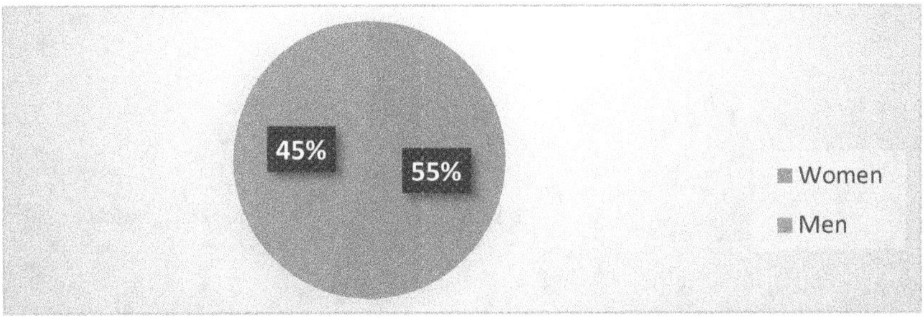

Figure 13.1. Distribution of Interviewees' Gender

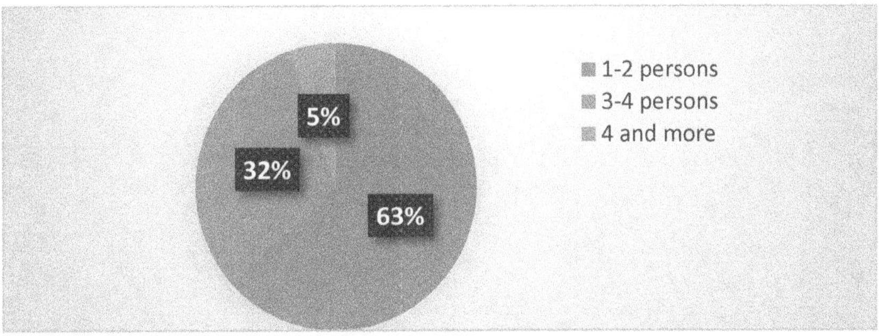

Figure 13.2. Number of people living in the households of the interviewees

The second group of questions sought information about the interviewees' working conditions while working remotely. The percentages in the following tables show that in Greece at least, remote work had not been applied massively before the COVID-19 pandemic. During the pandemic, most businesses followed a rotation schedule of working remotely and in situ to avoid overcrowding of the workspaces. In general, as it seems below, remote work was, in most cases, obligatory for the employees.

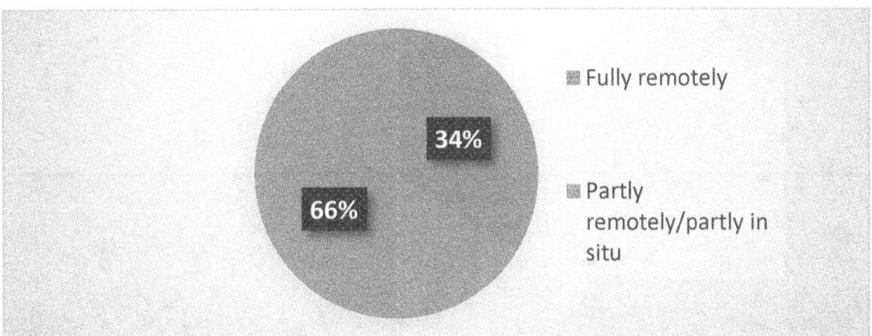

Figure 13.3. Do you work entirely remotely or partly remotely and partly in situ?

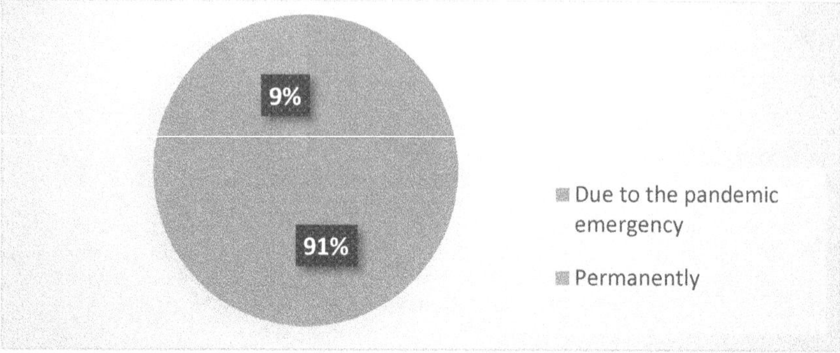

Figure 13.4. Do you work remotely due to the current pandemic or permanently?

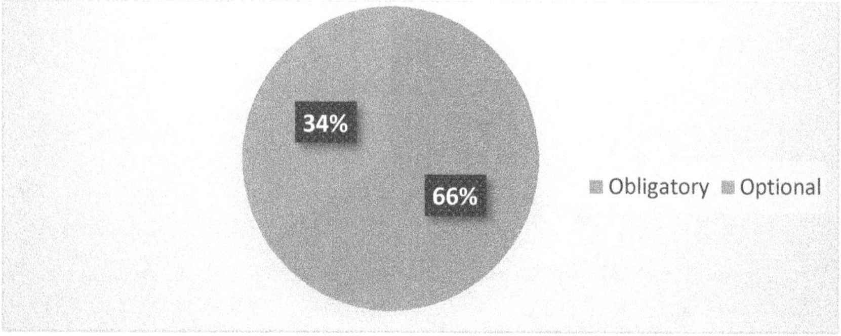

Figure 13.5. Has remote work been obligatory for you?

The third group of questions focused on the use of technological equipment and infrastructure within remote work.

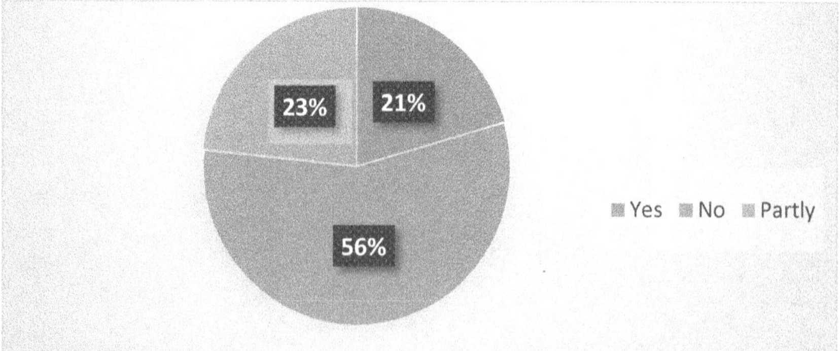

Figure 13.6. Has your employer provided you with the necessary equipment to work remotely?

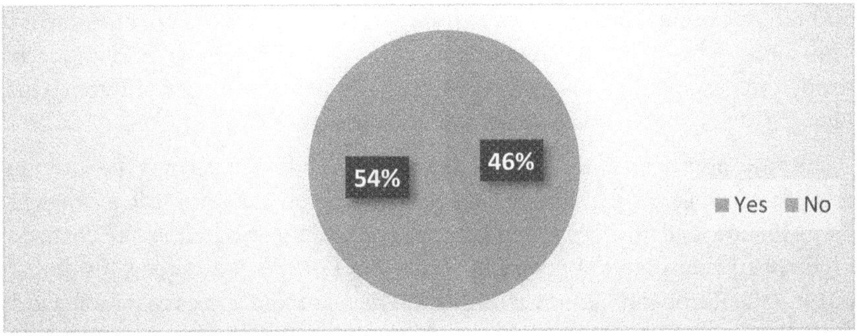

Figure 13.7. Did you or anyone in your household need to purchase or borrow technical equipment for either remote work or education?

As indicated in the tables above, despite remote working mainly being obligatory for the employees, most of them had to provide themselves with the equipment and infrastructure they used, either wholly or partly. About half of the interviewees had no alternative but to purchase or borrow equipment for themselves or a household member for remote work and education. Families with three, four, or even more members had to use similar equipment to work or attend classes simultaneously. At this point, we should note that Greece has been among the countries in which the economic affordability of many citizens had considerably diminished during the 2009 economic crisis. Moreover, among EU countries, Greece had the highest percentage of citizens who lost their jobs during the pandemic, and it was the third among the EU countries concerning the deterioration of the citizens' economic conditions (European Commission 2020; Ahrendt et al. 2021). Given these facts, the obligatory purchase of equipment for remote work and education, as shown in Tables 6 and 7, was problematic for many households.

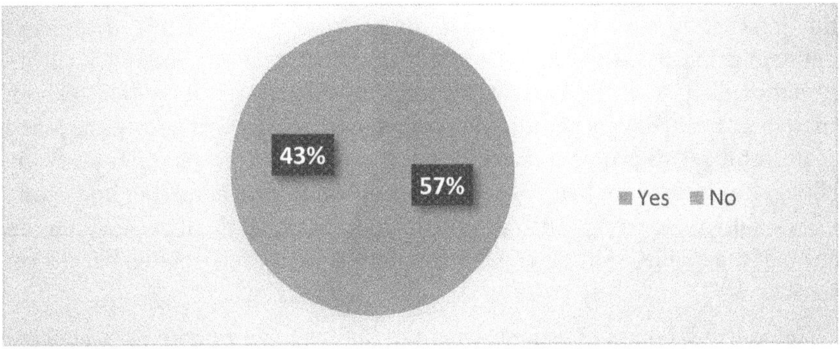

Figure 13.8. Have you been confronted with issues concerning internet connection while working remotely?

Another technological aspect of remote work concerns the use of the internet. The rapid growth of the demand for high-speed internet throughput in residential establishments caused severe problems with the internet QoS (Quality of Service) in many areas, as shown in Figure 13.8.

The table above indicates sociotechnical inequalities concerning the internet as a remote work infrastructure. Greece scores below the European average as far as the average internet speed is concerned. At the same time, the cost of a high-speed internet connection is above the European average (Eurofound 2019; 2020. European Commission 2020). The restricted access to remote work equipment and infrastructure (motility) has been related to various sociotechnical problems (Sheller 2018: 45-74; Kaufmann et al. 2004: 745-756). According to Greek media sources, many students had to attend classes using smartphones because their parents would use the household's computer(s) for remote work (Makridis 2020). Even worse, some students would use their smartphones at communal places or buildings where public WiFi connection was available because their home internet connection was poor, or their parents could not afford it. Moreover, as mentioned in similar sources in 2020 ("Remote Education" 2020; "Rethymno..." 2020), there had been cases of employees or students asking their neighbors not to use the internet during working hours to reduce regional overall internet traffic. During the first months of the pandemic, a press release of the Greek Ministry of Digital Administration asked citizens to avoid using the internet for reasons of entertainment during rush hours ("Lockdown: what we should avoid" 2020; "Urgent Call of the Government" 2020; "Danger of internet crash" 2020)

As shown in the STS bibliography, it is only during a crisis or breakdown of a sociotechnical system that people notice its very existence (Bijker and Law 1992: 1-16). The internet connection as a remote work infrastructure during the pandemic is exemplary. Never again had so many users realized massively that the internet is a network that can suffer from traffic, before the recent pandemic, in Greece at least. Due to this situation, the users had practically to be either thrifty with internet use or would eventually have to pay for faster and, therefore, more expensive internet connection services to earn their living uneventfully. This problem revealed that it was not self-evident that everyone could afford to pay for a high-speed internet connection at his/her house that s/he would use to work, which proved untrue. From that point of view, one can see that the quality of living of many employees partly deteriorated because of remote work.

The fourth group of questions sought to find whether remote work affected the quality of employees' living. Therefore, it included questions about the interviewees' work experience, as shown in the following figures.

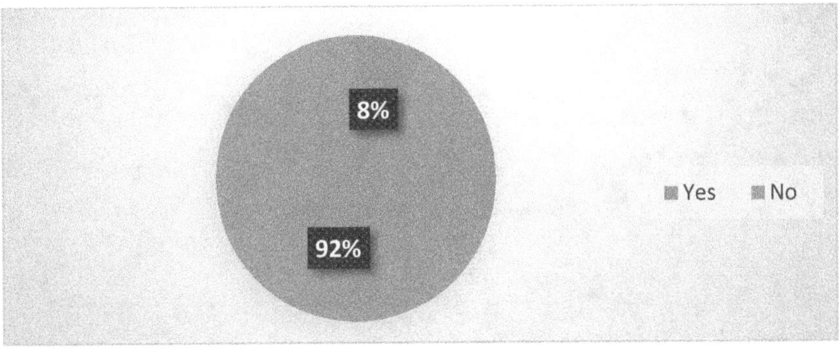

Figure 13.9. Has your salary been affected because of working remotely?

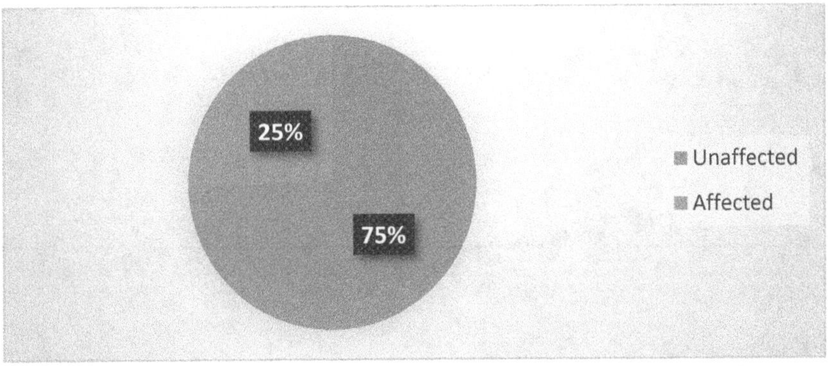

Figure 13.10. Has your feeling of professional security been affected because you work remotely?

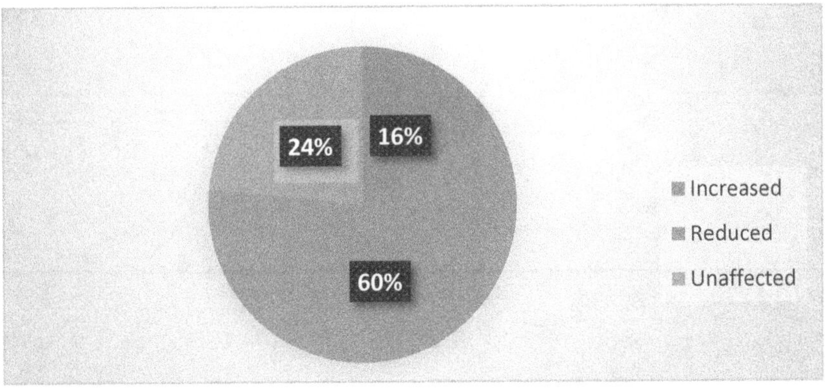

Figure 13.11. Has your ability to communicate with colleagues about common professional concerns increased or decreased while you worked remotely?

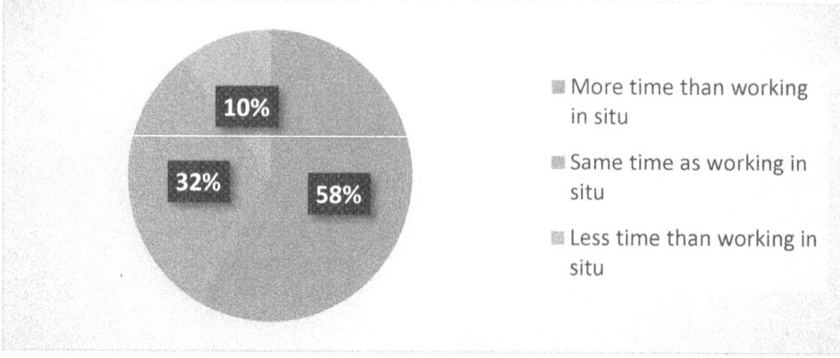

Figure 13.12. Has the duration of your labor time while you work remotely increased compared to your labor time in situ?

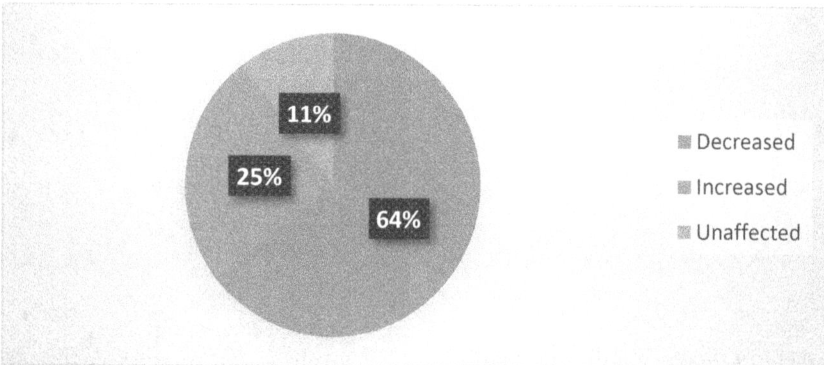

Figure 13.13. Has your ability to balance labor and family/leisure time increased or decreased while you worked remotely?

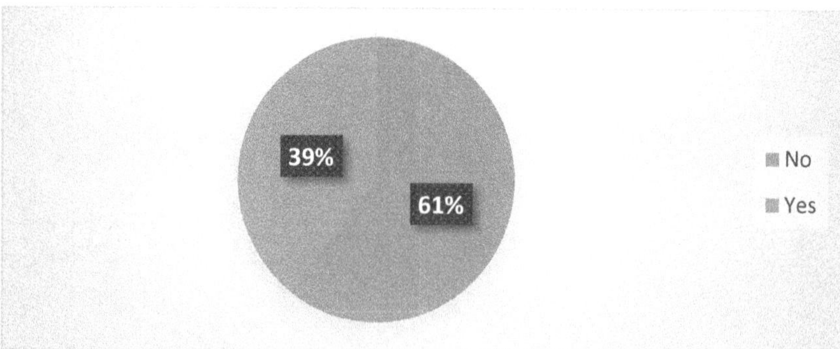

Figure 13.14. Do you think that your gender had a role in being burdened with more domestic tasks while working remotely?

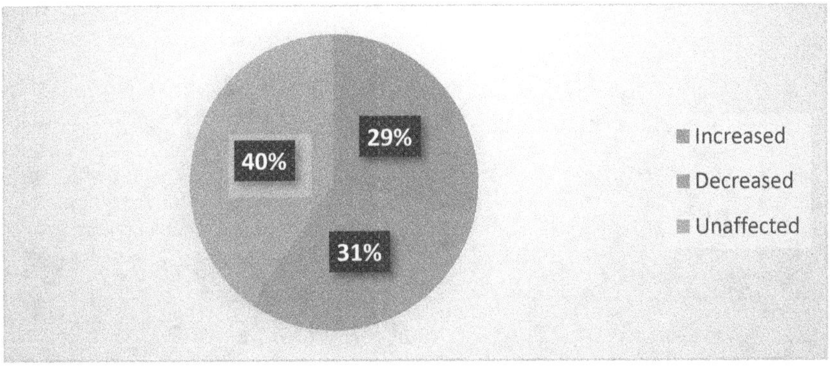

Figure 13.15. Have your creativity and productivity increased or decreased while working remotely?

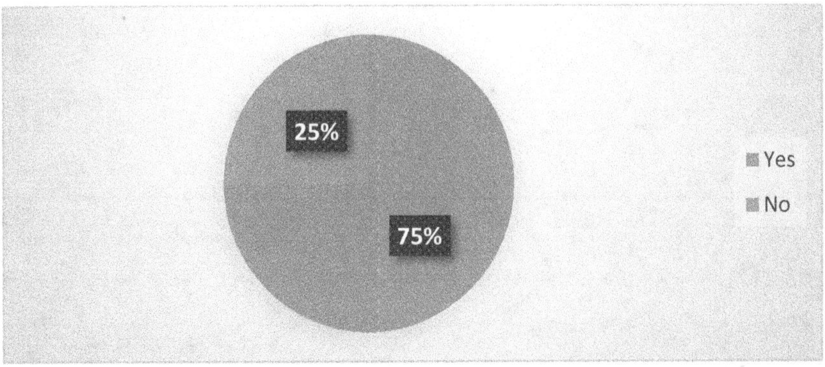

Figure 13.16. Have you obtained while working remotely new technical skills that might be useful for you while working in situ?

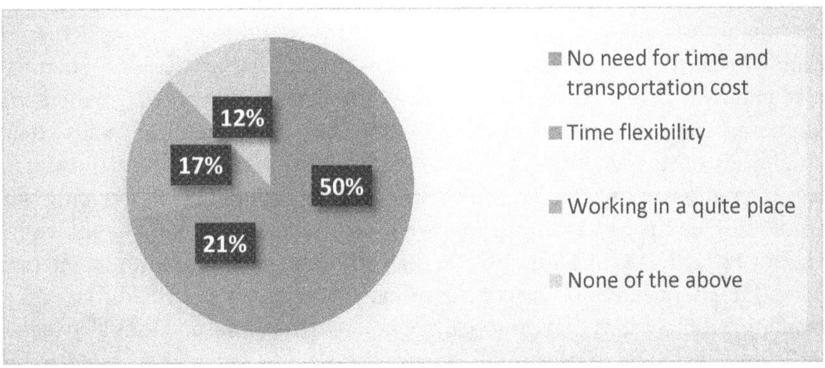

Figure 13.17. What do you think is the most crucial advantage of working remotely?

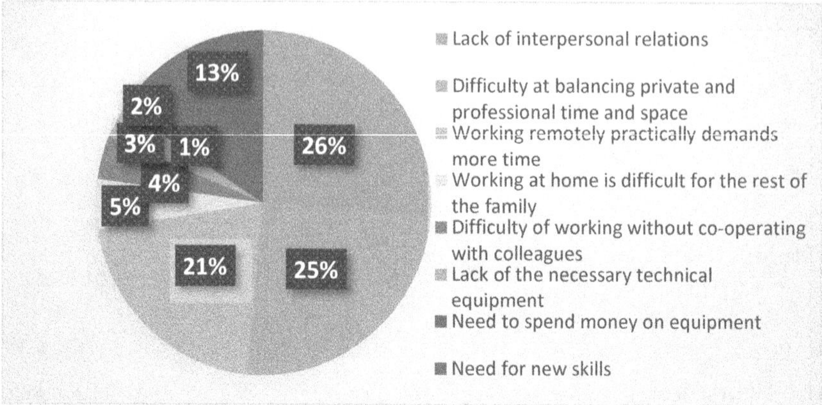

Figure 13.18. What do you think is the most crucial disadvantage of working remotely?

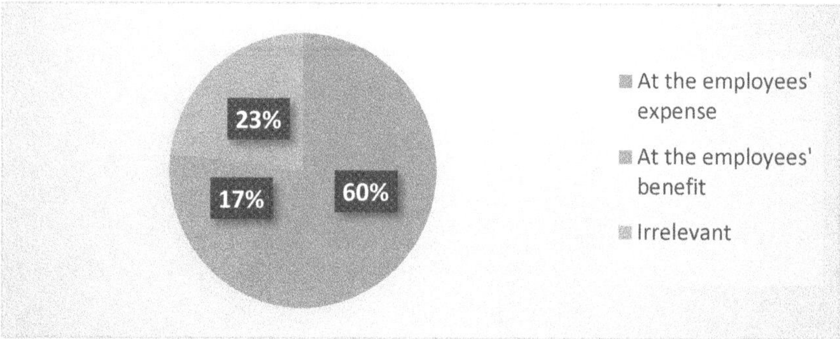

Figure 13.19. Do you think the remote work flexibility will be at the employees' labor rights expense in the long term?

As seen in the tables above, the parameters mostly affected by remote work did not mainly include the financial aspect. Most of the research participants claimed that neither their salary nor the feeling of their professional security was affected. Likewise, the employees' productivity remained unaffected. Moreover, remote work positively influenced the technical skills that employees developed to respond to new professional needs. On the other hand, remote work negatively affects the working experience by reducing the employees' ability to communicate with their co-workers. Finally, as shown in Figures 13.12, 13.13, 13.17, and 13.18, the parameter most notably affected has been time management. Most of the research participants claimed that while working remotely, they had to work more hours than before, and they also faced difficulty balancing professional and domestic life since they had to be available for employers or clients practically anytime. This fact can explain the pessimistic view many research participants express concerning the role of remote work. After all, they claim that reflexivity of working relations will be at the employees' expense in the long term.

The balance of professional and domestic time also concerns gender, as shown in Figure 13.14. Several recent case studies illustrate the issue of gender stereotypes' reinstatement in connection to remote work (Manzo and Minello 2020; Hardill and Green 2003; Nash and Churchill 2020). As previously mentioned, the flexibility of remote work practically abrogated the eight-hour labor, which was the cornerstone of 20th-century labor rights. At the same time, it privatized work aspects concerning social insurance, such as childcare. The growth of social inequalities has been unavoidably connected to re-establishing the so-called traditional gender roles within the domestic space. Despite employees of both genders having to work remotely, the fact that domestic tasks are traditionally considered female tasks burden women unequally. In the specific case study, the question of whether the interviewees thought their gender affected how much they were burdened with household chores was addressed to both men and women. That is because the experience of being burdened with tasks is subjective and concerns both genders in the first place. Even so, this is indicative of what really happens, up to an extent. The percentage of the interviewed women who replied positively to this question was 62%, while the respective percentage for men was 12%. In any case, in the long term, the fact that women are unequally burdened with balancing professional and domestic duties because of their gender might lead to the re-emergence of gender inequalities. As mentioned before, digital technologies are supposedly connected with deconstructing gender stereotypes. Therefore, it is noteworthy that the sociotechnical networks constructed within remote work status might relate to these stereotypes' potential reestablishment.

The fifth group of questions concerned net locality, namely the disconnection of employees' residence area from their work basis due to their ability to work remotely. The relevant answers are exposed in the tables below.

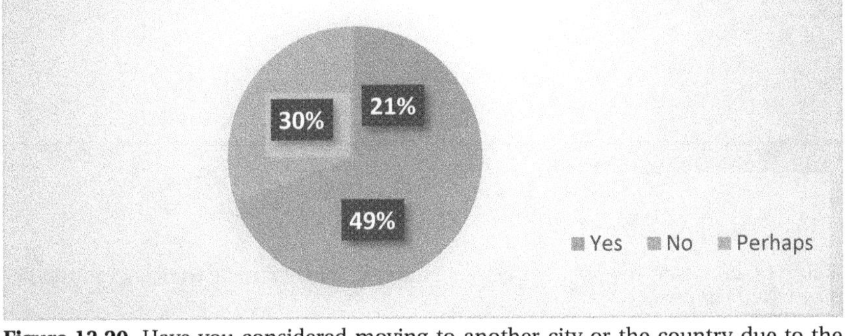

Figure 13.20. Have you considered moving to another city or the country due to the ability to work remotely?

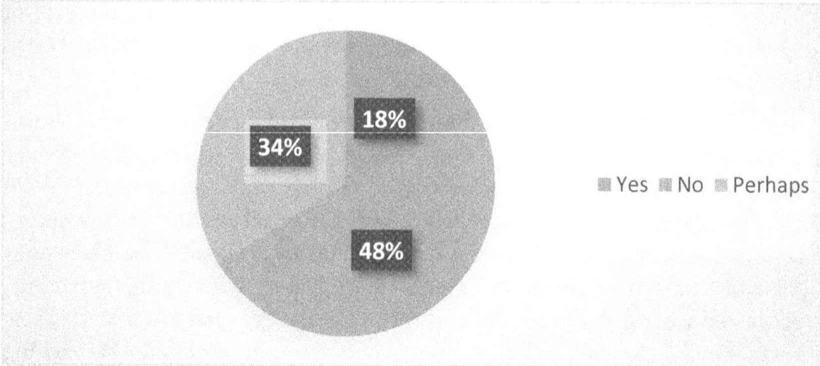

Figure 13.21. Have you considered seeking a job abroad while staying in Greece due to the ability to work remotely?

The interviewees' replies in the tables above show a potential geographical disconnection between the employees' home and work basis. As already analyzed in the third part of this chapter, the reconstruction of sociotechnical networks within remote work potentially entails the reconstruction of mobility networks. This situation can change geographic relations and the so-called economic centers and peripheries' roles. Hence, the massive application of remote work could lead to restructuring labor mobilities and the resignification of local and global both as geographic and notional categories. Still, it is too early to claim whether this trend will develop further.

Finally, the sixth category of questions focused on the interviewees' opinions about remote work, such as its impact on labor rights. Their replies are exposed below.

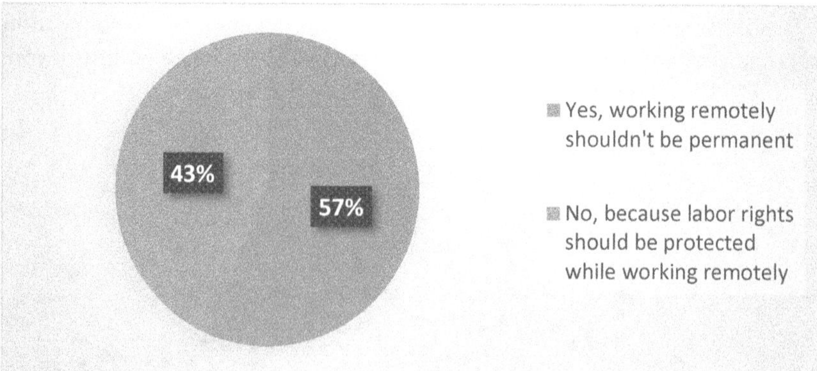

Figure 13.22. Would you oppose the legal establishment of remote work after the end of the current pandemic?

The opinions are almost equally divided concerning the legal establishment of remote work. The issue of remote work has been introduced within Greek legislation since 2006. However, even if remote work had been technologically and legally feasible for more than a decade, it was because of the social and economic conditions created during the COVID-19 pandemic that this was applied as a sociotechnical system. Moreover, the Greek parliament voted on a new labor bill in May 2021, which attempted to regulate remote work further while aiming at a more flexible context regarding the employers' and employees' relations. One of its most controversial changes concerned the conditional abolishment of the eight-hour labor, causing strong disagreements on behalf of the parliamentary opposition and the syndicates. The opinions of the interviewees concerning this legal bill appear in the following table.

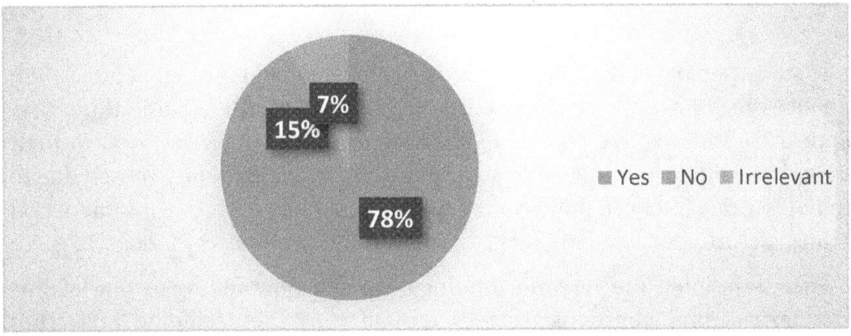

Figure 13.23. Do you think the reforms of the current legal bill voted in the Greek Parliament will be at the employees' expense?

Finally, the last two tables show the interviewees' overall expectations about the future of remote work. It appears that most of the research participants expect that, one way or another, remote work will bring permanent changes to the labor market despite it being temporarily applied due to COVID-19 urgency. Furthermore, they are open to working remotely permanently or occasionally.

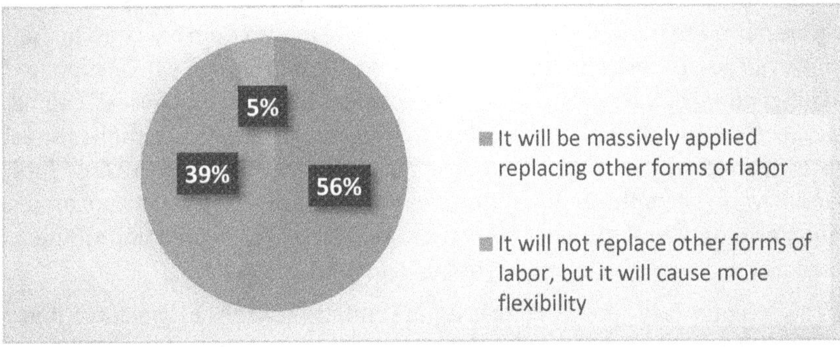

Figure 13.24. What do you think about the future of remote work?

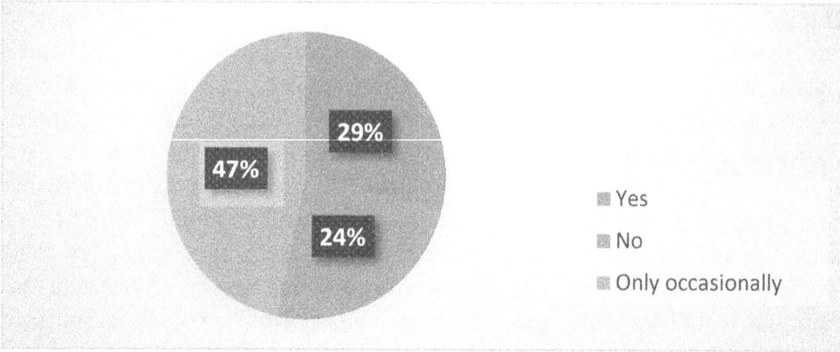

Figure 13.25. Would you like to continue working remotely after the end of the current pandemic?

To sum up, most of the research participants remain unaffected as far as their finances or their professional security is concerned. At the same time, they appear to find positive aspects of working remotely. However, they, in their majority, claim to have been negatively affected by the flexibility caused due to remote work. Moreover, they seemed to have unequal access to equipment and infrastructures, reproducing socio-economic and professional inequalities.

After describing the research findings, we can now return to one of this chapter's central questions: How is remote work connected to post-truth conditions? The following section discusses this issue.

Discussion on research findings

The overall assessment of the research findings shows that the rapid application of remote work in Greece during the COVID-19 pandemic has caused the immediate reconstruction of sociotechnical labor networks. The consequences of this reconstruction can be summarized into four key points.

First, on the macroscale level, the delocalization of work has created a decentralization trend since several citizens have moved from big cities to their home villages to work remotely. This trend is probably limited when compared to other countries. However, we should consider that Athens, Greece's capital, has been hydrocephalic for more than half a century (See, for example, Burgel 1978). Therefore, this trend might have a positive effect because it could limit the uneven geographical distribution of sociotechnical networks concerning labor, economy, and mobility infrastructure. Moreover, it could contribute to establishing a more sustainable working relationship model.

The second key point concerns the intensification of sociotechnical inequalities. As shown in the previous section, most of the employees had to provide on their own the technical infrastructure they used to work remotely. This infrastructure included essential equipment (e.g., computers, headsets, microphones) and the internet. At the same time, most of the interviewees faced severe problems with their internet connection. As previously discussed, Greece scores below the average internet speed and above the average internet cost of the European Union. On top of this, the purchasing power of Greek households has shrunk remarkably since the 2009 economic crisis, and it has further deteriorated during the pandemic. Consequently, the accessibility (or motility) of many Greek households –especially those with three or more members- to work and education has become unaffordable.

The third key point also concerns sociotechnical inequalities related to merging the domestic and professional spheres. Remote work has turned houses into multi-spaces; therefore, many collective labor rights such as eight-hour labor, parental or sick leaves, paid overtime work, and access to social insurance are considered personal problems that should be privately solved. Additionally, the domestic sphere has been historically connected with gender roles; thus, privatizing social insurance matters reproduces gender inequalities. These inequalities are often described as the 'natural' resultant of nuclear family arrangements, denying their sociocultural origins (Manzo and Minello 2020; Hardill and Green 2003; Nash and Churchill 2020). Moreover, the domination of a masculinist model of labor history (e.g., the history of the male working class) as the only model of history has reinforced a male concept of class or 'social.' Therefore, the gendered aspect of labor remains concealed (Betti et al. 2022).

Finally, the fourth key point concerns the scattering of professional collectivities. Most employees mentioned the limitation of their ability to share their common professional concerns with their colleagues as one of the main disadvantages of remote work. Therefore, the employees are supposed to face the abovementioned sociotechnical inequalities individually. Besides, according to most of the interviewees, the new labor bill that was voted on in the Greek Parliament in 2021 establishes new labor standards. These standards promote a neoliberal model of individualism and flexibility, which results not only from the legislation dynamics but also from the labor market. Within this model, the abovementioned inequalities are not recognized as such; on the contrary, they are described as 'challenges' that the employees must meet so that they will be 'competitive.' Of course, the urge to be competitive to meet new sociotechnical standards is not necessarily negative. Many interviewees pointed out that they developed new technical skills while working remotely. However, within the same context, having children, facing health issues, or being unaffordable to provide professional infrastructure (for which the employers are normally responsible) are also considered a lack of professional competitiveness. Moreover, the sociotechnical inequalities described above can potentially delocalize within remote work. Employers could seek 'competitive' employees worldwide within a digitalized and delocalized labor market. This situation might also result in the employees' willingness to accept more sociotechnical inequalities as an inevitable option. Unless the labor legislation aims to solve these problems, such sociotechnical inequalities could become global. After all, the global and local levels are multi-dimensionally networked in the digital communications era.

The above-described key can answer the question concerning the connection between remote work and the post-truth discourse. The sociotechnical labor inequalities potentially entail several 'post-truth' pathogeneses, such as the demonization of technology and the growing mistrust toward abstract conceptualization of science, democracy, or social progress. Technology has been strongly connected with social progress within the 20th-century public discourse. When people have more technology in their everyday lives in ways that deepen sociotechnical inequalities, as in remote work, that leads to a growing mistrust towards both technology and social progress values, like democracy.

In short, the reconstruction of sociotechnical labor networks entails the deterioration of many aspects of quality of life for most of the interviewees of the particular case study. The digitalization of labor has created instability because of unequal access to education, health, work, or natural resources. This situation can also be described as decreased access to mobility capitals (motility). Within the changing structure of the global labor market, the rise of precarious, informal, and unstable working relations has become increasingly frequent globally. Based on that, this chapter argues that the sociotechnical inequalities built within the remote work networks contribute to understanding post-truth aspects. The connection of these inequalities with the 'post-truth' phenomenon will be discussed in the next section.

Conclusions

After discussing the research results, one can return to the theoretical question that served as its starting point: how can the STS tools help understand the so-called post-truth age? Furthermore, what helps us understand remote work's connection with post-truth conditions?

As far as the concept of post-truth is concerned, this chapter argues that it is based on ahistorical presuppositions. However, it is helpful because it describes a context in which the persuasion method has a more significant influence than facts on public decisions with massive impact. This phenomenon has already been analyzed within the approach of the performative function of language and in literary theory (See, for example, Austin 1962; Foucault 1973; Barthes 1975). Moreover, the fact that the multiplicity of information sources and the social media explosion has escalated this phenomenon testifies to McLuhan's argument that the medium itself has a critical role in the message impact (McLuhan 1964: 7-13). Therefore, it has been argued that theories that have deconstructed the concept of truth and facts bear the burden of dealing with such ideas' consequences when they are misinterpreted in the public discourse.

Regarding this, the argument of this chapter concerning the first of the questions mentioned at the beginning of this part is that the STS paradigm provides a useful methodological framework to understand the 'post-truth' environment. The illustration of the heterogeneous character of this phenomenon creates a context for analyzing its sociotechnical structure. As mentioned in the Introduction of this chapter, both the sociotechnical networks of remote work and the 'post-truth' framework share common structural features, such as virtual communication and the digitalization of identities of the sociotechnical agents. From that point of view, it becomes evident that the post-truth problem is not only about the rationality or irrationality of discourses, as a literary theory approach would argue. It is mainly about sociotechnical networks in which heterogeneous actors interact through digital identities. Some of these identities are real, and some belong to 'trolls' or internet bots, which create a context of interaction among hybrids with fluid identities.

That brings us to the last of the research questions mentioned in the Introduction to this study, namely the connection of remote work with the post-truth phenomenon. This chapter argues that the discourses related to the post-truth phenomenon are re/produced within the same sociotechnical infrastructure in which remote work networks develop. The establishment of remote work further contributes to the sociotechnical reconstruction of the spatial and temporal context in which human or hybrid identities interact. The remote work system's heterogeneous structure comprises employers and employees, the technological equipment and infrastructure, the labor market, and labor legislation. The inequalities constructed within remote work do not only have to do with the system itself but with the interaction of the system's heterogeneous factors.

Technologies are usually connected with a progressive narrative; however, within remote work, they tend to introduce changes at the expense of employees. Therefore, they enforce power relations. Such an approach might explain post-truth phenomena, such as the demonization of technology or the growing mistrust toward the abstract conceptualization of science, democracy, or social progress. The fact that such ideas develop within the 'post-truth' context with the use of technology seems like an oxymoron; however, controversies are usually a structural element of ideologies. In any case, the focus is on the agents most impacted by these inequalities, which contributes to the illustration of the pathogeneses, which empower the appeal of demagoguery in broad audiences. While the resonance of populist rhetoric has often been considered a form of 'irrationality' over public opinion, post-truth phenomena have structural causes related to the fact that the heterogeneous sociotechnical character of labor inequalities is concealed.

In conclusion, the chapter argues that the study of the contemporary sociotechnical networks' inequalities is crucial for understanding the post-truth era. The very term 'post-truth' presupposes a fear of a digital society in which the opinions of both experts and the public are equally valid; consequently, core values of modernism, such as reason and the scientific method, are doubted. If that is the case, we might better understand the challenges of the 'post-truth' conditions for the STS, as illustrated in the case study of remote work. This case study shows that we need more participatory forms of science and technology to fight inequalities and work inclusively instead of discriminating the 'public' from the 'experts.' According to such an epistemological view, scientific methods and principles should not be merely produced within an introverted scientific community; on the contrary, they should be tuned to society to form science as a pluralistic inquiry that enforces democracy and equality.

References

Ahrendt, Daphne, Massimiliano Mascherini, Sanna Nivakoski and Eszter Sándor. 2021. "Living, working and COVID-19: Mental health and trust decline across EU as pandemic enters another year." (Luxemburg: European Foundation for the Improvement of the Living and Workings Conditions, Publications Office of the European Union. https://www.eurofound.europa.eu/publications/report/2020/living-working-and-covid-19 (Accessed, May 1st, 2021).

Austin, John Langshaw. 1962. *How to Do Things with Words. The William James Lectures Delivered at Harvard University in 1955.* London: Oxford University Press.

Barthes, Roland. 1975. *The Pleasure of the Text.* Trans. Richard Miller. New York: Hill and Wang.

Betti, Eloisa, Leda Papastefanaki, Marica Tolomelli, Susan Zimmermann. 2022. "Thinking the History of Women's Activism into Global Labor History." Pp. 1-34 in *Women, Work and Activism. Chapters of an Inclusive History of Labor in the Long Twentieth Century,* edited by Eloisa Betti, Leda Papastefanaki, Marica Tolomelli, Susan Zimmermann. Budapest: Central European University Press.

Bhambra, Gurminder K. 2017. "Brexit, Trump and 'methodological whiteness': on misrecognition of race and class." *British Journal of Sociology* 68: 214-232.

Bijker, Wiebe E. and John Law. 1992. "General Introduction." Pp. 1-16 in *Shaping Technology/Building Society* edited by Wiebe E. Bijker and John Law. Cambridge Massachusetts: The MIT Press.

Bonikowski, Bart. 2017. "Ethno-nationalist populism and the mobilization of collective resentment." *British Journal of Sociology* 68: 181–213.

Brynjolfsson, Erik, John J. Horton, Adam Ozimek, Daniel Rock, Garima Sharma, and Hong Yi TuYe. 2020. "COVID-19 and Remote Work: An Early Look at US Data." *NBER WORKING SERIES*, National Bureau of Economic Research. Working Paper 27344 (June 2020): 1–26. www.nber.org/papers/w27344.

Burgel, Guy. 1978. *Athens. The Development of a Mediterranean Capital City.* Athens: Exantas.

Castells, Manuel. 2004. "Informationalism, Networks, and the Network Society: A Theoretical Blueprint." Pp. 3-45 in *The Network Society. A Cross-Cultural Perspective* edited by M. Castells. Cheltenham – Northampton: Edward Elgar Publishing.

Castells, Manuel. 2010. *The Rise of the Network Society: The Information Age: Economy, Society, and Culture.* 2nd ed. . Malden-Oxford: Wiley.

Cresswell, Tim. 2006. *On the Move. Mobilities in the Modern Western World*, New York-Oxon: Routledge.

Cresswell, Tim. 2015. *Place: An Introduction.* Malden-Oxford: Wiley and Blackwell Publishing.

"Danger of internet crash in Greece". 2020. https://paradrasi.gr/koronoios-deka-odigies-gia-na-min-krasarei-to-diadiktyo/ (accessed August 16th, 2022).

De Sousa e Silva, Adriana and Eric Gordon. 2014. "Net Locality." Pp. 134-142 in *The Routledge Handbook of Mobilities* edited by Peter Adey, David Bissel, Kevin Hannam, Peter Merriman, and Mimi Sheller. Oxon-New York: Routledge.

Fari, Simone and Massimo Moraglio. 2016. "The Function of Flows Between Centers and Peripheries." Pp.10-17 in *Peripheral Flows. A Historical Perspective on Mobility Between Cores and Fringes* edited by Simone Fari and Massimo Moraglio. New Castle: Cambridge Scholars Publishing.

Eurofound. 2019. "European Union Mobile and Fixed Broadband Prices in Europe 2019." Luxemburg: Publications of Office of the European Union.

Eurofound. 2020. "Telework and ICT-based mobile work: Flexible working in the digital age, New forms of employment series." (Luxembourg: European Foundation for the Improvement of the Living and Workings Conditions, Publications Office of the European Union). https://www.eurofound.europa.eu/publications/report/2020/telework-and-ict-based-mobile-work-flexible-working-in-the-digital-age (Accessed, May 1st, 2021).

European Commission. 2020. "Shaping Europe's Digital Future." Luxembourg: European Foundation for the Improvement of the Living and Workings Conditions, Publications Office of the European Union). https://digital-strategy.ec.europa.eu/en/policies/desi-connectivity (Accessed, May 1st, 2021).

Foucault, Michel. 1973. *The Order of Things: An Archaeology of the Human Sciences.* New York: Vintage.

Hård, Mikael, and Thomas J. Misa. 2008. "Modernizing European Cities." Pp. 1-22 in *Urban Machinery. Inside Modern European Cities* edited by Mikael Hård and Thomas J. Misa. Cambridge Massachusetts: The MIT Press.

Hardill, Irene and Anna Green. 2003. "Remote Working: Altering the Spatial Contours of Work and Home in the New Economy." *New Economy, Work, and Technology* 18, (3): 212–222.

Herf, Jeffrey. 1984. *Reactionary Modernism. Technology, Culture, and Politics in Weimar and the Third Reich.* Cambridge: Cambridge University Press.

Hoffman, Steve G. 2018. "The Responsibilities and Obligations of STS in a Moment of Post-Truth Demagoguery." *Engaging Science. Technology and Society* 4: 444–452.

Iyengar, Shanto and Douglas S. Massey. 2019. "Scientific Communication in a Post-Truth Society." *PNAS* 116 (16): 7656-7661.

Jasanoff, Sheila, and Hilton R. Simmet. 2017. "No Funeral Bells: Public reason in a 'post-truth age." *Social Studies of Science* 47 (5): 751–770.

Journal of the Government of the Greek Republic, A 55/11.03.2020. B 4134/25.09.2020. B 1194/27.03.2021.

Kaufmann, Vincent, Manfred Max Bergman and Dominique Joye. 2004 "Motility: Mobility as Capital." *Journal of Urban and Regional Research.* 28 (4): 745-756.

Keeling, Trevor, Derek Clemens-Croome and Etienne Rosch. 2015. "The Effect of Agile Workspace and Remote Working on Experiences of Privacy and Satisfaction" *Buildings* 18: 880–898.

Kuhn, Thomas. 1962. *The Structure of Scientific Revolutions.* Chicago and London: The University of Chicago Press.

Kylili, Angeliki, Nicholas Afxentiou, Loucas Georgiou, Christiana Panteli, Phoebe Zoe Morskink-Georgali, Andri Panayidou, Constantinos Papouis, and Paris A. Fokaides. 2020. "The role of Remote Working in smart cities: lessons learnt from the COVID-19 pandemic." *Energy Sources, Part A: Recovery, Utilization, and Environmental Effects* https://doi.org/10.1080/15567036.2020.1831108 (Accessed, May 2021).

"Lockdown: What we should avoid so the internet will not crash during the quarantine." 2020. https://www.ethnos.gr/greece/article/131891/lockdown osaprepeinaapofeygoymegianamhnkrasareitodiadiktyosthnkarantina (accessed August 16th, 2022).

Lynch, Michael. 2017. "STS, symmetry and post-truth." *Social Studies of Science* 47 (4): 593–599.

Makridis, Yorgos. 2020. "Remote Education - 1st Keratsini General Highschool: 55% of students attend classes using mobile phones." https://www.alfavita.gr/ekpaideysi/339563_tilekpaideysi-1o-gel-keratsinioy-55-ton-mathiton-kanei-mathima-apo-kinito (Accessed May 25th, 2021).

Manzo Lidia Katia C. and Alexandra Minello. 2020. "Mothers, Childcare Duties, and Remote Working under COVID-19 Lockdown in Italy: Cultivating Communities of Care." *Dialogues in Human Geography* 10 (2): 120-123.

Marres, Noortje. 2018. "Why We Can't Have Our Facts Back." *Engaging Science, Technology, and Society* 4: 423-443.

McLuhan, Marshall. 1964. *Understanding Media: The Extensions of Man.* New York: McGraw Hill.

Mollino, Monica, Emanuela Ingusci, Fulvio Signore, Amelia Manuti, Maria Luisa Giancaspro, Vincenzo Russo, Margherita Zito and Claudio Cortese.

2020. "Wellbeing Costs of Technology Use During Covid-19 Remote Working. An Investigation Using the Italian Translation of the Technostress Creators Scale." *Sustainability* 12: 1-20.

Nash, Meredith and Brendan Churchill. 2020. "Caring During COVID-19: A Gendered Analysis of Australian University Responses to Managing Remote Working and Caring Responsibilities." *Gender Work and Organization*: 1–14.

Nowotny, Helga, Peter Scotts and Michael Gibbons. 2013. *Rethinking Science. Knowledge and the Public in an Age of Uncertainty*. Cambridge – Maiden: Polity (4th Edition).

Oldenziel, Ruth. 2009. "Exporting the American Cold-War Kitchen: Challenging Americanization, Technological Transfer, and Domestication." Pp. 315–340 in *The Cold-War Kitchen. Americanization, Technology and European User* edited by Ruth Oldenziel and Karin Zachmann,. Cambridge Massachusetts-London: The MIT Press.

Papazafeiropoulou, Alexia-Sophia. 2016. "The Affirmation of Semi-Periphery. A Case Study on Greek Automobility, 1930-2000." Pp. 28-50 in *Peripheral Flows. A Historical Perspective on Mobility Between Cores and Fringes* edited by Simone Fari and Massimo Moraglio. New Castle: Cambridge Scholars Publishing.

Papazafeiropoulou, Alexia-Sophia. 2021. Questionnaire. Research on the remote work conditions in Greece during the COVID-19 pandemic, 13/05/2021-14/06/2021. https://docs.google.com/forms/d/143D2claimruom PNBmfdHlXTTyVgq2HbTnqgPJ0l5MAU/edit.

Prasad, KVD, Mruthyanjaya Rao Mangypudi, Rajesh W. Vaidya, Budumuru Muralidhar. 2020. "Organizational Climate Opportunities, Challenges and Psychological Wellbeing of the Remote Working Employees during Covid-19 Pandemic: A General Linear Model Approach with Reference to Information Technology Industry in Hyderabad." *International Journal of Advanced Research in Engineering and Technology* 11, (4): 372-389.

"Remote Education: Students and their parents denounce network problems and resource deficits." 2020. https://www.megatv.com/2020/11/19/tilekpaideysi -provlimata-sto-systima-kai-elleipsi-meson-kataggelloun-goneis-kai-mathites/. (Accessed May 25, 2021).

"Rethymno…2020: Remote Education for two young girls on their building entrance" 2020. https://www.newsit.gr/topikes-eidhseis/rethymno-2020-tilekpaideysi-sto-pezouli-tis-polykatoikias-gia-dyo-mikra-koritsia/3150337/ (Accessed May 25th, 2021).

Sheller, Mimi. 2018. *Mobility Justice. The Politics of Movement in an Age of Extremes*, London-Brooklyn: Verso.

Sismondo, Sergio.2017. "Post-truth?" *Social Studies of Science*. 47 (1): 3–6.

"Urgent call of the government: Do not use the internet…" 2020. .https://www.athensmagazine.gr/article/news/486405-ektakth-proeidopoihsh-ths-kybernhshs -mhn-mpainete-sto-internet (accessed August 16th, 2022).

Wajcman, Judy. 2004. *Technofeminism*, Cambridge-Malden: Polity.

Wajcman, Judy. 2009. "Feminist Theories of Technology," *Cambridge Journal of Economics* 34: 1–10.

Author biographies

Janine Aldous Arantes is a Senior Lecturer in the College of Arts, Business, Law, Education and Information Technology at Victoria University. With over 20 years of experience in various educational contexts, Janine currently teaches into initial teach education programs and holds a Victoria University Research Fellowship with the Institute of Sustainable Industries. Her research interests include the implications of integrating digital technologies in schools, AI technologies, remote and online teaching and learning, and teachers' rights alongside the shifting nature of teachers' identities in digital classrooms. Janine is an Early Career Researcher, having completed her PhD in 'Big Data, Black Boxes and Bias: The Algorithmic Identity and Educational Practice' from the University of Newcastle. Her passion lies in critical data studies and digital humanities.

Malcolm Ashmore is the author of *The Reflexive Thesis* (University of Chicago Press, 1989) and co- author of *Health and Efficiency* (Open University Press, 1989). Within STS, he has researched reflexivity, the debunking of scientific fraud, the false/recovered memory controversy, and, with Olga Restrepo Forero, the ironies of state documentary practices, particularly in Colombia. He is Honorary Fellow in the Department of Social Sciences, Loughborough University, UK; and is now (almost) retired.

Lynda Dunlop is the Director of Education for Environmental Sustainability at York, and a senior lecturer in science education. Her research focuses on environmental interventions and scientific controversies, their impacts on young people, and the educational implications arising from these. She works in interdisciplinary teams to investigate the social impacts of fracking, climate interventions, biofuel production and human spaceflight. Her public engagement work has focused on the use of philosophical inquiry to examine questions about contemporary science and technology and develop critical thinking and sensitive communication about science, technology and society.

Rachan Daimary is an Assistant Professor in the Department of Journalism and Mass Communication, Faculty of Arts, Manipal University, Jaipur, Rajasthan, India. Rachan is an STS researcher working on science communication, biodiversity, and wildlife conservation. His doctoral work investigated the relationship between conservation scientists and laypeople for effective wildlife conservation management in India. Currently, he is interested in working in the research area of human-nature relationships to further his understanding of science communication.

Javier Guerrero-C. obtained his PhD in Social Studies of Science and Technology from the University of Edinburgh (UK) and his MSc in Science and Technology Studies from the University of Edinburgh (UK). Interested in understanding the dynamics of participation and interactions in online social networks, datafication processes, and the consequences of digital infrastructures, platforms and algorithms. He has analyzed conspiracy theories online in Colombian politics, with results published in: *You the readers will complete the list. The Castrochavismo conspiracy theory.*

Michael Kilburn is a Professor of Political Science and International Studies at Endicott College in Beverly, MA. A graduate of Emory University's interdisciplinary Institute for the Liberal Arts, his 2001 dissertation, "The Merry Ghetto: The Czech Underground in the time of Normalization" was one of the first academic studies of underground culture under late socialism in Czechoslovakia. His current research interests include international relations, human rights, political and critical theory, anarchism, oral history, narratology, and geography, with a critical focus on irony and system dysfunction as an analytical tool.

Hemant Kumar teaches at the Centre for Studies in Science, Technology and Innovation Policy, School of Social Sciences, Central University of Gujarat, Gandhinagar, Gujarat, India. He has spent more than a decade in India working on innovations in the informal sector with an emphasis on grassroots innovations. Other areas of interest for him include STS studies, marginalization in STEM, science, technology, and innovation diplomacy, as well as the history and philosophy of science and technology.

Kyunghwan Lee is a doctoral candidate in Sociology at the University of Southern California (USC). He obtained his MA and BA from Yonsei University, Seoul, Korea. His research grapples with how knowledge and ideas travel globally and interact with the object and the world. His dissertation project, entitled 'Making the Economy through the National Income in South Korea,' attempts to trace how the national economy in South Korea has been constituted as a quantified entity through a collection of statistical practices, economic knowledge, and policy-making technologies.

Karen McGregor Richmond is a postdoctoral research fellow at iCourts, the Danish Centre of Excellence in International Courts. Karen holds an LLB (Edinburgh), an LLM in Advocacy, and a PhD in Law. Her work is situated at the boundary of law and science. Karen's doctoral monograph used STS perspectives to chart the influence of marketization on forensic science provision, whilst her postdoctoral work focuses on developing methods to measure the probative value of evidence. She is a member of the Faculty of Evidence-Based Health Care at Oxford University.

Ezgi Pehlivanli holds a PhD from Middle East Technical University (METU), Department of Sociology. She taught Gender and Social Theory at the Research Center for Science and Technology Policy Studies, METU, between 2016-2022. She was a postdoc fellow at Lund University, Department of Gender Studies in 2018. Pehlivanli Kadayifci is currently double affiliated as a postdoc fellow at the University of Stavanger, Department of Media, Culture and Social Sciences and Middle East Technical University. Her research interests are feminist science and technology studies, gender, body, science discourse, political sociology, discursive politics, and populism.

Alexia Sofia Papazafeiropoulou teaches History and Sociology of Technology and Urban Civilization at the National Technical University of Athens (NTUA). She conducts postdoctoral research at the same university, titled: "The Notion of the Subject within the Context of Mobilities. A study of the bibliographical current as a potentially new paradigm." The title of her Ph.D. thesis is "Car Culture and Road Networks in Greece, 1930-1980." She has published papers on Mobility and Science and Technology Studies in international academic journals and chapters in collective volumes.

Tino Plümecke is a postdoctoral researcher and co-investigator at the Institute of Sociology at the University of Freiburg, Germany, within the project "Human Diversity in the New Life Sciences: Social and Scientific Effects of Biological Differentiations." His research interests are in science and technology studies, theories of discrimination and critical race studies with a specific focus on the history and the recent developments in genetics and post/genomics.

Olga Restrepo Forero is a Professor at Departamento de Sociología, Universidad Nacional de Colombia, Bogotá. She has published on Darwinism in Colombia and the historiography of Darwinism in Latin America, on scientific writing and scientific rhetoric. She has also researched, with Malcolm Ashmore, institutional settings that produce and negotiate trust relationships, such as notaries and identification cards, and their uses in Colombian society. From 2010 to 2013, she directed and published an enormous research project called *'Ensamblado en Colombia: producción de saberes y contrucción de ciudadanías'* ('Assembling Colombia: producing knowledge and constructing citizenships'). Most of her previously published work is in Spanish.

Joshua Stubbs is a PhD of Education student in the Psychology in Education Research Centre at the University of York. His research focuses on education, health and social justice. He has contributed to research on how to establish more inclusive social and academic environments within education; how to support students' transition from one stage of education to another; and how to incorporate philosophical inquiry into the curriculum. He is also interested in helping young people to identify logical fallacies, construct better arguments, and become more politically literate and engaged.

Laura Schnieder is a medical sociologist by training. Currently, she is employed as a project coordinator in clinical research. Her social scientific research focuses on questions at the intersection of the production of biomedical knowledge and feminist theory.

Pedro Alves da Veiga is a Portuguese transdisciplinary artist and researcher with a degree in Computer Science and earned a PhD in Digital Media Art. He is a Professor at Aberta University in Lisbon, Portugal, the subdirector of the Digital Media Arts PhD programme and a member of the Scientific Council of the Research Centre for Arts and Communication (CIAC). He developed his research at the intersection of art, science and technology; focused on interactive and multimedia generative systems, new media artivism, curatorship, and a/r/cography, a novel creative research methodological approach.

Matjaz Vidmar is a Lecturer in Engineering Management at the University of Edinburgh and Deputy Director of the Institute for the Study of Science, Technology and Innovation. He is researching innovation processes, R&D (eco)systems, and future strategies and design, especially within the space industry, artificial intelligence, and data-driven economy. He leads interdisciplinary projects spanning arts, science, and civil society, he is involved in several start-up companies, and he delivers an extensive public engagement programme. More at www.blogs.ed.ac.uk/vidmar

Index

L

M

N

O

P

R

S

T

W

www.ingramcontent.com/pod-product-compliance
Lightning Source LLC
LaVergne TN
LVHW010722210926
848485LV00007B/1421/J
* 9 7 8 1 6 4 8 8 9 7 8 8 7 *